SAMPLING, ANALYSIS,& MONITORING METHODS

A Guide to EPA Requirements

C. C. Lee, Ph. D.

Research Program Manager
National Risk Management Research Laboratory
U. S. Environmental Protection Agency
Cincinnati, Ohio 45268

Chairman of Executive Steering Committee
National Technical Workgroup on Mixed Waste Thermal Treatment

Founder of the International Congress on Toxic Combustion Byproducts
Washington, D.C. 20036

Member of Policy Review Group
Center for Clean Technology
University of California, Los Angeles (UCLA)
Los Angeles, California 90024

Adjunct Professor
Civil and Environmental Engineering Department
University of Cincinnati
Cincinnati, Ohio 45221

Government Institutes, Inc.
Rockville, Maryland

Government Institutes, Inc., 4 Research Place, Suite 200,
Rockville, Maryland 20850

ISBN: 0-86587-477-8

Printed in the United States of America

TABLE OF CONTENTS

SECTION 1. ENVIRONNMENTAL LAWS AND REGULATIONS

1.1 LIST OF ENVIRONMENTAL LAWS AND REGULATIONS 1-3

SECTION 2. CHEMICAL CROSS-REFERENCE

2.1 CHEMICAL CROSS-REFERENCE FOR SAMPLING, ANALYSIS,
MONITORING AND RISK ASSESSMENT 2-2

SECTION 3. SOLID WASTE PROGRAMS

3.1. SUMMARY OF SOLID WASTE SAMPLING, ANALYSIS AND MONITORING METHODS 3-2
    TABLE 3.IA. TEST METHODS--ALPHABETICAL ORDER BY TEST PARAMETER 3-2
    TABLE 3.IB. TEST METHODS--NUMBER ORDER BY METHOD 3-11

3.2. MUNICIPAL WASTE (40CFR258) 3-20
    TABLE 3.2. CONSTITUENTS FOR DETECTION MONITORING (40CFR258, APPENDIX 1) 3-20
    TABLE 3.3. ANALYSIS METHODS FOR MUNICIPAL WASTE CONSTITUENTS 3-23

3.3. HAZARDOUS WASTE 3-32
    TABLE 3.4. ANALYSIS METHODS FOR HAZARDOUS WASTE CONSTITUENTS 3-32

3.4. GROUND-WATER 3-53
    TABLE 3.5. GROUND-WATER MONITORING LIST (40CFR264, APPENDIX IX) 3-53

3.5. WASTE ANALYSIS 3-61

3.6. REFERENCES (40CFR260.11.a) 3-61

SECTION 4. DRINKING WATER PROGRAMS

4. I. MONITORING, SAMPLING AND ANALYTICAL REQUIREMENTS IN 40CFR 4-2
    4.1.1. COLIFORM SAMPLING (40CFR141.21) 4-2
    4.1.2. TURBIDITY SAMPLING AND ANALYTICAL REQUIREMENTS (40CFR141.22) 4-2
    4.1.3. INORGANIC CHEMICALS 4-2
    TABLE 4.1. DETECTION LIMITS FOR INORGANIC CONTAMINANTS (40CFR141.23.a) 4-2
    TABLE 4.2. ANALYTICAL METHODS FOR THE LISTED INORGANIC CONTAMINANTS
        (40CFR141.23.k) 4-4
    4.1.4. ORGANIC CHEMICALS 4-6
    TABLE 4.3. ORGANIC CONTAMINANTS AND MCL LIMITS 4-6
    TABLE 4.4. SYNTHETIC ORGANIC CONTAMINANTS AND MCL LIMITS 4-7
    4.1.5. ANALYTICAL METHODS FOR RADIOACTIVITY (40CFR141.25 AND TABLE 4.3) 4-9
    TABLE 4.5. MCL STANDARDS AND ANALYTICAL METHODS FOR RADIOACTIVITY 4-9
    4.1.6. TOTAL TRIHALOMETHANES SAMPLING, ANALYTICAL
        AND OTHER REQUIREMENTS (40CFR141.30) 4-10
    4.1.7. ANALYSIS OF TRIHALOMETHANES (40CFR141, APPENDIX C) 4-10
    4.1.8. MICROBIOLOGICAL CONTAMINANTS 4-10

4.2. MONITORING, SAMPLING AND ANALYTICAL METHODS IN EPA PUBLICATIONS 4-10
    4.2.1. METHODS FOR THE DETERMINATION OF INORGANIC SUBSTANCES
        IN ENVIRONMENTAL SAMPLES 4-10
    4.2.2. METHODS FOR THE DETERMINATION OF ORGANIC COMPOUNDS IN DRINKING WATER 4-10
    TABLE 4.6. ANALYTE - METHOD CROSS-REFERENCE (EPA-91/7) 4-12

4.2.3. METHODS FOR CHEMICAL ANALYSIS OF WATER AND WASTES 4-25
TABLE 4.7A. TEST METHODS--ALPHABETICAL ORDER BY PARAMETER 4-25
TABLE 4.7B. TEST METHODS--NUMBER ORDER BY METHOD 4-31

4.3. REFERENCES 4-37

SECTION 5. WATER PROGRAMS

5.1. NPDES PERMIT APPLICATION TESTING REQUIREMENTS (40CFR122, APPENDIX D) 5- 2
TABLE 5.1. TESTING REQUIREMENTS FOR ORGANIC TOXIC POLLUTANTS 5- 2
TABLE 5.2. ORGANIC TOXIC POLLUTANTS IN EACH OF FOUR FRACTIONS 5- 4
TABLE 5.3. OTHER TOXIC POLLUTANTS (METALS AND CYANIDE) AND TOTAL PHENOLS 5-6
TABLE 5.4. CONVENTIONAL AND NONCONVENTIONAL POLLUTANTS
REQUIRED TO BE TESTED 5-6
TABLE 5.5. TOXIC POLLUTANTS AND HAZARDOUS SUBSTANCES
REQUIRED TO BE IDENTIFIED 5-7

5.2. IDENTIFICATION OF TEST PROCEDURES (40CFR136.3) 5-10
TABLE 5.6. LIST OF APPROVED BIOLOGICAL TEST PROCEDURES (40CFR136.3 TABLE IA) 5-10
TABLE 5.7. LIST OF APPROVED BIOLOGICAL TEST PROCEDURES (40CFR136.3 TABLE IB) 5-11
TABLE 5.8. LIST OF APPROVED TEST PROCEDURES
FOR NON-PESTICIDE ORGANIC COMPOUNDS 5-25
TABLE 5.9. LIST OF APPROVED TEST PROCEDURES FOR PESTICIDES 5-29
TABLE 5.10. LIST OF APPROVED RADIOLOGIC TEST PROCEDURES (40CFR136.3 TABLE IE) 5-33
TABLE 5.11. REQUIRED CONTAINERS, PRESERVATION TECHNIQUES, AND HOLDING TIMES . 5-34

5.3. METHODS FOR ORGANICS CHEMICAL ANALYSIS OF MUNICIPAL AND
INDUSTRIAL WASTEWATERS (40CFR136, APPENDIX A) 5-38

5.4. REFERENCES (40CFR136.3.b) 5-38

SECTION 6. AIR PROGRAMS

6.l. NATIONAL PRIMARY AND SECONDARY AMBIENT AIR QUALITY STANDARDS (40CFR50) 6- 2
6.1.1. NATIONAL STANDARDS 6- 2
6.1.2. REFERENCE METHODS FOR CRITERIA POLLUTANTS 6-2
TABLE 6.lA. REFERENCE METHODS--ALPHABETICAL ORDER BY CRITERIA POLLUTANTS 6-2
TABLE 6.lB. REFERENCE METHODS--NUMBER ORDER BY METHOD 6-3

6.2. CONTINUOUS EMISSION MONITORS FOR S0₂ (40CFR52, APPENDIX D) 6-4

6.3. NATIONAL STANDARDS FOR NEW STATIONARY SOURCES 6-4
TABLE 6.2A. NATIONAL STANDARDS--ALPHABETICAL ORDER BY INDUSTRY CATEGORY 6-4
TABLE 6.2B. NATIONAL STANDARDS--40CFR60 ORDER BY SUBPART 6-8

6.4. REFERENCE METHODS FOR STATIONARY SOURCE STANDARDS 6-12
TABLE 6.3A. REFERENCE METHODS--ALPHABETICAL ORDER BY METHOD PARAMETER 6-12
TABLE 6.3B. REFERENCE METHODS--NUMBER ORDER BY METHOD 6-15

6.5 NATIONAL EMISSION STANDARDS FOR HAZARDOUS AIR POLLUTANTS (HAPS) (40CFR60) 6-18
TABLE 6.4. HAZARDOUS AIR POLLUTANTS AND THEIR STANDARDS 6-18
TABLE 6.5A TEST METHODS FOR HAPS--ALPHABETICAL ORDER BY POLLUTANT 6-25
TABLE 6.5B. TEST METHODS FOR HAPS--NUMBER ORDER BY METHOD 6-26

6.6. NATIONAL EMISSION STANDARDS FOR HAPs FOR SOURCE CATEGORIES (40CFR63) 6-27
TABLE 6.6. TEST METHODS FOR HAPs 6-27

6.7. TESTS METHODS FOR FUELS AND FUEL ADDITIVES (40CFR80, APPENDIXES A-G) 6-27

SECTION 7. RISK ASSESSMENT PROGRAMS

7.1. INTRODUCTION 7- 2

7.2. INTEGRATED RISK INFORMATION SYSTEM (IRIS) 7- 2
      TABLE 7.1. INTEGRATED RISK INFORMATION SYSTEM COMPOUNDS 7- 2
      TABLE 7.2. REFERENCE AIR CONCENTRATIONS (RAC) (40CFR266, APPENDIX IV) 7-23
      TABLE 7.3. RISK SPECIFIC DOSES (RsD) (10-5) (40CFR266, APPENDIX V) 7-26
      TABLE 7.4. PRODUCTS OF INCOMPLETE COMBUSTION (PICs) FOUND IN STACK EFFLUENTS . 7-29

ACRONYMS AND REFERENCES

ACRONYMS   8-2
REFERENCES   8-4

PREFACE

This book is a guide to obtaining testing information for chemicals that require sampling, analysis, and monitoring under the environmental measurement requirements of various environmental laws and regulations. The book also provides citations for health-related information needed for conducting risk assessment.

An environmental measurement is, qualitatively and/or quantitatively, the determination of chemical species that may cause adverse effects on the environment or human health. It includes sampling, analysis, monitoring, quality control and quality assurance. Risk assessment is defined as: (1) the determination of the kind and degree of hazard posed by an agent (such as a harmful substance); (2) the extent to which a particular group of people has been or may be exposed to the agent; and (3) the present or potential health risk that exists due to the agent. It is the probability of injury, disease, or even death (under specific circumstances) and is a cornerstone of environmental decision-making. Both environmental measurement and risk assessment require the identification of chemical species and their characteristics.

It is well recognized that environmental laws and regulations are the driving forces behind environmental protection. One of the key elements of successful environmental protection is the conduct of environmental measurement and risk assessment studies. However, thousands of chemicals have been regulated by many very complicated environmental laws and regulations. Furthermore, the specific requirements for sampling, analysis and monitoring of these chemicals are randomly distributed throughout the entire 40 CFR (Code of Federal Regulations) from Part 1 to Part 1517. In addition, measurement and risk assessment techniques are either published by EPA or developed by other professional societies. The huge volume of the CFRs and other publications make information searches extremely difficult. To help ease this difficulty, this guide was created. The major areas covered in this book are summarized below:

(1) Section 1 is a summary of the environmental laws and their corresponding regulations in 40 CFR.

(2) Section 2 is a chemical cross-reference for the sampling, analysis, monitoring, and risk assessment regulations under the various major environmental laws.

(3) Sections 3, 4, 5, and 6 list chemicals, their emission standards, and their measurement methods under the requirements of the Resource Conservation and Recovery Act (RCRA), the Safe Drinking Water Act (SDWA), the Clean Water Act (CWA), and the Clean Air Act (CAA), respectively. The requirements for the RCRA, SDWA, CWA, and CAA are also applicable to other environmental laws such as the Superfund Act, the Toxic Substances Control Act, and the Federal Insecticide, Fungicide, and Rodenticide Act.

(4) Section 7 provides sources from which health-related information can be obtained for performing risk assessment calculations.

Many tables have been developed for this book. At the end of each Table, notes are provided to help readers understand the table contents. This book is intended to be a reference book for those who are involved in environmental protection activities. It is an essential tool and will make many environmental jobs much easier!

ABOUT THE AUTHOR

Dr. C.C. Lee is a Research Program Manager at the National Risk Management Research Laboratory of the U.S. Environmental Protection Agency in Cincinnati, Ohio. In addition, he is currently a member of the Policy Review Group to the Center for Clean Technology at the University of California, Los Angeles (UCLA). He is also the Chairman of the Sponsoring Committee to the International Congress on Toxic Combustion Byproducts (ICTCB). He initiated the ICTCB and served as the Chairman of the First and Second Congresses which were held in 1989 and 1991, respectively.

Dr. Lee has more than 20 years experience in conducting various engineering and research projects which often involve multi-environmental issues ranging from clean air and clean water control to solid waste disposal. He has been recognized as a worldwide expert in the thermal treatment of medical and hazardous wastes, witness his leading discussions on medical waste disposal technologies at a Meeting conducted by the Congressional Office of Technology Assessment. Also, at the initiation of the U.S. State Department, he served as head of the U.S. delegation to the Conference on "National Focal Points for the Low- and Non-Waste Technology" (it was sponsored by the United Nations and held in Geneva, Switzerland on August 28-30, 1978). He has been invited to lecture on various issues regarding solid waste disposal in numerous national and international conferences, and he has published more than 100 papers and reports in various environmental areas.

He received a B.S. from the National Taiwan University in 1964, and a M.S. and Ph.D. from the North Carolina State University in 1968 and 1972, respectively. Before joining EPA in 1974, he was an Assistant Professor at the North Carolina State University.

NOTICE

This book was written and edited by Dr. C. C. Lee in his private capacity. No official support or endorsement by the U.S. Environmental Protection Agency is intended nor should be inferred.

Section 1

LIST OF
ENVIRONMENTAL LAWS AND REGULATIONS

SECTION 1. MAJOR ENVIRONMENTAL LAWS AND ENVIRONMENTAL REGULATIONS	
Environmental Laws	Environmental Regulations (CFR Cites)
• National Environmental Education Act Grants	Chapter I, Subchapter C • 40CFR47
• Clean Air Act (CAA) of 1970 • Clean Air Act Amendments (CAAA) of 1990	Chapter I, Subchapter C - Air Programs • 40CFR50-99
• Federal Transit Act	Chapter I, Subchapter C - Air Programs • 40CFR51.390-51.464
• Safe Drinking Water Act (SDWA) of 1974	Chapter I, Subchapter D - Water Programs • 40CFR141-149
• Federal Insecticide, Fungicide, and Rodenticide Act (FIFRA)	Chapter I, Subchapter E - Pesticide Programs • 40CFR150-189
• Atomic Energy Act (AEC) of 1954	Chapter I, Subchapter F - Radiation Protection Programs • 40CFR190-192 also in • 10CFR20.3; 30.4; 40.4 & 70.4
• Noise Control Act (NCA) 1972	Chapter I, Subchapter G - Noise Abatement Programs • 40CFR201-211
• Marine Protection, Research, and Sanctuaries Act (MPRSA) of 1972	Chapter I, Subchapter H - Ocean Dumping • 40CFR220-233
• Solid Waste Disposal Act (SWDA) of 1965 • Resource Conservation Recovery Act (RCRA) of 1976 • Hazardous and Solid Waste Act (HSWA) of 1984	Chapter I, Subchapter I - Solid Wastes • 40CFR240-299
• Comprehensive Environmental Response, Compensation, and Liability Act of 1980 (CERCLA) • Superfund Amendments and Reauthorization Act (SARA) of 1986	Chapter I, Subchapter J - Superfund, Emergency Planning, and Community Right-To-Know Programs • 40CFR300-374
• Federal Water Pollution Control Act (FWPCA) of 1972 • Clean Water Act (CWA) of 1977	Chapter I, Subchapter D - Water Programs • 40CFR100-140 Chapter I, Subchapter N - Effluent Guidelines and Standards • 40CFR400-699
• Toxic Substances Control Act (TSCA) of 1976	Chapter I, Subchapter R - Toxic Substances Control Act • 40CFR700-799
• National Environmental Policy Act (NEPA) 1969 • Environmental Quality Improvement Act (EQIA) 1970	Chapter I, Subchapter V - Council on Environmental Quality • 40CFR1500-1517
• Occupational Safety and Health Act (OSHA) of 1970	Chapter XVII - Occupational Safety and Health Administration Department of Labor • 29CFR1900-1910

Section 2

CHEMICAL CROSS-REFERENCE
FOR SAMPLING, ANALYSIS, MONITORING
AND RISK ASSESSMENT

Alphabetical Order

SAMPLING, ANALYSIS, AND MONITORING METHODS: A GUIDE TO EPA REQUIREMENTS

CAS RN	Alphabetic Order	type[1]	122[2]	136[3]	141[4]	EPA.[5]	258[6]	261[7]	264[8]	266[9]	hap[10]	iris[11]	pic[12]
83-329	Acenaphthene		122-AD-T2/1B	136.3-T1C/1; 136.3-T2			258-A2		264-A9			iris	
208-968	Acenaphthylene		122-AD-T2/2B	136.3-T1C/2; 136.3-T2		91/7	258-A2		264-A9			iris	
30560-191	Acephate											iris	
75-070	Acetaldehyde		122-AD-T5							266-A4	hap	iris	
60-355	Acetamide										hap	iris	
34256-821	Acetochlor											iris	
67-641	Acetone						258-A1/16; A2		264-A9			iris	
75-058	Acetonitrile; Methyl cyanide	v					258-A2	261-A8	264-A9	266-A4	hap	iris	
129-066	Acetonylbenzyl-4-hydroxycoumarin(3-alpha-)	sv						261-A8					
98-862	Acetophenone	sv					258-A2	261-A8	264-A9	266-A4	hap	iris	
75-365	Acetyl chloride	v						261-A8				iris	
591-082	Acetyl-2-thiourea(-1)	sv						261-A8					
53-963	Acetylaminofluorene(2-); 2-AAF	sv					258-A2	261-A8	264-A9		hap		
	Acidity			136.3-T1B/1; 136.3-T2									
	Acifluorfen					91/7						iris	
62476-599	Acifluorfen, sodium											iris	
107-028	Acrolein	v	122-AD-T2/1V	136.3-T1C/3; 136.3-T2			258-A2	261-A8	264-A9	266-A4	hap	iris	
79-061	Acrylamide	v						261-A8		266-A5	hap	iris	
79-107	Acrylic acid	v									hap	iris	
107-131	Acrylonitrile	v	122-AD-T2/2V	136.3-T1C/4; 136.3-T2			258-A1/17; A2	261-A8	264-A9	266-A5	hap*	iris	
	Adhesive and sealants		122-AD-T1										
111-693	Adiponitrile											iris	
1402-682	Aflatoxins							261-A8					
15972-608	Alachlor				141.61.c	91/7						iris	
1596-845	Alar											iris	
116-063	Aldicarb				141.61.c	91/7		261-A8		266-A4		iris	

SECTION 2. CHEMICAL CROSS-REFERENCE FOR SAMPLING, ANALYSIS, MONITORING AND RISK ASSESSMENT - ALPHABETICAL ORDER

CAS RN	Alphabetic Order	type[1]	122[2]	136[3]	141[4]	EPA-[5]	258[6]	261[7]	264[8]	266[9]	hap[10]	iris[11]	pic[12]
1646-884	Aldicarb sulfone				141.61.c	91/7						iris	
1646-873	Aldicarb sulfoxide				141.61.c	91/7							
309-002	Aldrin	sv	122-AD-T2/1P	136.3-T1D/1; 136.3-T2		91/7	258-A2	261-A8	264-A9	266-A5		iris	
	Alkalinity			136.3-T1B/2; 136.3-T2	141.89								
74223-646	Ally											iris	
107-186	Allyl alcohol	v	122-AD-T5					261-A8		266-A4		iris	
107-051	Allyl chloride		122-AD-T5				258-A2	261-A8	264-A9		hap	iris	
	alpha emitters											iris	
	alpha-Counting error, pCi per liter			136.3-T1E/2; 136.3-T2									
	alpha-Emitting radium isotopes												
	alpha-Total, pCi per liter			136.3-T1E/1; 136.3-T2									
7429-905	Aluminum		122-AD-T1									iris	
	Aluminum forming												
20859-738	Aluminum phosphide	solid						261-A8		266-A4		iris	
	Aluminum, total		122-AD-T4	136.3-T1B/3; 136.3-T2									
67485-294	Amdro											iris	
834-128	Ametryn			136.3-T1D/2; 136.3-T2		91/7						iris	
	Amino(2-)-1-methylbenzene (o-Toluidine)												
	Amino(4-)-1-methylbenzene (p-Toluidine)												
92-671	Aminobiphenyl(4-)	sv					258-A2	261-A8	264-A9		hap		
	Aminocarb			136.3-T1D/3; 136.3-T2									
2763-964	Aminomethyl-3-isoxazolol(5-)	sv						261-A8					
504-245	Aminopyridine(4-)							261-A8				iris	
33089-611	Amitraz											iris	
61-825	Amitrole	sv						261-A8					

SAMPLING, ANALYSIS, AND MONITORING METHODS: A GUIDE TO EPA REQUIREMENTS

CAS RN	Alphabetic Order	type[1]	122[2]	136[3]	141[4]	EPA-[5]	258[6]	261[7]	264[8]	266[9]	hap[10]	iris[11]	pic[12]
7664-417	Ammonia											iris	
	Ammonia (as N)			136.3-T1B/4; 136.3-T2									
631-618	Ammonium acetate											iris	
16325-476	Ammonium methacrylate											iris	
7773-060	Ammonium sulfamate											iris	
7803-556	Ammonium vanadate							261-A8					
	Amyl acetate		122-AD-T5										
62-533	Aniline	sv	122-AD-T5					261-A8	264-A9	266-A5	hap	iris	
90-040	Anisidine(ortho-)										hap	iris	
120-127	Anthracene		122-AD-T2/3B	136.3-T1C/5; 136.3-T2		91/7	258-A2		264-A9			iris	
7440-360	Antimony									266-A4		iris	
(Total)	Antimony		122-AD-T3	136.3-T1B/5; 136.3-T2			258-A1/1; A2		264-A9				
7440-360	Antimony and compounds, N.O.S.	metal						261-A8			hap		
1309-644	Antimony trioxide											iris	
74115-245	Apollo											iris	
140-578	Aramite	sv						261-A8	264-A9			iris	
	Arcolein	v						261-A8					
	Aroclor (General screen)					91/7							
12674-112	Aroclor 1016					91/7						iris	
	Aroclor 1221					91/7							
	Aroclor 1232					91/7							
	Aroclor 1242					91/7							
12672-296	Aroclor 1248					91/7						iris	
11097-691	Aroclor 1254					91/7						iris	
	Aroclor 1260					91/7							
7440-382	Arsenic									266-A5			
(Total)	Arsenic		122-AD-T3	136.3-T1B/6; 136.3-T2			258-A1/2; A2		264-A9				

2 - 4

SECTION 2. CHEMICAL CROSS-REFERENCE FOR SAMPLING, ANALYSIS, MONITORING AND RISK ASSESSMENT - ALPHABETICAL ORDER

CAS RN	Alphabetic Order	type[1]	122[2]	136[3]	141[4]	EPA[5]	258[6]	261[7]	264[8]	266[9]	hap[10]	iris[11]	pic[12]
7778-394	Arsenic acid HAsO$_4$	metal						261-A8					
7440-382	Arsenic and compounds, N.O.S.	metal						261-A8			hap*		
7440-382	Arsenic, inorganic											iris	
1303-282	Arsenic pentoxide As$_2$O$_5$	metal						261-A8					
1327-533	Arsenic trioxide As$_2$O$_3$	metal						261-A8					
7784-421	Arsine											iris	
1332-214	Asbestos		122-AD-T5		141.23.a; 141.23.k						hap*	iris	
81405-858	Assert											iris	
76578-148	Assure											iris	
3337-711	Asulam											iris	
	Atraton			136.3-T1D/4; 136.3-T2		91/7							
1912-249	Atrazine			136.3-T1D/5; 136.3-T2	141.61.c	91/7						iris	
492-808	Auramine	sv						261-A8					
	Auto and other laundries		122-AD-T1										
65195-553	Avermectin B1											iris	
115-026	Azaserine	sv						261-A8					
	Azinphos methyl			136.3-T1D/6; 136.3-T2								iris	
103-333	Azobenzene												
	Bacteria: coliform (fecal) in presence of chlorine			136.3-T1A/2; 136.3-T2									
	Bacteria: coliform (fecal)			136.3-T1A/1; 136.3-T2									
	Bacteria: coliform, total in presence of chlorine			136.3-T1A/4; 136.3-T2									
	Bacteria: coliform, total			136.3-T1A/3; 136.3-T2									
	Bacteria: fecal streptococci			136.3-T1A/5; 136.3-T2									
	Barban			136.3-T1D/7; 136.3-T2									
7440-393	Barium									266-A4		iris	

SAMPLING, ANALYSIS, AND MONITORING METHODS: A GUIDE TO EPA REQUIREMENTS

CAS RN	Alphabetic Order	type[1]	122[2]	136[3]	141[4]	EPA-[5]	258[6]	261[7]	264[8]	266[9]	hap[10]	iris[11]	pic[12]
(Total)	Barium		122-AD-T4	136.3-T1B/7; 136.3-T2	141.23.a; 141.23.k		258-A1/3; A2		264-A9				
7440-393	Barium and compounds, N.O.S.	metal						261-A8					
542-621	Barium cyanide	metal						261-A8		266-A4		iris	
	Battery manufacturing		122-AD-T1										
55179-312	Baycor											iris	
114-261	Baygon					91/7						iris	
43121-433	Bayleton											iris	
55219-653	Baytan											iris	
68359-375	Baythroid											iris	
1861-401	Benefin											iris	
17804-352	Benomyl											iris	
25057-890	Bentazon					91/7						iris	
71-432	Benxene									266-A5			
56-553	Benz(a)anthracene	sv				91/7		261-A8				iris	
56-553	Benz(a)anthracene							261-A8		266-A5			
225-514	Benz(c)acridine	sv						261-A8					
98-873	Benzal chloride							261-A8					
100-527	Benzaldehyde											iris	
71-432	Benzene	v	122-AD-T2/3V	136.3-T1C/6; 136.3-T2	141.61.a	91/7	258-A1/18; A2	261-A8	264-A9			iris	pic
71-432	Benzene (including benzene from gasoline)										hap*		
	Benzene, 2-amino-1-methyl							261-A8					
	Benzene, 4-amino-1-methyl							261-A8					
98-873	Benzene, dichloromethyl-	sv						261-A8					
98-055	Benzenearsonic acid	metal						261-A8					
108-985	Benzenethiol	sv						261-A8				iris	
92-875	Benzidine	sv	122-AD-T2/4B	136.3-T1C/7; 136.3-T2				261-A8		266-A5	hap	iris	

SECTION 2. CHEMICAL CROSS-REFERENCE FOR SAMPLING, ANALYSIS, MONITORING AND RISK ASSESSMENT - ALPHABETICAL ORDER

CAS RN	Alphabetic Order	type[1]	122[2]	136[3]	141[4]	EPA-[5]	258[6]	261[7]	264[8]	266[9]	hap[10]	iris[11]	pic[12]
	Benzo(a)anthracene		122-AD-T2/5B	136.3-T1C/8; 136.3-T2									
56-553	Benzo(a)anthracene; Benzanthracene						258-A2	261-A8	264-A9			iris	
50-328	Benzo(a)pyrene	sv	122-AD-T2/6B	136.3-T1C/9; 136.3-T2	141.61.c	91/7	258-A2	261-A8	264-A9			iris	
50-328	Benzo(a)pyrene	266-A5											
205-992	Benzo(b)fluoranthene	sv		136.3-T1C/10; 136.3-T2		91/7	258-A2	261-A8	264-A9			iris	
192-972	Benzo(e)pyrene											iris	
191-242	Benzo(ghi)perylene		122-AD-T2/8B	136.3-T1C/11; 136.3-T2		91/7	258-A2		264-A9			iris	
205-823	Benzo(j)fluoranthene	sv						261-A8				iris	
207-089	Benzo(k)fluoranthene		122-AD-T2/9B	136.3-T1C/12; 136.3-T2		91/7	258-A2	261-A8	264-A9			iris	
	Benzofluoranthene(3,4-)		122-AD-T2/7B										
65-850	Benzoic acid											iris	
	Benzonitrile		122-AD-T5										
106-514	Benzoquinone(p-)	sv						261-A8					
98-077	Benzotrichloride	sv						261-A8			hap	iris	
100-516	Benzyl alcohol						258-A2		264-A9				
	Benzyl butyl phthalate			136.3-T1C/14; 136.3-T2									
100-447	Benzyl chloride	sv	122-AD-T5	136.3-T1C/13; 136.3-T2				261-A8			hap	iris	
(Total)	Beryllium		122-AD-T3	136.3-T1B/8; 136.3-T2			258-A1/4; A2		264-A9				
7440-417	Beryllium									266-A5		iris	
7440-417	Beryllium and compounds, N.O.S.	sv						261-A8			hap*		
13510-491	Beryllium sulfate											iris	
	Beta-counting error, pCi			136.3-T1E/4; 136.3-T2									
	Beta-total, pCi per liter			136.3-T1E/3; 136.3-T2									
319-846	BHC(alpha-)		122-AD-T2/2P	136.3-T1D/8; 136.3-T2			258-A2		264-A9				
319-857	BHC(beta-)		122-AD-T2/3P	136.3-T1D/9; 136.3-T2			258-A2		264-A9				
319-868	BHC(delta-)		122-AD-T2/5P	136.3-T1D/10; 136.3-T2			258-A2		264-A9				

SAMPLING, ANALYSIS, AND MONITORING METHODS: A GUIDE TO EPA REQUIREMENTS

CAS RN	Alphabetic Order	type[1]	122[2]	136[3]	141[4]	EPA-[5]	258[6]	261[7]	264[8]	266[9]	hap[10]	iris[11]	pic[12]
58-899	BHC(gamma-); Lindane		122-AD-T2/4P	136.3-T1D/11; 136.3-T2			258-A2		264-A9				
141-662	Bidrin											iris	
	Biochemical oxygen demand (BOD5)			136.3-T1B/9; 136.3-T2									
82657-043	Biphenthrin											iris	
92-524	Biphenyl										hap		
92-524	Biphenyl(1,1-)											iris	
108-601	Bis(2-chloro-1-methyl-ethyl) ether; 2,2'-dichlorodiisopropyl ether						258-A2		264-A9				
111-911	Bis(2-chloroethoxy) methane	sv	122-AD-T2/10B	136.3-T1C/15; 136.3-T2			258-A2	261-A8	264-A9			iris	
	Bis(2-chloroethoxymethane)												
111-444	Bis(2-chloroethyl) ether		122-AD-T2/11B	136.3-T1C/16; 136.3-T2			258-A2	261-A8	264-A9				
494-031	Bis(2-chloroethyl)-2-naphthylamine(n,n-)	sv						261-A8					
111-444	Bis(2-chloroethyl)ether	sv						261-A8		266-A5			
108-601	Bis(2-chloroisopropyl)ether	sv	122-AD-T2/12B					261-A8					
39638-329	Bis(2-chloroisopropyl) ether											iris	
	bis(2-Ethylhexyl)adipate					91/7							
117-817	Bis(2-ethylhexyl)phthalate (DEHP)	sv	122-AD-T2/13B	136.3-T1C/17; 136.3-T2		91/7	258-A2	261-A8	264-A9	266-A5	hap		pic
111-444	Bis(chloroethyl)ether										hap	iris	
542-881	Bis(chloromethyl)ether	sv						261-A8		266-A5		iris	
80-057	Bisphenol A.											iris	
35400-432	Bolstar											iris	
7440-428	Boron (Boron and Borates only)											iris	
	Boron, total		122-AD-T4	136.3-T1B/10; 136.3-T2									
	Bromacil					91/7							
	Bromide		122-AD-T4	136.3-T1B/11; 136.3-T2									

CAS RN	Alphabetic Order	type[1]	122[2]	136[3]	141[4]	EPA.[5]	258[6]	261[7]	264[8]	266[9]	hap[10]	iris[11]	pic[12]
	Brominated dibenzofurans												
598-312	Bromoacetone	v						261-A8				iris	
	Bromobenzene					91/7							pic
74-975	Bromochloromethane; Chlorobromomethane					91/7	258-A1/19; 258-A2					iris	pic
75-274	Bromodichloromethane; Dibromochloromethane			136.3-T1C/18; 136.3-T2		91/7	258-A1/20; A2		264-A9			iris	
1511-622	Bromodifluoromethane											iris	
101-553	Bromodiphenyl ether(p-)											iris	
74-964	Bromoethane											iris	
75-252	Bromoform; Tribromomethane	v	122-AD-T2/5V	136.3-T1C/19; 136.3-T2		91/7	258-A1/21; A2	261-A8	264-A9		hap	iris	pic
74-839	Bromomethane	v		136.3-T1C/20; 136.3-T2		91/7		261-A8		266-A4		iris	pic
101-553	Bromophenyl phenyl ether(4-)	sv	122-AD-T2/14B	136.3-T1C/21; 136.3-T2			258-A2	261-A8	264-A9			iris	
75-627	Bromotrichloromethane											iris	
1689-845	Bromoxynil											iris	
1689-992	Bromoxynil octanoate											iris	
357-573	Brucine	sv						261-A8				iris	
	Butachlor					91/7							
106-990	Butadiene(1,3-)									266-A5	hap*	iris	
71-363	Butanol(n-)											iris	
	Butanone peroxide(2-)	sv						261-A8					
	Butyl acetate		122-AD-T5										
85-687	Butyl benzyl phthalate; Benzyl butyl phthalate	sv					258-A2	261-A8	264-A9			iris	pic
	Butyl(n-) alcohol												
	Butyl-4,6-dinitrophenol(2-sec-)	sv	122-AD-T5					261-A8					
	Butylamine												

SAMPLING, ANALYSIS, AND MONITORING METHODS: A GUIDE TO EPA REQUIREMENTS

CAS RN	Alphabetic Order	type[1]	122[2]	136[3]	141[4]	EPA-[5]	258[6]	261[7]	264[8]	266[9]	hap[10]	iris[11]	pic[12]
2008-415	Butylate					91/7						iris	
	Butylbenzene(n-)					91/7							
	Butylbenzene(sec-)					91/7							
	Butylbenzene(tert-)					91/7							
85-687	Butylbenzyl phthalate	sv	122-AD-T2/15B					261-A8					
	Butylbenzylphthalate					91/7							
507-200	Butylchloride(t-)											iris	
85-701	Butylphthalyl butylglycolate											iris	
75-605	Cacodylic acid							261-A8				iris	
(Total)	Cadmium		122-AD-T3	136.3-T1B/12; 136.3-T2	141.23.a; 141.23.k; 141.89		258-A1/5; A2		264-A9				
7440-439	Cadmium									266-A5		iris	
7440-439	Cadmium and compounds, N.O.S.	metal						261-A8			hap*		
104098-499	Cadre											iris	
13765-190	Calcium chromate H_2CrO_4	metal						261-A8					
156-627	Calcium cyanamide										hap		
592-018	Calcium cyanide $Ca(CN)_2$	cyanide						261-A8		266-A4		iris	
105-602	Caprolactam										hap	iris	
2425-061	Captafol											iris	
133-062	Captan		122-AD-T5	136.3-T1D/12; 136.3-T2							hap	iris	
63-252	Carbaryl		122-AD-T5	136.3-T1D/13; 136.3-T2		91/7					hap	iris	
1563-662	Carbofuran		122-AD-T5		141.61.c	91/7					hap	iris	
75-150	Carbon disulfide	v	122-AD-T5				258-A1/22; A2	261-A8	264-A9	266-A4	hap	iris	
353-504	Carbon oxyfluoride	v						261-A8				iris	
56-235	Carbon tetrachloride	v	122-AD-T2/6V	136.3-T1C/22; 136.3-T2	141.61.a	91/7	258-A1/23; A2	261-A8	264-A9	266-A5	hap*	iris	pic
	Carbonaceous biochemical oxygen demand (CBOD5)			136.3-T1B/14; 136.3-T2									

SECTION 2. CHEMICAL CROSS-REFERENCE FOR SAMPLING, ANALYSIS, MONITORING AND RISK ASSESSMENT - ALPHABETICAL ORDER

CAS RN	Alphabetic Order	type[1]	122[2]	136[3]	141[4]	EPA.[5]	258[6]	261[7]	264[8]	266[9]	hap[10]	iris[11]	pic[12]
353-504	Carbonyl fluoride											iris	
463-581	Carbonyl sulfide										hap	iris	
	Carbophenothion			136.3-T1D/14; 136.3-T2									
55285-148	Carbosulfan											iris	
5234-684	Carboxin					91/7						iris	
120-809	Catechol										hap	iris	
	Chemical oxygen demand (COD)			136.3-T1B/15; 136.3-T2									
	Chloraform			136.3-T1C/27; 136.3-T2									
75-876	Chloral	v						261-A8		266-A4		iris	
302-170	Chloral hydrate											iris	
133-904	Chloramben					91/7					hap	iria	
305-033	Chlorambucil	sv						261-A8					
57-749	Chlordane				141.61.c								
57-749	Chlordane (alpha and gamma isomers)	sv	122-AD-T2/6P	136.3-T1D/15; 136.3-T2			258-A2	261-A8	264-A9	266-A5	hap	iris	
	Chlordane (Technical)					91/7							
	Chlordane-alpha					91/7							
	Chlordane-gamma					91/7							
	Chloride			136.3-T1B/16; 136.3-T2									
90982-324	Chlorimuron-ethyl											iris	
	Chlorinated benzenes, N.O.S.	v						261-A8					
	Chlorinated biphenyls												
	Chlorinated dibenzo-p-dioxins												
	Chlorinated dibenzodioxins												
	Chlorinated dibenzofurans												
	Chlorinated ethane, N.O.S.	v						261-A8					

CAS RN	Alphabetic Order	type[1]	122[2]	136[3]	141[4]	EPA_[5]	258[6]	261[7]	264[8]	266[9]	hap[10]	iris[11]	pic[12]
	Chlorinated fluorocarbons, N.O.S.	v						261-A8					
	Chlorinated naphthalene, N.O.S.	sv						261-A8					
	Chlorinated phenol, N.O.S.	sv						261-A8					
7782-505	Chlorine										hap	iris	
	Chlorine (free)									266-A4			
506-774	Chlorine cyanide											iris	
10049-044	Chlorine dioxide											iris	
7758-192	Chlorine, total residual		122-AD-T4	136.3-T1B/17; 136.3-T2									
	Chlorite					91/7							
	Chlormeb											iris	
75-683	Chloro-1,1-difluoroethane(1-)											iris	
2837-890	Chloro-1,1,1,2-tetrafluoroethane(2-)											iris	
126-998	Chloro-1,3-butadiene(2-)							261-A8		266-A4		iris	
106-898	Chloro-2,3-epoxypropane(1-)	v						261-A8					
	Chloro-3-methylphenol(4-)			136.3-T1C/23; 136.3-T2									
59-507	Chloro-m-cresol(p-); 4-Chloro-3-methylphenol	sv	122-AD-T2/8A				258-A2	261-A8	264-A9			iris	
107-200	Chloroacetaldehyde	v						261-A8					
79-118	Chloroacetic acid										hap		
532-274	Chloroacetophenone(2-)										hap	iris	
	Chloroalkyl ethers, N.O.S.	v						261-A8					
	Chloroanaphthalene(2-)								264-A9				
106-478	Chloroaniline(p-)	sv					258-A2	261-A8	264-A9			iris	
108-907	Chlorobenzene	v	122-AD-T2/7V	136.3-T1C/24; 136.3-T2		91/7	258-A1/24; A2	261-A8	264-A9		hap	iris	pic
510-156	Chlorobenzilate										hap	iris	
510-156	Chlorobenzilate	sv				91/7	258-A2	261-A8	264-A9				

SECTION 2. CHEMICAL CROSS-REFERENCE FOR SAMPLING, ANALYSIS, MONITORING AND RISK ASSESSMENT - ALPHABETICAL ORDER

CAS RN	Alphabetic Order	type[1]	122[2]	136[3]	141[4]	EPA-[5]	258[6]	261[7]	264[8]	266[9]	hap[10]	iris[11]	pic[12]
	Chlorobiphenyl(2-)					91/7							
109-693	Chlorobutane(1-)											iris	
78-864	Chlorobutane(2-)											iris	
41851-507	Chlorocyclopentadiene											iris	
75-456	Chlorodibromomethane		122-AD-T2/8V									iris	
75-003	Chloroethane; Ethyl chloride		122-AD-T2/9V	136.3-T1C/25; 136.3-T2		91/7	258-A1/25; A2		264-A9				
110-758	Chloroethyl vinyl ether(2-)	v	122-AD-T2/10V	136.3-T1C/26; 136.3-T2				261-A8					
67-663	Chloroform; Trichloromethane	v	122-AD-T2/11V			91/7	258-A1/26; A2	261-A8	264-A9	266-A5	hap*	iris	pic
74-873	Chloromethane	v		136.3-T1C/28; 136.3-T2		91/7		261-A8				iris	
74-873	Chloromethane	266-A5											
107-302	Chloromethyl methyl ether	v						261-A8			hap	iris	
91-587	Chloronaphthalene(2-)	sv	122-AD-T2/16B	136.3-T1C/29; 136.3-T2			258-A2	261-A8					
91-587	Chloronaphthalene(beta-)							261-A8				iris	
95-578	Chlorophenol(2-)	sv	122-AD-T2/1A	136.3-T1C/30; 136.3-T2			258-A2	261-A8	264-A9			iris	
95-578	Chlorophenol(o-)	sv						261-A8					pic
122-883	Chlorophenoxyacetic acid(4-)											iris	
934-736	Chlorophenyl methyl sulfoxide(p-)											iris	
98-571	Chlorophenyl methyl sulfone(p-)											iris	
123-091	Chlorophenyl methyl sulfide(p-)											iris	
7005-72-3	Chlorophenyl phenyl ether(4-)						258-A2						
	Chlorophenyl thiourea(1-)							261-A8					
934-736	Chlorophenyl((1-)o-)thiourea	sv						261-A8					
95-578	Chlorophenyl(1-(o-))							261-A8					
5344-821	Chlorophenyl(4-) phenyl ether		122-AD-T2/17B	136.3-T1C/31; 136.3-T2					264-A9				

SAMPLING, ANALYSIS, AND MONITORING METHODS: A GUIDE TO EPA REQUIREMENTS

CAS RN	Alphabetic Order	type[1]	122[2]	136[3]	141[4]	EPA-[5]	258[6]	261[7]	264[8]	266[9]	hap[10]	iris[11]	pic[12]
126-998	Chloroprene						258-A2	261-A8	264-A9		hap*	iris	
	Chloropropene(3-)							261-A8					
	Chloropropham			136.3-T1D/16; 136.3-T2									
542-767	Chloropropionitrile(3-)	sv						261-A8					
1897-456	Chlorothalonil					91/7						iris	
	Chlorotoluene(2-)					91/7							
	Chlorotoluene(4-)					91/7							
95-498	Chlorotoluene(o-)											iris	
101-213	Chlorpropham					91/7						iris	
2921-882	Chlorpyrifos		122-AD-T5									iris	
5598-130	Chlorpyrifos-methyl											iris	
64902-723	Chlorsulfuron											iris	
(Total)	Chromium		122-AD-T3	136.3-T1B/19; 136.3-T2	141.23.a; 141.23.k		258-A1/6; A2						
7440-473	Chromium and compounds, N.O.S.	metal						261-A8			hap*		
16065-831	Chromium III									266-A4			
7440-473	Chromium VI							261-A8		266-A5			
	Chromium VI dissolved			136.3-T1B/18; 136.3-T2									
16065-831	Chromium(III), insoluble salts											iris	
	Chromium(III), soluble salts											iris	
18540-299	Chromium(VI)											iris	
218-019	Chrysene	sv	122-AD-T2/18B	136.3-T1C/32; 136.3-T2		91/7	258-A2	261-A8	264-A9			iris	
6358-538	Citrus red No. 2	sv						261-A8					
	Coal mining		122-AD-T1										
8007-452	Coal tar creosote	sv						261-A8				iris	
(Total)	Cobalt		122-AD-T4	136.3-T1B/20; 136.3-T2			258-A1/7; A2		264-A9				
7440-484	Cobalt										hap	iris	

SECTION 2. CHEMICAL CROSS-REFERENCE FOR SAMPLING, ANALYSIS, MONITORING AND RISK ASSESSMENT - ALPHABETICAL ORDER

CAS RN	Alphabetic Order	type[1]	122[2]	136[3]	141[4]	EPA-[5]	258[6]	261[7]	264[8]	266[9]	hap[10]	iris[11]	pic[12]
	Coil coating		122-AD-T1										
8007-452	Coke oven emissions										hap*	iris	
	Coliform				141.21								
	Color		122-AD-T4										
	Color platinum cobalt units or dominant wavelength, hue, luminance purity			136.3-T1B/21; 136.3-T2									
	Colorimetric, automated ascorbic acid				141.89								
	Conductance, specific conducance			136.3-T1B/64; 136.3-T2									
	Conductivity				141.89								
7440-508	Copper											iris	
(Total)	Copper		122-AD-T3	136.3-T1B/22; 136.3-T2	141.89		258-A1/8; A2		264-A9				
544-923	Copper cyanide	cyanide						261-A8		266-A4		iris	
	Copper forming		122-AD-T1										
56-724	Coumaphos		122-AD-T5									iris	
8001-589	Creosote	sv						261-A8				iris	
1319-773	Cresol		122-AD-T5							266-A4			
1319-773	Cresol (Cresylic acid)	sv						261-A8					
108-394	Cresol(m-); 3-methylphenol						258-A2		264-A9		hap		
95-487	Cresol(o-); 2-methylphenol						258-A2		264-A9		hap		
106-445	Cresol(p-); 4-methylphenol						258-A2		264-A9		hap		
1319-773	Cresols/cresylic acid (isomers and mixture)										hap		
4170-303	Crotonaldehyde	sv	122-AD-T5					261-A8					
123-739	Crotonaldehyde									266-A4		iris	
98-828	Cumene										hap	iris	
21725-462	Cyanazine											iris	
57-125	Cyanide						258-A2						

SAMPLING, ANALYSIS, AND MONITORING METHODS: A GUIDE TO EPA REQUIREMENTS

CAS RN	Alphabetic Order	type[1]	122[2]	136[3]	141[4]	EPA.[5]	258[6]	261[7]	264[8]	266[9]	hap[10]	iris[11]	pic[12]
	Cyanide amendable to chlorination			136.3-T1B/24; 136.3-T2									
	Cyanide compounds										hap		
57-125	Cyanide, free									266-A4		iris	
	Cyanide, total		122-AD-T3	136.3-T1B/23; 136.3-T2					264-A9				
	Cyanides (soluble salts and complexes) N.O.S.	cyanide						261-A8					
460-195	Cyanogen	gas						261-A8		266-A4		iris	
506-683	Cyanogen bromide	gas						261-A8		266-A4		iris	
506-774	Cyanogen chloride	gas						261-A8					
14901-087	Cycasin							261-A8					
1134-232	Cycloate					91/7						iris	
108-941	Cyclohexane		122-AD-T5										
	Cyclohexanone											iris	
131-895	Cyclohexyl-4,6-dinitrophenol(2-)	sv						261-A8					
108-918	Cyclohexylamine											iris	
50-180	Cyclophosphamide							261-A8					
68085-858	Cyhalothrin/Karate											iris	
52315-078	Cypermethrin											iris	
66215-278	Cyromazine											iris	
94-757	D(2,4-)		122-AD-T5	136.3-T1D/17; 136.3-T2	141.61.c			261-A8			hap		
	D(2,4-), salts, esters							261-A8					
94-757	D(2,4-); 2,4-Dichlorophenoxyacetic acid						258-A2		264-A9				
1861-321	Dacthal											iris	
75-990	Dalapon				141.61.c	91/7							
75-990	Dalapon, sodium salt											iris	
39515-418	Danitol											iris	

SECTION 2. CHEMICAL CROSS-REFERENCE FOR SAMPLING, ANALYSIS, MONITORING AND RISK ASSESSMENT - ALPHABETICAL ORDER

CAS RN	Alphabetic Order	type[1]	122[2]	136[3]	141[4]	EPA[5]	258[6]	261[7]	264[8]	266[9]	hap[10]	iris[11]	pic[12]
20830-813	Daunomycin	sv						261-A8					
	DCPA mono and diacid metabolites					91/7							
72-548	DDD	sv						261-A8					
	DDD(1,1-dichloro-2,2-bis(p-chlorophenyl)ethane)(4,4'-)		122-AD-T2/9P	136.3-T1D/18; 136.3-T2		91/7							
72-548	DDD(4,4'-)								264-A9				
72-548	DDD(4,4-)						258-A2						
3547-044	DDE	sv									hap		
72-559	DDE	sv				91/7		261-A8					
	DDE(1,1-dichloro-2,2-bis(p-chlorophenyl)ethylene)(4,4'-)		122-AD-T2/8P	136.3-T1D/19; 136.3-T2									
	DDE(4,4'-)								264-A9				
72-559	DDE(4,4-)						258-A2						
50-293	DDT	sv						261-A8		266-A5			
	DDT(1,1,1-trichloro-2,2-bis(p-chlorophenyl)ethane)(4,4'-)					91/7							
	DDT(4,4'-)		122-AD-T2/7P	136.3-T1D/20; 136.3-T2					264-A9				
50-293	DDT(4,4-)						258-A2						
1163-195	Decabromodiphenyl ether				141.61.c							iris	
52918-635	Deltamethrin				141.61.c							iris	
	Dementon-o			136.3-T1D/21; 136.3-T2									
	Dementon-s			136.3-T1D/22; 136.3-T2									
8065-483	Demeton											iris	
103-231	Di(2-ethylhexyl)adipate				141.61.c							iris	
117-817	Di(2-ethylhexyl)phthalate				141.61.c							iris	
84-742	Di-n-butyl phthalate	sv	122-AD-T2/26B	136.3-T1C/51; 136.3-T2		91/7	258-A2	261-A8	264-A9	266-A4			
117-840	Di-n-octyl phthalate	sv	122-AD-T2/29B	136.3-T1C/52; 136.3-T2			258-A2	261-A8	264-A9			iris	
621-647	Di-n-propylnitrosamine	sv						261-A8					

SAMPLING, ANALYSIS, AND MONITORING METHODS: A GUIDE TO EPA REQUIREMENTS

CAS RN	Alphabetic Order	type[1]	122[2]	136[3]	141[4]	EPA.[5]	258[6]	261[7]	264[8]	266[9]	hap[10]	iris[11]	pic[12]
2303-164	Diallate	sv					258-A2	261-A8	264-A9				
95-807	Diaminotoluene(2,4-)											iris	
333-415	Diazinon		122-AD-T5	136.3-T1D/23; 136.3-T2		91/7						iris	
334-883	Diazomethane										hap	iris	
226-368	Dibenz(ah)acridine	sv						261-A8					
53-703	Dibenz(ah)anthracene	sv				91/7	258-A2	261-A8	264-A9			iris	
224-420	Dibenz(aj)acridine	sv						261-A8					
5385-751	Dibenzo(ae)fluoranthene											iris	
192-654	Dibenzo(ae)pyrene	sv						261-A8					
53-703	Dibenzo(ah)anthracene		122-AD-T2/19B	136.3-T1C/33; 136.3-T2						266-A5			
189-640	Dibenzo(ah)pyrene	sv						261-A8					
189-559	Dibenzo(ai)pyrene	sv						261-A8					
194-592	Dibenzo(cg)carbazole(7h-)	sv						261-A8					
132-649	Dibenzofuran	v					258-A2		264-A9		hap	iris	
96-128	Dibromo-3-chloropropane(1,2-); DBCP					91/7	258-A1/28; A2	261-A8	264-A9	266-A5	hap	iris	
106-376	Dibromobenzene(1,4-)											iris	
124-481	Dibromochloromethane; Chlorodibromomethane			136.3-T1C/34; 136.3-T2		91/7	258-A1/27; A2		264-A9			iris	
96-128	Dibromochloropropane				141.61.c								
594-183	Dibromodichloromethane											iris	
2050-477	Dibromodiphenyl ether(p,p'-)											iris	
106-934	Dibromoethane(1,2-); Ethylene dibromide	v				91/7	258-A1/29; A2	261-A8	264-A9	266-A5		iris	
74-953	Dibromomethane	v				91/7		261-A8					
74-953	Dibromomomethane										hap		
84-742	Dibutyl phthalate							261-A8			hap	iris	
1918-009	Dicamba		122-AD-T5	136.3-T1D/24; 136.3-T2		91/7						iris	
	Dichlobenil		122-AD-T5										

CAS RN	Alphabetic Order	type[1]	122[2]	136[3]	141[4]	EPA[5]	258[6]	261[7]	264[8]	266[9]	hap[10]	iris[11]	pic[12]
	Dichlofenthion			136.3-T1D/25; 136.3-T2									
	Dichlone		122-AD-T5										pic
	Dichloran			136.3-T1D/26; 136.3-T2									
1717-006	Dichloro-1-fluoroethane(1,1-)											iris	
764-410	Dichloro-2-butene(1,4-)	v						261-A8					
	Dichloro-2-butene(cis-1,4-)	v											pic
110-576	Dichloro-2-butene(trans-1,4-)						258-A1/32; A2		264-A9				
306-832	Dichloro-2,2,2-trifluoroethane(1,1-)											iris	
79-436	Dichloroacetic acid	v						261-A8				iris	
25321-226	dichlorobenzene, N.O.S.	v											
	Dichlorobenzene (meta, ortho and para isomers)	v						261-A8					
95-501	Dichlorobenzene(1,2-)		122-AD-T2/20B	136.3-T1C/35; 136.3-T2		91/7						iris	
541-731	Dichlorobenzene(1,3-)		122-AD-T2/21B	136.3-T1C/36; 136.3-T2		91/7						iris	
106-467	Dichlorobenzene(1,4-)		122-AD-T2/22B	136.3-T1C/37; 136.3-T2		91/7					hap	iris	
541-731	Dichlorobenzene(m-); 1,3-Dichlorobenzene	sv					258-A2	261-A8	264-A9				pic
95-501	Dichlorobenzene(o-); 1,2-Dichlorobenzene	sv			141.61.a		258-A1/30; A2	261-A8	264-A9	266-A4			pic
106-467	Dichlorobenzene(p-); 1,4-Dichlorobenzene	sv			141.61.a		258-A1/31; A2	261-A8	264-A9	266-A4			pic
91-941	Dichlorobenzidine(3,3'-)	sv	122-AD-T2/23B	136.3-T1C/38; 136.3-T2			258-A2	261-A8	264-A9		hap	iris	
	Dichlorobenzoic acid(3,5-)					91/7							
	Dichlorobiphenyl(2,3-)					91/7							
	Dichlorobromomethane		122-AD-T2/12V										
75-718	Dichlorodifluoromethane	v		136.3-T1C/39; 136.3-T2		91/7	258-A2	261-A8	264-A9	266-A4		iris	
72-548	Dichlorodiphenyl dichloroethane(p,p'-)											iris	

SAMPLING, ANALYSIS, AND MONITORING METHODS: A GUIDE TO EPA REQUIREMENTS

CAS RN	Alphabetic Order	type[1]	122[2]	136[3]	141[4]	EPA-[5]	258[6]	261[7]	264[8]	266[9]	hap[10]	iris[11]	pic[12]
72-559	Dichlorodiphenyldichloroethylene(p,p'-)											iris	
50-293	Dichlorodiphenyltrichloroethane(p,p'-)											iris	
75-343	Dichloroethane(1,1-); Ethylidene chloride	v	122-AD-T2/14V	136.3-T1C/40; 136.3-T2		91/7	258-A1/33; A2	261-A8	264-A9	266-A5		iris	
107-062	Dichloroethane(1,2-); Ethylele dichloride	v	122-AD-T2/15V	136.3-T1C/41; 136.3-T2	141.61.a	91/7	258-A1/34; A2	261-A8	264-A9	266-A5		iris	
	Dichloroethene(1,1-)			136.3-T1C/42; 136.3-T2		91/7							
	Dichloroethene(cis-1,2-)					91/7							
	Dichloroethene(trans-1,2-)	v	122-AD-T2/26V	136.3-T1C/43; 136.3-T2		91/7		261-A8					
111-444	Dichloroethyl ether							261-A8			hap		
25323-302	Dichloroethylene, N.O.S.	v						261-A8					
75-354	Dichloroethylene(1,1-); 1,1-Dichloroethene; Vinylidene chloride	v	122-AD-T2/16V		141.61.a	91/7	258-A1/35; A2	261-A8	264-A9	266-A5		iris	
156-605	Dichloroethylene(1,2-)							261-A8					
156-592	Dichloroethylene(cis-1,2-); cis-1,2-Dichloroethene				141.61.a		258-A1/36; A2					iris	
156-605	Dichloroethylene(trans-1,2-)				141.61.a		258-A1/37; A2		264-A9			iris	
108-601	Dichloroisopropyl ether							261-A8					
75-092	Dichloromethane	v			141.61.a			261-A8				iris	
111-911	Dichloromethoxy ethane							261-A8					
542-881	Dichloromethyl ether							261-A8					
120-832	Dichlorophenol(2,4-)	sv	122-AD-T2/2A	136.3-T1C/44; 136.3-T2			258-A2	261-A8	264-A9	266-A4		iris	
87-650	Dichlorophenol(2,6-)	sv					258-A2	261-A8	264-A9				
94-826	Dichlorophenoxy(4-(2,4-))butyric acid; 2,4-DB					91/7						iris	
94-757	Dichlorophenoxyacetic acid(2,4-); 2,4-D					91/7						iris	
	Dichlorophenoxyacetic acid												
696-286	Dichlorophenylarsine	metal						261-A8					

CAS RN	Alphabetic Order	type[1]	122[2]	136[3]	141[4]	EPA-[5]	258[6]	261[7]	264[8]	266[9]	hap[10]	iris[11]	pic[12]
120-365	Dichloroprop											iris	
26638-197	Dichloropropane, N.O.S.	v						261-A8					
78-875	Dichloropropane(1,2-); Propylene dichloride	v	122-AD-T2/17V	136.3-T1C/45; 136.3-T2	141.61.a	91/7	258-A1/38; A2	261-A8	264-A9			iris	
142-289	Dichloropropane(1,3-); Trimethylene dichloride					91/7	258-A2						
542-756	Dichloropropane(1,3-) N.O.S.							261-A8					
594-207	Dichloropropane(2,2-); Isopropylidene chloride					91/7	258-A2						
	Dichloropropanol												
26545-733	Dichloropropanol, N.O.S.	sv						261-A8				iris	
616-239	Dichloropropanol(2,3-)												
26952-238	Dichloropropene, N.O.S.	v						261-A8					
563-586	Dichloropropene(1,1-)					91/7	258-A2						
542-756	Dichloropropene(1,3-)	v						261-A8		266-A5	hap	iris	
10061-015	Dichloropropene(cis-1,3-)						258-A1/39						
10061-015	Dichloropropene(cis-1,3-)			136.3-T1C/46; 136.3-T2		91/7	258-A2		264-A9				
10061-026	Dichloropropene(trans-1,3-)			136.3-T1C/47; 136.3-T2		91/7	258-A1/40; A2		264-A9				
	Dichloropropionic(2,2-) acid		122-AD-T5										
	Dichloropropylene(1,3-)		122-AD-T2/18V			91/7							
	Dichlorprop					91/7					hap	iris	
62-737	Dichlorvos		122-AD-T5			91/7							
	Dichlrophenoxyacetic(2,4-) acid	sv		136.3-T1D/27; 136.3-T2				261-A8					
115-322	Dicofol	sv	122-AD-T2/10P	136.3-T1D/28; 136.3-T2		91/7	258-A2	261-A8	264-A9	266-A5		iris	
60-571	Dieldrin	sv						261-A8				iris	
1464-535	Diepoxybutane(1,2:3,4-)	sv						261-A8					
Not found	Diesel engine emissions											iris	
111-422	Diethanolamine										hap		

SAMPLING, ANALYSIS, AND MONITORING METHODS: A GUIDE TO EPA REQUIREMENTS

CAS RN	Alphabetic Order	type[1]	122[2]	136[3]	141[4]	EPA-[5]	258[6]	261[7]	264[8]	266[9]	hap[10]	iris[11]	pic[12]
	Diethyl amine		122-AD-T5										
297-972	Diethyl o-pyrazinyl phosphorothioate(o,o-)	sv						261-A8					
84-662	Diethyl phthalate	sv	122-AD-T2/24B	136.3-T1C/48; 136.3-T2			258-A2	261-A8	264-A9	266-A4		iris	pic
	Diethyl s-methyl ester of phosphorodithioic acid(o,o-)	sv						261-A8					
64-675	Diethyl sulfate										hap	iris	
121-697	Diethyl(n,n-) aniline										hap		
297-972	Diethyl-o-2-pyrazinyl phosphorothioate(o,o-); Thionazin						258-A2		264-A9				
297-972	Diethyl-o-pyrazinyl(o,o-)											iris	
311-455	Diethyl-p-nitrophenylphosphate											iris	
692-422	Diethylarsine	metal						261-A8					
693-210	Diethylene glycol dinitrate											iris	
112-345	Diethylene glycol monobutyl ether											iris	
124-174	Diethylene glycol monobutyl ether acetate											iris	
123-911	Diethyleneoxide(1,4-)							261-A8					
1615-801	Diethylhydrazine(1,2-)											iris	
1615-801	Diethylhydrazine(n,n'-)	sv						261-A8					
	Diethylphosphoric acid, o-p-nitrophenyl ester	sv						261-A8					
	Diethylphthalate					91/7							
56-531	Diethylstilbesterol	sv						261-A8		266-A5			
43222-486	Difenzoquat											iris	
35367-385	Diflubenzuron											iris	
75-376	Difluoroethane(1,1-)											iris	
94-586	Dihydrosafrole	sv						261-A8					

SECTION 2. CHEMICAL CROSS-REFERENCE FOR SAMPLING, ANALYSIS, MONITORING AND RISK ASSESSMENT - ALPHABETICAL ORDER

CAS RN	Alphabetic Order	type[1]	122[2]	136[3]	141[4]	EPA.[5]	258[6]	261[7]	264[8]	266[9]	hap[10]	iris[11]	pic[12]
55-914	Diisopropylfluorophosphate (DFP)	sv						261-A8					
1445-756	Diisopropyl methylphosphonate											iris	
81777-891	Dimethazone											iris	
55290-647	Dimethipin											iris	
60-515	Dimethoate	sv					258-A2	261-A8	264-A9			iris	
119-904	Dimethoxybenzidine(3,3'-)	sv						261-A8		266-A4	hap		
70-382	Dimethrin											iris	
	Dimethyl amine		122-AD-T5										
60-117	Dimethyl aminoazobenzene										hap	iris	
79-447	Dimethyl carbamoyl chloride										hap		
68-122	Dimethyl formamide										hap		
756-796	Dimethyl methylphosphonate											iris	
131-113	Dimethyl phthalate	sv	122-AD-T2/25B	136.3-T1C/50; 136.3-T2			258-A2	261-A8	264-A9		hap	iris	pic
77-781	Dimethyl sulfate	sv						261-A8			hap	iris	
120-616	Dimethyl terephthalate											iris	
57-147	Dimethyl(1,1-) hydrazine					91/7					hap		
119-937	Dimethyl(3,3'-) benzidine										hap		
	Dimethyl(3,3-)-1-(methylthio)-2-butanone, o-((methylamino)carbonyl)oxi me, see thiofanox							261-A8					
	Dimethyl(7,12-)benz(a)anthracene								264-A9				
	Dimethyl(alpha,alpha-)phenethylamine								264-A9				
	Dimethyl-2,3,5,6-tetrachloroterephthalate; DCPA; Dacthal												
124-403	Dimethylamine											iris	

2 - 23

SAMPLING, ANALYSIS, AND MONITORING METHODS: A GUIDE TO EPA REQUIREMENTS

CAS RN	Alphabetic Order	type[1]	122[2]	136[3]	141[4]	EPA-[5]	258[6]	261[7]	264[8]	266[9]	hap[10]	iris[11]	pic[12]
	Dimethylamino(p-)azobenzene								264-A9				
60-117	Dimethylaminoazobenzene(p-)	sv					258-A2	261-A8					
57-97-6	Dimethylbenz(a)anthracene(7,12-)						258-A2						
57-976	Dimethylbenz(a)anthracene(7,12-)	sv						261-A8					
119-937	Dimethylbenzidine(3,3'-)	sv					258-A2	261-A8	264-A9			iris	
79-447	Dimethylcarbamoyl chloride							261-A8				iris	
68-122	Dimethylformamide(n,n-)											iris	
	Dimethylhydrazine(1,1-) (UDMH)												
540-738	Dimethylhydrazine(1,2-)											iris	
57-147	Dimethylhydrazine(1,1-)	sv						261-A8				iris	
540-738	Dimethylhydrazine(1,2-)	sv						261-A8					
62-759	Dimethylnitrosamine									266-A5			
122-098	Dimethylphenethylamine(alpha, alpha-)	sv						261-A8					
105-679	Dimethylphenol(2,4-); m-Xylenol	sv	122-AD-T2/3A	136.3-T1C/49; 136.3-T2			258-A2	261-A8	264-A9			iris	pic
576-261	Dimethylphenol(2,6-)											iris	
95-658	Dimethylphenol(3,4-)											iris	
	Dimethylphthalate					91/7							
	Dinitrotoluene(2,4-)			136.3-T1C/54; 136.3-T2									
	Dinitro(4,6-)-o-cresol								264-A9				
	Dinitro(4,6-)-o-cresol, and salts										hap		
534-52-1	Dinitro-o-cresol 4,6-Dinitro-2-methylphenol(4,6-)						258-A2						
	Dinitro-o-cresol salts(4,6-)	sv						261-A8					
534-521	Dinitro-o-cresol(4,6-)	sv	122-AD-T2/4A					261-A8				iris	

SECTION 2. CHEMICAL CROSS-REFERENCE FOR SAMPLING, ANALYSIS, MONITORING AND RISK ASSESSMENT - ALPHABETICAL ORDER

CAS RN	Alphabetic Order	type[1]	122[2]	136[3]	141[4]	EPA[5]	258[6]	261[7]	264[8]	266[9]	hap[10]	iris[11]	pic[12]
131-895	Dinitro-o-cyclohexyl phenol(4,6-)											iris	
	Dinitrobenzene												
25154-545	Dinitrobenzene, N.O.S.	sv						261-A8				iris	
99-650	Dinitrobenzene(m-)						258-A2		264-A9			iris	
528-290	Dinitrobenzene(o-)											iris	
	Dinitrophenol(2,3-)			136.3-T1C/53; 136.3-T2									
51-285	Dinitrophenol(2,4-)	sv	122-AD-T2/5A				258-A2	261-A8	264-A9	266-A4	hap	iris	
	Dinitrotoluene mixture, 2,4-/2,6-											iris	
121-142	Dinitrotoluene(2,4-)	sv	122-AD-T2/27B				258-A2	261-A8	264-A9	266-A5	hap	iris	
606-202	Dinitrotoluene(2,6-)	sv	122-AD-T2/28B	136.3-T1C/55; 136.3-T2			258-A2	261-A8	264-A9			iris	
88-857	Dinoseb; DNBP; 2-sec-Butyl-4,6-dinitrophenol				141.61.c	91/7	258-A2	261-A8	264-A9	266-A4			
	Dintrobenzene		122-AD-T5										
123-911	Dioxane(1,4-)	v					258-A2	261-A8	264-A9	266-A5	hap	iris	
	Dioxathion			136.3-T1D/29; 136.3-T2									
957-517	Diphenamid					91/7							
122-394	Diphenylamine	sv					258-A2	261-A8	264-A9	266-A4		iris	
122-667	Diphenylhydrazine(1,2-)	sv						261-A8		266-A5	hap	iris	
	Diphenylhydrazine(1,2-) (as azobenzene)		122-AD-T2/30B										
85-007	Diquat		122-AD-T5		141.61.c							iris	
1937-377	Direct black 38											iris	
2602-462	Direct blue 6											iris	
16071-866	Direct brown 95											iris	
298-044	Disulfoton	sv	122-AD-T5	136.3-T1D/30; 136.3-T2		91/7	258-A2	261-A8	264-A9			iris	
	Disulfoton sulfone					91/7							
	Disulfoton sulfoxide					91/7							

SAMPLING, ANALYSIS, AND MONITORING METHODS: A GUIDE TO EPA REQUIREMENTS

CAS RN	Alphabetic Order	type[1]	122[2]	136[3]	141[4]	EPA-[5]	258[6]	261[7]	264[8]	266[9]	hap[10]	iris[11]	pic[12]
505-293	Dithiane(1,4-)											iris	
541-537	Dithiobiuret							261-A8					
	Dithiobiuret(2,4-)	sv						261-A8					
330-541	Diuron		122-AD-T5	136.3-T1D/31; 136.3-T2								iris	
2439-103	Dodine											iris	
95465-999	Ebufos											iris	
	Electric and electronic components		122-AD-T1										
	Electroplating		122-AD-T1										
115-297	Endosulfan	sv						261-A8		266-A4		iris	
959-988	Endosulfan I			136.3-T1D/32; 136.3-T2		91/7	258-A2		264-A9				
33213-659	Endosulfan II			136.3-T1D/33; 136.3-T2		91/7	258-A2		264-A9				
1031-078	Endosulfan sulfate		122-AD-T2/13P	136.3-T1D/34; 136.3-T2		91/7	258-A2		264-A9				
	Endosulfan(alpha-)		122-AD-T2/11P										
	Endosulfan(beta-)		122-AD-T2/12P										
145-733	Endothall				141.61.c								
72-208	Endrin	sv	122-AD-T2/14P	136.3-T1D/35; 136.3-T2	141.61.c	91/7	258-A2	261-A8	264-A9	266-A4		iris	
7421-934	Endrin aldehyde		122-AD-T2/15P	136.3-T1D/36; 136.3-T2		91/7	258-A2		264-A9				
	Endrin metabolites	sv						261-A8					
Not found	Environmental Tobacco Smoke											iris	
106-898	Epichlorohydrin		122-AD-T5	136.3-T1C/56; 136.3-T2				261-A8		266-A5	hap*	iris	
51-434	Epinephrine							261-A8					
106-887	Epoxybutane(1,2-)										hap	iris	
6108-107	Epsilon-Hexachlorocyclohexane											iris	
16672-870	Ethephon											iris	
563-122	Ethion		122-AD-T5	136.3-T1D/37; 136.3-T2								iris	
64529-562	Ethiozin											iris	

SECTION 2. CHEMICAL CROSS-REFERENCE FOR SAMPLING, ANALYSIS, MONITORING AND RISK ASSESSMENT - ALPHABETICAL ORDER

CAS RN	Alphabetic Order	type[1]	122[2]	136[3]	141[4]	EPA.[5]	258[6]	261[7]	264[8]	266[9]	hap[10]	iris[11]	pic[12]
80844-071	Ethofenprox											iris	
	Ethoprop					91/7							
110-805	Ethoxyethanol(2-)					91/7						iris	
	Ethyl dipropylthiocarbamate(s-); EPTC											iris	
759-944	Ethyl dipropylthiocarbamate(s-)											iris	
141-786	Ethyl acetate											iris	
140-885	Ethyl acrylate										hap	iris	
100-414	Ethyl benzene										hap	iris	
51-796	Ethyl carbamate; urethane	sv						261-A8			hap		
75-003	Ethyl chloride							261-A8			hap	iris	
107-120	Ethyl cyanide	cyanide						261-A8				iris	
60-297	Ethyl ether												
97-632	Ethyl methacrylate	sv					258-A2	261-A8	264-A9			iris	
62-500	Ethyl methanesulfonate	sv					258-A2	261-A8	264-A9				
2104-645	Ethyl p-nitrophenyl phenylphosphorothioate											iris	
100-414	Ethylbenzene		122-AD-T2/19V	136.3-TIC/57; 136.3-T2	141.61.a	91/7	258-A1/41; A2		264-A9			iris	
107-153	Ethylene diamine		122-AD-T5									iris	
106-934	Ethylene dibromide		122-AD-T5		141.61.c			261-A8		266-A5	hap		
107-062	Ethylene dichloride							261-A8			hap*		
107-211	Ethylene glycol										hap	iris	
111-762	Ethylene glycol monobutyl ether											iris	
111-159	Ethylene glycol monoethyl ether acetate											iris	
110-805	Ethylene glycol monoethyl ether							261-A8					
151-564	Ethylene imine										hap		

SAMPLING, ANALYSIS, AND MONITORING METHODS: A GUIDE TO EPA REQUIREMENTS

CAS RN	Alphabetic Order	type[1]	122[2]	136[3]	141[4]	EPA[5]	258[6]	261[7]	264[8]	266[9]	hap[10]	iris[11]	pic[12]
75-218	Ethylene oxide	v						261-A8		266-A5	hap*		
96-457	Ethylene thiourea										hap	iris	
	Ethylenebisdithiocarbamic acid, salts and esters							261-A8					
111-546	Ethylenebisdithiocarbamic acid							261-A8					
151-564	Ethyleneimine	sv						261-A8				iris	
96-457	Ethylenethiourea	sv						261-A8					
	Ethylhexyl(n-(2-))bicyclo(2.2.1)-5-heptene-2,3-dicarboximide (MGK-264)					91/7							
75-343	Ethylidene dichloride							261-A8			hap		
84-720	Ethylphthalyl ethylglycolate											iris	
	Etridiazole					91/7							
101200-480	Explosives manufacturing		122-AD-T1										
	Express											iris	
52-857	Famphur	sv					258-A2	261-A8	264-A9			iris	
	Fecal coliform		122-AD-T4										
22224-926	Fenamiphos					91/7						iris	
	Fenarimol					91/7							
122-145	Fenitrothion			136.3-T1D/38; 136.3-T2								iris	
	Fenuron			136.3-T1D/39; 136.3-T2									
	Fenuron-TCA												
	Fine mineral fibers										hap		
2164-172	Fluometuron		122-AD-T2/31B									iris	
206-440	Fluoranthene	sv	122-AD-T2/32B	136.3-T1C/58; 136.3-T2			258-A2	261-A8	264-A9			iris	pic
86-737	Fluorene			136.3-T1C/59; 136.3-T2		91/7	258-A2		264-A9			iris	
	Fluoride, total		122-AD-T4	136.3-T1B/25; 136.3-T2									

CAS RN	Alphabetic Order	type[1]	122[2]	136[3]	141[4]	EPA-[5]	258[6]	261[7]	264[8]	266[9]	hap[10]	iris[11]	pic[12]
7782-414	Fluorine	gas						261-A8		266-A4			
7782-414	Fluorine (soluble fluoride)											iris	
640-197	Fluoroacetamide							261-A8					
640-197	Fluoroacetamide(2-)							261-A8				iris	
62-748	Fluoroacetic acid, sodium salt	sv						261-A8					
59756-604	Fluridone					91/7						iris	
56425-913	Flurprimidol											iris	
66332-965	Flutolanil											iris	
69409-945	Fluvalinate											iris	
133-073	Folpet											iris	
72178-020	Fomesafen											iris	
944-229	Fonofos											iris	
50-000	Formaldehyde	v	122-AD-T5					261-A8		266-A5	hap	iris	
23422-539	Formetanate hydrochloride											iris	
64-186	Formic acid							261-A8		266-A4		iris	
39148-248	Fosetyl-al											iris	
	Foundries		122-AD-T1										
110-009	Furan											iris	
98-011	Furfural		122-AD-T5									iris	
60568-050	Furmecyclox											iris	
69806-504	Fusilade											iris	
77182-822	Glufosinate-ammonium											iris	
765-344	Glycidaldehyde	sv										iris	
765-344	Glycidylaldehyde							261-A8		266-A4		iris	
	Glycol ethers										hap		
1071-836	Glyphosate				141.61.c							iris	
	Gold, total			136.3-T1B/26; 136.3-T2									

2 - 29

SAMPLING, ANALYSIS, AND MONITORING METHODS: A GUIDE TO EPA REQUIREMENTS

CAS RN	Alphabetic Order	type[1]	122[2]	136[3]	141[4]	EPA-[5]	258[6]	261[7]	264[8]	266[9]	hap[10]	iris[11]	pic[12]
	Gross alpha and gross beta												
	Gum and wood chemicals		122-AD-T1										
86-500	Guthion		122-AD-T5									iris	
100784-201	Halomethanes, N.O.S.	v						261-A8				iris	
	Halosulfuron methyl											iris	
69806-402	Haloxyfop-methyl												
	Hardness, total			136.3-T1B/27; 136.3-T2									
79277-273	Harmony											iris	
76-448	Heptachlor	sv	122-AD-T2/16P	136.3-T1D/40; 136.3-T2	141.61.c	91/7	258-A2	261-A8	264-A9	266-A5	hap	iris	
1024-573	Heptachlor epoxide		122-AD-T2/17P	136.3-T1D/41; 136.3-T2	141.61.c	91/7	258-A2	261-A8	264-A9	266-A5		iris	
	Heptachlor epoxide (alpha, beta, and gamma isomers)	sv						261-A8					
	Heptachlorobiphenyl(2,2',3,3',4,4',6-)					91/7							
	Heptachlorodibenzo-p-dioxins							261-A8					
	Heptachlorodibenzofurans							261-A8					
431-890	Heptafluoropropane(1,1,1,2,3,3,3-)											iris	
142-825	Heptane(n-)											iris	
87-821	Hexabromobenzene											iris	
36483-600	Hexabromodiphenyl ether											iris	
	Hexachloro-1,4,4a,5,8,8a-hexahydro-1,4:5,8-endo,endo-dimethanonaphthalene(1,2,3,4,10,10-)	sv						261-A8					
118-741	Hexachlorobenzene	sv	122-AD-T2/33B	136.3-T1C/60; 136.3-T2	141.61.c	91/7	258-A2	261-A8	264-A9	266-A5	hap	iris	pic
	Hexachlorobiphenyl(2,2',4,4',5,6'-)					91/7							
87-683	Hexachlorobutadiene	sv	122-AD-T2/34B	136.3-T1C/61; 136.3-T2		91/7	258-A2	261-A8	264-A9	266-A5	hap	iris	

SECTION 2. CHEMICAL CROSS-REFERENCE FOR SAMPLING, ANALYSIS, MONITORING AND RISK ASSESSMENT - ALPHABETICAL ORDER

CAS RN	Alphabetic Order	type[1]	122[2]	136[3]	141[4]	EPA-[5]	258[6]	261[7]	264[8]	266[9]	hap[10]	iris[11]	pic[12]
	Hexachlorocyclo-hexane, Technical									266-A5			
319-868	Hexachlorocyclohexane(delat -); delta-BHC; HCH-delta					91/7						iris	
	Hexachlorocyclohexane(all isomers)	sv						261-A8					
319-846	Hexachlorocyclohexane(alpha -); alpha-BHC; HCH-alpha					91/7				266-A5		iris	
319-857	Hexachlorocyclohexane(beta-); beta-BHC; HCH-beta					91/7				266-A5		iris	
58-899	Hexachlorocyclohexane(gam ma-); Lindane; gamma-BHC; HCH-gamma					91/7				266-A5		iris	
77-474	Hexachlorocyclopentadiene	sv	122-AD-T2/35B	136.3-T1C/62; 136.3-T2	141.61.c	91/7	258-A2	261-A8	264-A9	266-A4	hap*	iris	
	Hexachlorodibenzo-p-dioxin(1,2-Mixture)									266-A5			
	Hexachlorodibenzo-p-dioxins	sv						261-A8				iris	
19408-743	Hexachlorodibenzo-p-dioxin, mixture												
	Hexachlorodibenzofurans	sv						261-A8					
67-721	Hexachloroethane	v	122-AD-T2/36B	136.3-T1C/63; 136.3-T2			258-A2	261-A8	264-A9	266-A5	hap	iris	
70-304	Hexachlorophene	sv						261-A8	264-A9	266-A4		iris	
1888-717	Hexachloropropene	v					258-A2	261-A8	264-A9				
757-584	Hexaethyl tetraphosphate	sv						261-A8					
121-824	Hexahydro-1,3,5-trinitro-1,3,5-triazine											iris	
822-060	Hexamethylene diisocyanate(1,6-)										hap	iris	
680-319	Hexamethylphosphoramide										hap	iris	
110-543	Hexane										hap		
110-543	Hexane(n-)											iris	
591-786	Hexanone(2-); Methyl butyl ketone						258-A1/42; A2		264-A9			iris	

SAMPLING, ANALYSIS, AND MONITORING METHODS: A GUIDE TO EPA REQUIREMENTS

CAS RN	Alphabetic Order	type[1]	122[2]	136[3]	141[4]	EPA.[5]	258[6]	261[7]	264[8]	266[9]	hap[10]	iris[11]	pic[12]
51235-042	Hexazinone					91/7						iris	
302-012	Hydrazine	v						261-A8		266-A5	hap		
302-012	Hydrazine/Hydrazine sulfate											iris	
302-012	Hydrazine sulfate									266-A5			
7647-010	Hydrochloric acid										hap		
74-908	Hydrocyanic acid	gas						261-A8		266-A4			
	Hydrofluoric acid	anion						261-A8				iris	
7647-010	Hydrogen chloride									266-A4		iris	
74-908	Hydrogen cyanide							261-A8			hap	iris	
7664-393	Hydrogen fluoride							261-A8			hap	iris	
	Hydrogen ion (pH), pH units			136.3-T1B/28; 136.3-T2									
7783-064	Hydrogen sulfide	gas						261-A8		266-A4	hap	iris	
123-319	Hydroquinone										hap	iris	
	Hydroxycarbofuran(3-)					91/7							
	Hydroxydicamba(5-)					91/7							
	Hydroxydimethylarsine oxide	metal						261-A8					
35554-440	Imazalil											iris	
81335-377	Imazaquin											iris	
193-395	Indeno(1,2,3-cd)pyrene	sv	122-AD-T2/37B	136.3-T1C/64; 136.3-T2		91/7	258-A2	261-A8	264-A9				
193-395	Indeno(1,2,3-cd)pyrene		122-AD-T1									iris	
	Inorganic chemicals manufacturing		122-AD-T1										
	Iodomethane	v						261-A8					
	Ionizing radiation												
36734-197	Iprodione											iris	
	Iridium, total			136.3-T1B/29; 136.3-T2								iris	
	Iron and steel manufacturing		122-AD-T1										
	Iron dextran (complex)							261-A8					

CAS RN	Alphabetic Order	type[1]	122[2]	136[3]	141[4]	EPA[5]	258[6]	261[7]	264[8]	266[9]	hap[10]	iris[11]	pic[12]
	Iron, total		122-AD-T4	136.3-T1B/30; 136.3-T2									
42509-808	Isazophos											iris	
	Isobutanol												
78-831	Isobutyl alcohol	v					258-A2	261-A8	264-A9	266-A4		iris	
	Isocyanic acid, methyl ester	v						261-A8					
465-736	Isodrin			136.3-T1D/42; 136.3-T2			258-A2	261-A8	264-A9			iris	
78-591	Isophorone		122-AD-T2/38B	136.3-T1C/65; 136.3-T2			258-A2		264-A9		hap	iris	
	Isoprene		122-AD-T5									iris	
33820-530	Isopropalin		122-AD-T5										
	Isopropanolamine dodecylbenzenesulfonate											iris	
1832-548	Isopropyl methyl phosphonic acid												
	Isopropylbenzene					91/7							
	Isopropyltoluene(4-)					91/7							
120-581	Isosafrole	sv					258-A2	261-A8	264-A9			iris	
82558-507	Isoxaben		122-AD-T5									iris	
	Kelthane		122-AD-T5										
143-500	Kepone	sv	122-AD-T5				258-A2	261-A8	264-A9				
	Kjeldahl nitrogen, total, (as N)			136.3-T1B/31; 136.3-T2									
77501-634	Lactofen											iris	
303-341	Lasiocarpine							261-A8					
(Total)	Lead		122-AD-T3	136.3-T1B/32; 136.3-T2	141.89		258-A1/9; A2		264-A9				
7439-921	Lead	metal								266-A4			
301-042	Lead acetate	metal						261-A8				iris	
7439-921	Lead and compounds (inorganic)											iris	
7439-921	Lead and compounds, N.O.S.	metal						261-A8			hap		

CAS RN	Alphabetic Order	type[1]	122[2]	136[3]	141[4]	EPA[5]	258[6]	261[7]	264[8]	266[9]	hap[10]	iris[11]	pic[12]
7446-277	Lead phosphate	metal						261-A8					
1335-326	Lead subacetate	metal						261-A8					
	Leather and tanning finishing		122-AD-T1										
5989-275	Limonene(d-)											iris	
58-899	Lindane				141.61.c			261-A8			hap		
58-899	Lindane (all isomers)												
330-552	Linuron			136.3-T1D/43; 136.3-T2								iris	
83055-996	Londax											iris	
1702-176	Lontrel											iris	
12427-382	Magnesium, total		122-AD-T4	136.3-T1B/33; 136.3-T2								iris	
121-755	Malathion		122-AD-T5	136.3-T1D/44; 136.3-T2								iris	
108-316	Maleic anhydride	sv						261-A8		266-A4	hap	iris	
123-331	Maleic hydrazide	sv						261-A8			hap	iris	
109-773	Malononitrile	sv						261-A8					
12427-382	Maneb											iris	
7439-965	Manganese											iris	
	Manganese compounds										hap*		
	Manganese, total		122-AD-T4	136.3-T1B/34; 136.3-T2									
	Mechanical products manufacturing		122-AD-T1										
148-823	Melphalan	sv						261-A8					
24307-264	Mepiquat chloride											iris	
	Mercaptodimethur		122-AD-T5										
7487-947	Mercuric chloride											iris	
(Total)	Mercury		122-AD-T3	136.3-T1B/35; 136.3-T2	141.23.a; 141.23.k		258-A2		264-A9				
7439-976	Mercury	metal								266-A4			
7439-976	Mercury and compounds, N.O.S.	metal						261-A8			hap		

SECTION 2. CHEMICAL CROSS-REFERENCE FOR SAMPLING, ANALYSIS, MONITORING AND RISK ASSESSMENT - ALPHABETICAL ORDER

CAS RN	Alphabetic Order	type[1]	122[2]	136[3]	141[4]	EPA.[5]	258[6]	261[7]	264[8]	266[9]	hap[10]	iris[11]	pic[12]
7439-976	Mercury, elemental											iris	
628-864	Mercury fulminate	metal						261-A8				iris	
150-505	Merphos					91/7						iris	
78-488	Merphos oxide											iris	
2032-657	Mesurol											iris	
57837-191	Metalaxyl											iris	
126-987	Methacrylonitrile	sv					258-A2	261-A8	264-A9	266-A4		iris	
10265-926	Methamidophos											iris	
74-931	Methanethiol	v						261-A8				iris	
67-561	Methanol										hap	iris	
91-805	Methapyrilene	sv					258-A2	261-A8	264-A9				
950-378	Methidathion											iris	
	Methiocarb	sv		136.3-T1D/45; 136.3-T2		91/7							
16752-775	Methomyl	sv				91/7		261-A8		266-A4		iris	
72-435	Methoxychlor	sv	122-AD-T5	136.3-T1D/46; 136.3-T2		91/7	258-A2	261-A8	264-A9	266-A4	hap	iris	
109-864	Methoxyethanol(2-)				141.61.c						hap	iris	
79-209	Methyl acetate											iris	
96-333	Methyl acrylate											iris	
74-839	Methyl bromide; Bromomethane		122-AD-T2/20V				258-A1/43; A2	261-A8	264-A9		hap		
74-873	Methyl chloride; Chloromethane		122-AD-T2/21V				258-A1/44; A2	261-A8	264-A9		hap		
79-221	Methyl chlorocarbonate									266-A4		iris	
71-556	Methyl chloroform							261-A8			hap*		
1338-234	Methyl ethyl ketone peroxide							261-A8					
1338-234	Methyl ethyl ketone peroxide											iris	
78-933	Methyl ethyl ketone; MEK; 2-Butanone	v					258-A1/47; A2	261-A8	264-A9	266-A4	hap	iris	pic
60-344	Methyl hydrazine	v						261-A8		266-A5	hap		

SAMPLING, ANALYSIS, AND MONITORING METHODS: A GUIDE TO EPA REQUIREMENTS

CAS RN	Alphabetic Order	type[1]	122[2]	136[3]	141[4]	EPA.[5]	258[6]	261[7]	264[8]	266[9]	hap[10]	iris[11]	pic[12]
74-884	Methyl iodide; Iodomethane						258-A1/48; A2	261-A8	264-A9		hap	iris	
108-101	Methyl isobutyl ketone										hap	iris	
624-839	Methyl isocyanate							261-A8			hap	iris	
	Methyl mercaptan		122-AD-T5										
80-626	Methyl methacrylate	sv	122-AD-T5				258-A2	261-A8	264-A9		hap	iris	
66-273	Methyl methanesulfonate	sv					258-A2	261-A8	264-A9				
	Methyl paraoxon					91/7							
298-000	Methyl parathion	sv	122-AD-T5				258-A2	261-A8	264-A9	266-A4		iris	
25013-154	Methyl styrene											iris	
1634-044	Methyl tert-butyl ether							261-A8			hap	iris	
	Methyl-2-(methylthio)propionaldehyde-o-(methylcarbonyl)oxime(2-)							261-A8					
	Methyl-2-pentanone(4-); Methyl isobutyl ketone								264-A9				
108-101	Methyl-2-pentanone(4-); Methyl isobutyl ketone						258-A1/49; A2						
94-746	Methyl-4-chlorophenoxyacetic acid(2-)	sv										iris	
93-652	Methyl-4-chlorophenoxy(2-(2-))propionic acid	sv						261-A8				iris	
94-815	Methyl-4-chlorophenoxy(4-(2-)) butyric acid							261-A8				iris	
	Methyl-4,6-dinitrophenol(2-)			136.3-T1C/67; 136.3-T2									
	Methyl-n'-nitro-n-nitroso-guanidine(n-)	sv						261-A8					
	Methylaziridine(2-)	sv						261-A8					
79-221	Methylchlorocarbonate							261-A8					
56-495	Methylcholanthrene(3-)	sv					258-A2	261-A8	264-A9	266-A5			
101-144	Methylene bis (2-chloroaniline)(4,4'-)											iris	

CAS RN	Alphabetic Order	type[1]	122[2]	136[3]	141[4]	EPA.[5]	258[6]	261[7]	264[8]	266[9]	hap[10]	iris[11]	pic[12]
101-611	Methylene bis(N,N'-dimethyl)aniline(4,4'-)											iris	
74-953	Methylene bromide; Dibromomethane	v					258-A1/45; A2	261-A8	264-A9				pic
75-092	Methylene chloride; Dichloromethane	v	122-AD-T2/22V	136.3-T1C/66; 136.3-T2		91/7	258-A1/46; A2	261-A8	264-A9	266-A5	hap*		pic
101-779	Methylene dianiline(4,4'-)											iris	
101-688	Methylene diphenyl diisocyanate										hap		
101-688	Methylene diphenyl isocyanate											iris	
101-144	Methylene-bis(2-chloroaniline)(4,4'-)	sv						261-A8		266-A5	hap	iris	
101-779	Methylenedianiline(4,4'-)										hap		
75-865	Methyllactonitrile(2-)	sv						261-A8				iris	
22967-926	Methylmercury											iris	
91-576	Methylnaphthalene(2-)						258-A2		264-A9				
95-487	Methylphenol(2-)											iris	
108-394	Methylphenol(3-)											iris	
106-445	Methylphenol(4-)											iris	
56-042	Methylthiouracil	sv						261-A8					
51218-452	Metolachlor					91/7						iris	
21087-649	Metribuzin					91/7						iris	
	Mevinphos		122-AD-T5			91/7							
2385-855	Mexacarbate		122-AD-T5	136.3-T1D/47; 136.3-T2									
	Mirex			136.3-T1D/48; 136.3-T2								iris	
50-077	Mitomycin C	sv						261-A8					
70-257	MNNG (Guanidine, n-methyl-n'-nitro-n-nitroso-)							261-A8					
2212-671	Molinate					91/7						iris	
7439-987	Molybdenum											iris	

SAMPLING, ANALYSIS, AND MONITORING METHODS: A GUIDE TO EPA REQUIREMENTS

CAS RN	Alphabetic Order	type[1]	122[2]	136[3]	141[4]	EPA-[5]	258[6]	261[7]	264[8]	266[9]	hap[10]	iris[11]	pic[12]
	Molybdenum, total		122-AD-T4	136.3-T1B/36; 136.3-T2									
10599-903	Monochloramine											iris	
108-907	Monochlorobenzene				141.61.a							iris	
6923-224	Monocrotophos		122-AD-T5									iris	
	Monoethyl amine		122-AD-T5										
	Monomethyl amine		122-AD-T5										
60-344	Monomethylhydrazine											iris	
	Mononitrobenzene	sv											pic
	Monuron			136.3-T1D/49; 136.3-T2									
	Monuron-TCA			136.3-T1D/50; 136.3-T2									
505-602	Mustard gas	v						261-A8					
121-697	N-Dimethylaniline(N-)											iris	
142-596	Nabam											iris	
300-765	Naled		122-AD-T5									iris	
	Naphthalamine(1-)								264-A9				
91-203	Naphthalene	sv	122-AD-T2/39B	136.3-T1C/68; 136.3-T2		91/7	258-A2	261-A8	264-A9		hap	iris	pic
130-154	Naphthoquinone(1,4-)	sv					258-A2		264-A9			iris	
	Naphthyl-2-thiourea(1-)	sv						261-A8					
134-327	Naphthylamine(1-)	sv					258-A2	261-A8					
91-598	Naphthylamine(2-)	sv						261-A8	264-A9			iris	
134-327	Naphthylamine(alpha-)							261-A8					
91-598	Naphthylamine(beta-)							261-A8					
86-884	Naphthylthiourea(alpha-)							261-A8				iris	
15299-997	Napropamide					91/7						iris	
	Napthenic acid		122-AD-T5										
7440-020	Nickel									266-A5			
(Total)	Nickel		122-AD-T3	136.3-T1B/37; 136.3-T2			258-A1/10; A2		264-A9				

2 - 38

SECTION 2. CHEMICAL CROSS-REFERENCE FOR SAMPLING, ANALYSIS, MONITORING AND RISK ASSESSMENT - ALPHABETICAL ORDER

CAS RN	Alphabetic Order	type[1]	122[2]	136[3]	141[4]	EPA-[5]	258[6]	261[7]	264[8]	266[9]	hap[10]	iris[11]	pic[12]
7440-020	Nickel and compounds, N.O.S.	metal						261-A8			hap*		
13463-393	Nickel carbonyl	metal						261-A8				iris	
557-197	Nickel cyanide	metal						261-A8		266-A4		iris	
7440-020	Nickel refinery dust									266-A5		iris	
Various	Nickel, soluble salts											iris	
12035-722	Nickel subsulfide									266-A5		iris	
54-115	Nicotine	sv						261-A8					
	Nicotine salts	sv						261-A8					
1929-824	Nitrapyrin											iris	
14797-558	Nitrate	sv										iris	
	Nitrate (as N)			136.3-T1B/38; 136.3-T2	141.23.a; 141.23.k								
	Nitrate-nitrite (as N)		122-AD-T4	136.3-T1B/39; 136.3-T2									
10102-439	Nitric oxide	gas						261-A8		266-A4		iris	
14797-650	Nitrite							261-A8				iris	
	Nitrite (as N)			136.3-T1B/40; 136.3-T2	141.23.a; 141.23.k								
99-558	Nitro-o-toluidine(5-)	sv					258-A2	261-A8	264-A9				
88-744	Nitroaniline(2-)											iris	
99-092	Nitroaniline(m-); 3-Nitroanile						258-A2		264-A9				
88-744	Nitroaniline(o-); 2-Nitroaniline						258-A2		264-A9				
100-016	Nitroaniline(p-); 4-Nitroaniline	sv					258-A2	261-A8	264-A9				
98-953	Nitrobenzene	sv	122-AD-T2/40B	136.3-T1C/69; 136.3-T2			258-A2	261-A8	264-A9	266-A4	hap	iris	
92-933	Nitrobiphenyl(4-)										hap	iris	
10102-440	Nitrogen dioxide	gas						261-A8				iris	
	Nitrogen mustard, hydrochloride salt	v						261-A8					

2 - 39

SAMPLING, ANALYSIS, AND MONITORING METHODS: A GUIDE TO EPA REQUIREMENTS

CAS RN	Alphabetic Order	type[1]	122[2]	136[3]	141[4]	EPA-[5]	258[6]	261[7]	264[8]	266[9]	hap[10]	iris[11]	pic[12]
51-752	Nitrogen mustard	v						261-A8					pic
	Nitrogen mustard, n-oxide, hydrochloride salt	v						261-A8					
126-852	Nitrogen mustard n-oxide	v						261-A8					
	Nitrogen, total organic		122-AD-T4										
55-630	Nitroglycerin	sv						261-A8					
556-887	Nitroguanidine											iris	
	Nitrophenol(2-)		122-AD-T2/6A	136.3-T1C/70; 136.3-T2									
100-027	Nitrophenol(4-)	sv	122-AD-T2/7A	136.3-T1C/71; 136.3-T2		91/7		261-A8			hap		
88-755	Nitrophenol(o-); 2-Nitrophenol	sv					258-A2		264-A9				pic
100-027	Nitrophenol(p-); 4-Nitrophenol						258-A2	261-A8	264-A9			iris	
79-469	Nitropropane(2-)							261-A8		266-A5	hap	iris	
	Nitroquinoline-1-oxide(4-)	sv						261-A8	264-A9			iris	
35576-911	Nitrosamines, N.O.S.	sv						261-A8					
684-935	Nitroso-n-methylurea(n-)										hap	iris	
924-163	Nitroso-di-n-butylamine(n-)									266-A5		iris	
924-163	Nitroso-n-butylamine(n-)									266-A5			
759-739	Nitroso-n-ethylurea(n-)	sv						261-A8				iris	
10595-956	Nitroso-n-methylethylamine(n-)											iris	
684-935	Nitroso-n-methylurea(n-)											iris	
684-935	Nitroso-n-methylurea(n-)	sv						261-A8					
684-935	Nitroso-n-methylurea(n-)									266-A5			
615-532	Nitroso-n-methylurethane(n-)	sv						261-A8					
	Nitrosodi-n-butyl-amine(n-)								264-A9				
924-163	Nitrosodi-n-butylamine(n-)	sv					258-A2	261-A8					
621-647	Nitrosodi-n-propylamine(n-)		122-AD-T2/42B	136.3-T1C/73; 136.3-T2								iris	

SECTION 2. CHEMICAL CROSS-REFERENCE FOR SAMPLING, ANALYSIS, MONITORING AND RISK ASSESSMENT - ALPHABETICAL ORDER

CAS RN	Alphabetic Order	type[1]	122[2]	136[3]	141[4]	EPA-[5]	258[6]	261[7]	264[8]	266[9]	hap[10]	iris[11]	pic[12]
1116-547	Nitrosodiethanolamine(n-)											iris	
55-185	Nitrosodiethylamine(n-)	sv					258-A2	261-A8	264-A9	266-A5		iris	
62-759	Nitrosodimethylamine(n-)	sv	122-AD-T2/41B	136.3-T1C/72; 136.3-T2			258-A2	261-A8	264-A9		hap	iris	
86-306	Nitrosodiphenylamine(n-)	sv	122-AD-T2/43B	136.3-T1C/74; 136.3-T2			258-A2		264-A9			iris	
	Nitrosodipropylamine(n-)								264-A9				
621-64-7	Nitrosodipropylamine(n-); n-Nitroso-N-dipropylamine; Di-n-propylnitrosamine						258-A2						
10595-956	Nitrosomethylethylamine(n-)	sv					258-A2	261-A8	264-A9				
4549-400	Nitrosomethylvinylamine(n-)	sv						261-A8					
59-892	Nitrosomorpholine(n-)	sv						261-A8	264-A9		hap		
16543-558	Nitrosonornicotine(n-)	sv					258-A2	261-A8	264-A9				
100-754	Nitrosopiperidine(n-)	sv					258-A2	261-A8	264-A9				
930-552	Nitrosopyrrolidine(n-)	sv					258-A2	261-A8	264-A9	266-A5		iris	
13256-229	Nitrososarcosine(n-)	sv						261-A8					
	Nitrotoluene		122-AD-T5										
63936-561	Nonabromodiphenyl ether											iris	
	Nonachlor(cis-)					91/7							
	nonachlor(trans-)					91/7							
	Nonferrous metals manufacturing		122-AD-T1										
27314-132	Norflurazon					91/7						iris	
	Nuburon			136.3-T1D/51; 136.3-T2									
85509-199	NuStar											iris	
32536-520	Octabromodiphenyl ether											iris	
	Octachlorobiphenyl(2,2',3,3',4,5',6,6')					91/7							
2691-410	Octahydro-1,3,5,7-tetranitro-1,3,5,7-tetrazocine											iris	

CAS RN	Alphabetic Order	type[1]	122[2]	136[3]	141[4]	EPA.[5]	258[6]	261[7]	264[8]	266[9]	hap[10]	iris[11]	pic[12]
152-169	Octamethylpyro-phosphoramide	sv						261-A8				iris	
	Oil and grease, total recoverable		122-AD-T4	136.3-T1B/41; 136.3-T2									
	Ore mining		122-AD-T1										
	Organic carbon, total (TOC)			136.3-T1B/42; 136.3-T2									
	Organic chemicals manufacturing		122-AD-T1										
	Organic Constituents:												
	Organic nitrogen (as N)			136.3-T1B/43; 136.3-T2									
	Orthophosphate (as P)			136.3-T1B/44; 136.3-T2									
	Orthophosphate, unfiltered, no digestion or hydrolysis				141.89								
19044-883	Oryzalin											iris	
20816-120	Osmium tetroxide	metal						261-A8				iris	
	Osmium, total			136.3-T1B/45; 136.3-T2									
	Oxabicyclo(2.2.1)heptane-2,3-dicarboxylic acid	sv						261-A8				iris	
19666-309	Oxadiazon											iris	
77732-093	Oxadixyl											iris	
23135-220	Oxamyl; Vydate				141.61.c	91/7						iris	
	Oxybis(1-chloropropane)(2,2-)			136.3-T1C/75; 136.3-T2									
42874-033	Oxyfluorfen											iris	
	Oxygen dissolved			136.3-T1B/46; 136.3-T2									
76738-620	Paclobutrazol											iris	
	Paint and ink formulation		122-AD-T1										
	Palladium, total			136.3-T1B/47; 136.3-T2									
123-637	Paraldehyde	v						261-A8				iris	
	Paraldehyde (trimer of acetaldehyde)												

SECTION 2. CHEMICAL CROSS-REFERENCE FOR SAMPLING, ANALYSIS, MONITORING AND RISK ASSESSMENT - ALPHABETICAL ORDER

CAS RN	Alphabetic Order	type[1]	122[2]	136[3]	141[4]	EPA-[5]	258[6]	261[7]	264[8]	266[9]	hap[10]	iris[11]	pic[12]
1910-425	Paraquat											iris	
56-382	Parathion	sv	122-AD-T5				258-A2	261-A8	264-A9		hap	iris	
	Parathion ethyl			136.3-T1D/53; 136.3-T2									
	Parathion methyl			136.3-T1D/52; 136.3-T2									
	PCB-1016		122-AD-T2/24P	136.3-T1C/76; 136.3-T2									
	PCB-1221		122-AD-T2/20P	136.3-T1C/77; 136.3-T2									
	PCB-1232		122-AD-T2/21P	136.3-T1C/78; 136.3-T2									
	PCB-1242		122-AD-T2/18P	136.3-T1C/79; 136.3-T2									
	PCB-1248		122-AD-T2/22P	136.3-T1C/80; 136.3-T2									
	PCB-1254		122-AD-T2/19P	136.3-T1C/81; 136.3-T2									
	PCB-1260		122-AD-T2/23P	136.3-T1C/82; 136.3-T2									
1336-363	PCBs									266-A5			
	Pebulate					91/7							
40487-421	Pendimethalin											iris	
32534-819	Pentabromodiphenyl ether											iris	
608-935	Pentachlorobenzene	sv					258-A2	261-A8	264-A9	266-A4		iris	
	Pentachlorobenzofurans	sv						261-A8					
	Pentachlorobiphenyl(2,2',3,4,6-)				141.61.c	91/7							
25329-355	Pentachlorocyclopentadiene											iris	
	Pentachlorodibenzo-p-dioxins	sv						261-A8					
76-017	Pentachloroethane	sv						261-A8	264-A9			iris	
82-688	Pentachloronitrobenzene; PCNB	sv		136.3-T1D/54; 136.3-T2			258-A2	261-A8	264-A9	266-A5	hap	iris	
87-865	Pentachlorophenol	sv	122-AD-T2/9A	136.3-T1C/83; 136.3-T2			258-A2	261-A8	264-A9	266-A4	hap	iris	pic
	Pentachlorophenol; PCP					91/7							
354-336	Pentafluoroethane											iris	
355-259	Perfluorobutane											iris	

SAMPLING, ANALYSIS, AND MONITORING METHODS: A GUIDE TO EPA REQUIREMENTS

CAS RN	Alphabetic Order	type[1]	122[2]	136[3]	141[4]	EPA.[5]	258[6]	261[7]	264[8]	266[9]	hap[10]	iris[11]	pic[12]
355-420	Perfluorohexane											iris	pic
52645-531	Permethrin											iris	
	Permethrin(cis-)					91/7							
	Permethrin(trans-)					91/7							
	Perthane			136.3-T1D/55; 136.3-T2									
	Pesticides		122-AD-T1										
	Petroleum refining		122-AD-T1										
	pH				141.89								
	Pharmaceutical preparations		122-AD-T1										
62-442	Phenacetin	sv					258-A2	261-A8	264-A9				
85-018	Phenanthrene		122-AD-T2/44B	136.3-T1C/84; 136.3-T2		91/7	258-A2	261-A8	264-A9			iris	
13684-634	Phenmedipham											iris	
108-952	Phenol	sv	122-AD-T2/10A	136.3-T1C/85; 136.3-T2			258-A2	261-A8	264-A9	266-A4	hap*	iris	pic
	Phenolics												
	Phenols, total		122-AD-T3	136.3-T1B/48; 136.3-T2									
	Phenolsulfanate		122-AD-T5										
25265-763	Phenylenediamine	sv						261-A8					
108-452	Phenylenediamine(m-)									266-A4		iris	
106-503	Phenylenediamine(p-)						258-A2		264-A9		hap		
62-384	Phenylmercuric acetate	metal								266-A4		iris	
62-384	Phenylmercury acetate							261-A8	264-A9				
103-855	Phenylthiourea							261-A8					
103-855	phenylthiourea(n-)											iris	
298-022	Phorate	sv					258-A2	261-A8	264-A9			iris	
2310-170	Phosalone											iris	
75-445	Phosgene	gas	122-AD-T5					261-A8			hap	iris	
732-116	Phosmet											iris	

SECTION 2. CHEMICAL CROSS-REFERENCE FOR SAMPLING, ANALYSIS, MONITORING AND RISK ASSESSMENT - ALPHABETICAL ORDER

CAS RN	Alphabetic Order	type[1]	122[2]	136[3]	141[4]	EPA-[5]	258[6]	261[7]	264[8]	266[9]	hap[10]	iris[11]	pic[12]
7803-512	Phosphine	v						261-A8		266-A4	hap	iris	
7664-382	Phosphoric acid											iris	
	Phosphorodithioic acid esters												
	Phosphorodithioic acid, o,o-diethyl s-((ethylthio)methyl) ester, see phorate							261-A8					
	Phosphorothioic acid, o,o-dimethyl o-(p-(dimethylamino)sulfonyl)phenyl ester, see famphur							261-A8					
7723-140	Phosphorus (elemental)			136.3-T1B/49; 136.3-T2							hap	iris	
	Phosphorus, total		122-AD-T4	136.3-T1B/50; 136.3-T2									
	Photographic equipment and supplies		122-AD-T1										
	photon emitters(beta-)											iris	
	Phthalic acid ester, N.O.S.	sv						261-A8					
85-449	Phthalic anhydride	sv						261-A8		266-A4	hap	iris	
1918-021	Picloram				141.61.c	91/7						iris	
109-068	Picoline(2-)	sv						261-A8	264-A9			iris	
80-568	Pinene(alpha-)											iris	
127-913	Pinene(beta-)											iris	
29232-937	Pirimiphos-methyl											iris	
	Plastic and synthetic materials manufacturing		122-AD-T1										
	Plastic processing		122-AD-T1										
	Platinum, total			136.3-T1B/51; 136.3-T2									
1336-363	Polychlorinated biphenyls, N.O.S.	sv			141.61.c			261-A8			hap		
1336-363	Polychlorinated biphenyls; PCBs; Aroclors						258-A2		264-A9			iris	
	Polychlorinated dibenzo-p-dioxins; PCDDs								264-A9				

SAMPLING, ANALYSIS, AND MONITORING METHODS: A GUIDE TO EPA REQUIREMENTS

CAS RN	Alphabetic Order	type[1]	122[2]	136[3]	141[4]	EPA-[5]	258[6]	261[7]	264[8]	266[9]	hap[10]	iris[11]	pic[12]
	Polychlorinated dibenzofurans; PCDFs								264-A9				
	Polychlorobiphenyls (General screen)					91/7							
	Polycyclic organic matter											iris	
	Polycyclic organic matter										hap*		
	Porcelain enameling		122-AD-T1										
7758-012	Potassium bromate											iris	
151-508	Potassium cyanide	cyanide						261-A8		266-A4		iris	
506-616	Potassium silver cyanide	metal						261-A8		266-A4		iris	
	Potassium, total			136.3-T1B/52; 136.3-T2									
	Printing and publishing		122-AD-T1										
67747-095	Prochloraz											iris	
	Prometon					91/7							
1610-180	Prometon											iris	
	Prometron			136.3-T1D/56; 136.3-T2									
7287-196	Prometryn			136.3-T1D/57; 136.3-T2		91/7						iris	
23950-585	Pronamide	sv				91/7	258-A2	261-A8	264-A9	266-A5		iris	
1918-167	Propachlor					91/7						iris	
1120-714	Propane sultone(1,3-)	sv						261-A8			hap	iris	
709-988	Propanil											iris	
2312-358	Propargite		122-AD-T5									iris	
107-197	Propargyl alcohol							261-A8				iris	
139-402	Propazine			136.3-T1D/58; 136.3-T2		91/7						iris	
31218-834	Propetamphos											iris	
122-429	Propham			136.3-T1D/59; 136.3-T2								iris	
60207-901	Propiconazole											iris	
57-578	Propiolactone(beta-)										hap	iris	

SECTION 2. CHEMICAL CROSS-REFERENCE FOR SAMPLING, ANALYSIS, MONITORING AND RISK ASSESSMENT - ALPHABETICAL ORDER

CAS RN	Alphabetic Order	type[1]	122[2]	136[3]	141[4]	EPA[5]	258[6]	261[7]	264[8]	266[9]	hap[10]	iris[11]	pic[12]
123-386	Propionaldehyde										hap	iris	
79-094	Propionic acid											iris	
107-120	Propionitrile; Ethyl cyanide						258-A2		264-A9				
114-261	Propoxur			136.3-T1D/60; 136.3-T2							hap		
107-108	Propylamine(n-)	sv						261-A8					
	Propylbenzene					91/7						iris	
	Propylbenzene(n-)					91/7							
78-875	Propylene dichloride							261-A8			hap		
57-556	Propylene glycol											iris	
52125-538	Propylene glycol monoethyl ether											iris	
107-982	Propylene glycol monomethyl ether											iris	
75-569	Propylene oxide		122-AD-T5								hap	iris	
75-558	Propyleneimine											iris	
75-558	Propylenimine(1,2-)							261-A8			hap		
51-525	Propylthiouracil	sv						261-A8					
	Propyn-1-ol(2-)	v						261-A8					
	Pulp and paper mills		122-AD-T1										
81335-775	Pursuit											iris	
51630-581	Pydrin											iris	
129-000	Pyrene		122-AD-T2/45B	136.3-T1C/86; 136.3-T2		91/7	258-A2		264-A9			iris	pic
	Pyrethrins		122-AD-T5										
110-861	Pyridine	sv						261-A8	264-A9			iris	
13593-038	Quinalphos									266-A4			
84087-014	Quinclorac											iris	
91-225	Quinoline		122-AD-T5								hap	iris	
106-514	Quinone										hap	iris	

2 - 47

SAMPLING, ANALYSIS, AND MONITORING METHODS: A GUIDE TO EPA REQUIREMENTS

CAS RN	Alphabetic Order	type[1]	122[2]	136[3]	141[4]	EPA-[5]	258[6]	261[7]	264[8]	266[9]	hap[10]	iris[11]	pic[12]
	Radioactivity		122-AD-T4										
	Radionuclides (including radon)										hap		
7440-144	Radium 226,228											iris	
	Radium-228												
	Radium-Total pCi per liter (a); Ra226, pCi per liter (b)			136.3-T1E/5; 136.3-T2									
	Radon (Inert gas only)											iris	
14859677	Radon 222											iris	
Not found	Refractory ceramic fibers											iris	
50-555	Reserpine	sv						261-A8		266-A5			
	Residue, filterable			136.3-T1B/54; 136.3-T2									
	Residue, nonfilterable			136.3-T1B/55; 136.3-T2									
	Residue, settleable			136.3-T1B/56; 136.3-T2									
	Residue, total			136.3-T1B/53; 136.3-T2									
	Residue, volatile			136.3-T1B/57; 136.3-T2									
10453-868	Resmethrin	sv										iris	
108-463	Resorcinol	sv						261-A8					
	Rhodium, total			136.3-T1B/58; 136.3-T2									
299-843	Ronnel											iris	
83-794	Rotenone											iris	
	Rubber processing		122-AD-T1										
	Ruthenium, total			136.3-T1B/59; 136.3-T2									
81-072	Saccharin	sv						261-A8					
	Saccharin salts	sv						261-A8					
94-597	Safrole	sv					258-A2	261-A8	264-A9				
78587-050	Savey											iris	
	Secbumeton			136.3-T1D/61; 136.3-T2									

SECTION 2. CHEMICAL CROSS-REFERENCE FOR SAMPLING, ANALYSIS, MONITORING AND RISK ASSESSMENT - ALPHABETICAL ORDER

CAS RN	Alphabetic Order	type[1]	122[2]	136[3]	141[4]	EPA-[5]	258[6]	261[7]	264[8]	266[9]	hap[10]	iris[11]	pic[12]
7783-008	Selenious acid									266-A4		iris	
(Total)	Selenium		122-AD-T3	136.3-T1B/60; 136.3-T2	141.23.a; 141.23.k		258-A1/I1; A2		264-A9				
7782-492	Selenium and compounds											iris	
7782-492	Selenium and compounds, N.O.S.	metal						261-A8			hap		
7783-008	Selenium dioxide							261-A8					
7446-346	Selenium sulfide											iris	
7488-564	Selenium sulfide	metal						261-A8					
630-104	Selenourea	metal						261-A8		266-A4		iris	
74051-802	Sethoxydim			136.3-T1D/62; 136.3-T2								iris	
	Siduron			136.3-T1B/61; 136.3-T2									
	Silica, dissolved				141.89								
	Silicon												
(Total)	Silver		122-AD-T3	136.3-T1B/62; 136.3-T2			258-A1/I2; A2		264-A9				
7440-224	Silver									266-A4		iris	
7440-224	Silver and compounds, N.O.S.	metal						261-A8					
506-649	Silver cyanide	metal						261-A8		266-A4		iris	
93-721	Silvex; 2,4,5-TP	sv					258-A2	261-A8	264-A9				
122-349	Simazine			136.3-T1D/63; 136.3-T2	141.61.c	91/7						iris	
	Simetryn					91/7							
	Soap and detergent manufacturing		122-AD-T1										
148-185	Sodium diethyldithiocarbamate											iris	
26628-228	Sodium azide											iris	
143-339	Sodium cyanide	cyanide						261-A8		266-A4		iris	
62-748	Sodium fluoroacetate											iris	
	Sodium, total			136.3-T1B/63; 136.3-T2									

SAMPLING, ANALYSIS, AND MONITORING METHODS: A GUIDE TO EPA REQUIREMENTS

CAS RN	Alphabetic Order	type[1]	122[2]	136[3]	141[4]	EPA[5]	258[6]	261[7]	264[8]	266[9]	hap[10]	iris[11]	pic[12]
	Steam electric power plants		122-AD-T1										
	Stirofos					91/7							
18883-664	Streptozotocin	sv						261-A8					
	Strobane		122-AD-T5	136.3-T1D/64; 136.3-T2									
7440-246	Strontium	metal										iris	
	Strontium sulfide							261-A8					
57-249	Strychnine	sv	122-AD-T5					261-A8		266-A4		iris	
	Strychnine salts	sv						261-A8					
100-425	Styrene		122-AD-T5		141.61.a	91/7	258-A1/50; A2				hap	iris	
96-093	Styrene oxide										hap		
	Sulfate		122-AD-T4	136.3-T1B/65; 136.3-T2									
	Sulfide		122-AD-T4	136.3-T1B/66; 136.3-T2									
18496-25-8	Sulfide						258-A2						
	Sulfite (as SO$_3$)		122-AD-T4	136.3-T1B/67; 136.3-T2									
	Surfactant		122-AD-T4	136.3-T1B/68; 136.3-T2									
	Swep			136.3-T1D/65; 136.3-T2									
99-35-4	sym-Trinitrobenzene						258-A2						
88671-890	Systhane	sv										iris	
93-765	T(2,4,5-)							261-A8					
	T(2,4,5-)		122-AD-T5	136.3-T1D/66; 136.3-T2					264-A9				
93-76-5	T(2,4,5-); 2,4,5-Trichlorophenoxyacetic acid						258-A2						
1746-016	TCDD(2,3,7,8-) (Dioxin)				141.61.c								
	TDE		122-AD-T5										
107534-963	Tebuconazole											iris	
34014-181	Tebuthiuron					91/7						iris	
608-731	technical Hexachlorocyclohexane											iris	

CAS RN	Alphabetic Order	type[1]	122[2]	136[3]	141[4]	EPA.[5]	258[6]	261[7]	264[8]	266[9]	hap[10]	iris[11]	pic[12]
	Temperature C.: Thermometric			136.3-T1B/69; 136.3-T2	141.89								pic
5902-512	Terbacil					91/7						iris	
13071-799	Terbufos					91/7						iris	
	Terbuthylazine			136.3-T1D/68; 136.3-T2									
886-500	Terbutryn					91/7						iris	
40088-479	Tetrabromodiphenyl ether											iris	
95-943	Tetrachlorobenzene(1,2,4,5-)						258-A2	261-A8	264-A9	266-A4		iris	
	Tetrachlorobiphenyl(2,2',4,4'-)					91/7							
695-772	Tetrachlorocyclopentadiene											iris	
1746-016	Tetrachlorodibenzo-p-dioxin(2,3,7,8-)	sv		136.3-T1C/87; 136.3-T2				261-A8	264-A9	266-A5	hap		
	Tetrachlorodibenzofurans	sv						261-A8					
25322-207	Tetrachloroethane, N.O.S.	v						261-A8					
79-345	Tetrachloroethane(1,1,2,2-)	v	122-AD-T2/23V	136.3-T1C/88; 136.3-T2		91/7	258-A1/52; A2	261-A8	264-A9	266-A5	hap	iris	
630-206	Tetrachloroethane(1,1,1,2-)					91/7	258-A1/51; A2	261-A8	264-A9			iris	
127-184	Tetrachloroethylene; Tetrachloroethene; Perchloroethylene	v	122-AD-T2/24V	136.3-T1C/89; 136.3-T2	141.61.a	91/7	258-A1/53; A2	261-A8	264-A9	266-A5	hap	iris	pic
58-902	Tetrachlorophenol(2,3,4,6-)	266-A4											
58-902	Tetrachlorophenol(2,3,4,6-)	sv					258-A2	261-A8	264-A9			iris	
961-115	Tetrachlorovinphos	sv										iris	
	Tetraethyl dithiopyrophosphate; Sulfotepp								264-A9				
78-002	Tetraethyl lead	metal						261-A8		266-A4		iris	
107-493	Tetraethyl pyrophosphate	sv						261-A8					
3689-245	Tetraethyldithiopyrophosphate	sv						261-A8				iris	

SAMPLING, ANALYSIS, AND MONITORING METHODS: A GUIDE TO EPA REQUIREMENTS

CAS RN	Alphabetic Order	type[1]	122[2]	136[3]	141[4]	EPA.[5]	258[6]	261[7]	264[8]	266[9]	hap[10]	iris[11]	pic[12]
107-493	Tetraethylpyrophosphate											iris	
811-972	Tetrafluoroethane(1,1,1,2-)											iris	
109-999	Tetrahydrofuran									266-A4		iris	
509-148	Tetranitromethane	v						261-A8				iris	
	Textile mills		122-AD-T1										
1314-325	Thallic oxide	metal	122-AD-T3					261-A8		266-A4		iris	
(Total)	Thallium			136.3-T1B/70; 136.3-T2			258-A1/13; A2		264-A9				
7440-280	Thallium	266-A4											
563-688	Thallium (I) acetate	metal						261-A8		266-A4		iris	
6533-739	Thallium (I) carbonate	metal						261-A8		266-A4		iris	
7791-120	Thallium (I) chloride	metal						261-A8		266-A4		iris	
10102-451	Thallium (I) nitrate	metal						261-A8		266-A4		iris	
7446-186	Thallium (I) sulfate	metal						261-A8		266-A4		iris	
7440-280	Thallium and compounds, N.O.S.	metal						261-A8					
12039-520	Thallium selenite	metal						261-A8		266-A4		iris	
62-555	Thioacetamide	sv						261-A8				iris	
28249-776	Thiobencarb											iris	
39196-184	Thiofanox							261-A8				iris	
74-931	Thiomethanol							261-A8					
23564-058	Thiophanate-methyl											iris	
108-985	Thiophenol							261-A8					
79-196	Thiosemicarbazide							261-A8				iris	
62-566	Thiourea	sv						261-A8		266-A5		iris	
137-268	Thiram	sv						261-A8		266-A4		iris	
	Timber products processing		122-AD-T1										
(Total)	Tin		122-AD-T4	136.3-T1B/71; 136.3-T2			258-A2		264-A9				
7550-450	Titanium tetrachloride										hap		

SECTION 2. CHEMICAL CROSS-REFERENCE FOR SAMPLING, ANALYSIS, MONITORING AND RISK ASSESSMENT - ALPHABETICAL ORDER

CAS RN	Alphabetic Order	type[1]	122[2]	136[3]	141[4]	EPA.[5]	258[6]	261[7]	264[8]	266[9]	hap[10]	iris[11]	pic[12]
	Titanium, total		122-AD-T4	136.3-T1B/72; 136.3-T2									
108-883	Toluene	v	122-AD-T2/25V	136.3-T1C/90; 136.3-T2	141.61.a	91/7	258-A1/54; A2	261-A8	264-A9	266-A4	hap*	iris	pic
26471-625	Toluene diisocyanate mixture(2,4-/2,6)											iris	
26471-625	Toluene diisocyanate							261-A8					
	Toluene diisocyanate(2,6-)	sv											pic
584-849	Toluene(2,4-) diisocyanate	hap									hap		
95-807	Toluene-2,4-diamine	sv						261-A8					
823-405	Toluene-2,6-diamine	sv						261-A8					
496-720	Toluene-3,4-diamine	sv						261-A8					
25376-458	Toluenediamine							261-A8					
	Toluenediamine, N.O.S.	sv						261-A8					
	Toluenediamine(2,4-)												
	Toluenediamine(2,6-)												
	Toluenediamine(3,4-)												
636-215	Toluidine hydrochloride(o-)							261-A8					
95-534	Toluidine(o-)						258-A2	261-A8	264-A9		hap		
106-490	Toluidine(p-)							261-A8					
	Tolylene diisocyanate	sv						261-A8					
	Total coliform												
	Total organic carbon												
	Total organic halides												
8001-352	Toxaphene	sv	122-AD-T2/25P	136.3-T1D/69; 136.3-T2	141.61.c		258-A2	261-A8	264-A9	266-A5	hap	iris	
	Toxaphene (Technical)					91/7							
93-721	TP(2,4,5-)		122-AD-T5	136.3-T1D/67; 136.3-T2	141.61.c								
66841-256	Tralomethrin											iris	
	Triademefon					91/7							
2303-175	Triallate											iris	

SAMPLING, ANALYSIS, AND MONITORING METHODS: A GUIDE TO EPA REQUIREMENTS

CAS RN	Alphabetic Order	type[1]	122[2]	136[3]	141[4]	EPA-[5]	258[6]	261[7]	264[8]	266[9]	hap[10]	iris[11]	pic[12]
82097-505	Triasulfuron												
615-543	Tribromobenzene(1,2,4-)											iris	
594-150	Tribromochloromethane											iris	
49690-940	Tribromodiphenyl ether											iris	
	Tribromomethane	v						261-A8					
56-359	Tributyltin oxide											iris	
	Trichloropropionic acid(2,4,5-)												
76-131	Trichloro-1,2,2-trifluoroethane(1,1,2-)											iris	
120-821	Trichloro-benzene(1,2,4-)				141.61.a								
79-005	Trichloro-ethane(1,1,2-)				141.61.a								
76-039	Trichloroacetic acid											iris	
120-821	Trichlorobenzene(1,2,4-)	sv	122-AD-T2/46B	136.3-T1C/91; 136.3-T2		91/7	258-A2	261-A8	264-A9	266-A4	hap	iris	pic
	Trichlorobenzene(1,2,3-)					91/7							
	Trichlorobiphenyl(2,4,5-)					91/7							
77323-843	Trichlorocyclopentadiene											iris	
	Trichloroethane		122-AD-T5										
71-556	Trichloroethane(1,1,1-)	v	122-AD-T2/27V	136.3-T1C/92; 136.3-T2	141.61.a	91/7	258-A1/55; A2	261-A8	264-A9			iris	pic
79-005	Trichloroethane(1,1,2-)	v	122-AD-T2/28V	136.3-T1C/93; 136.3-T2		91/7	258-A1/56; A2	261-A8	264-A9	266-A5	hap	iris	
	Trichloroethene			136.3-T1C/94; 136.3-T2		91/7							
79-016	Trichloroethylene; Trichloroethene	v	122-AD-T2/29V		141.61.a		258-A1/57; A2	261-A8	264-A9	266-A5	hap*	iris	pic
	Trichlorofan												
75-694	Trichlorofluoromethane; CFC-11			136.3-T1C/95; 136.3-T2		91/7	258-A1/58; A2		264-A9			iris	
75-707	Trichloromethanithiol	sv						261-A8					
75-694	Trichloromonofluoromethane	v						261-A8		266-A4			
95-954	Trichlorophenol(2.4.5-)	266-A4											
95-954	Trichlorophenol(2,4,5-)	sv					258-A2	261-A8	264-A9		hap	iris	

CAS RN	Alphabetic Order	type[1]	122[2]	136[3]	141[4]	EPA-[5]	258[6]	261[7]	264[8]	266[9]	hap[10]	iris[11]	pic[12]
88-062	Trichlorophenol(2,4,6-)	sv	122-AD-T2/11A	136.3-T1C/96; 136.3-T2			258-A2	261-A8	264-A9	266-A5	hap	iris	pic
93-721	Trichlorophenoxy(2-(2,4,5-))propionic acid; 2,4,5-TP; Silvex					91/7						iris	
93-765	Trichlorophenoxyacetic acid(2,4,5-); 2,4,5-T					91/7		261-A8				iris	
	Trichlorophenoxypropionic(2,4,5-) acid; TP(2,4,5-); silver							261-A8					
25735-299	Trichloropropane, N.O.S.	v						261-A8					
598-776	Trichloropropane(1,1,2-)											iris	
96-184	Trichloropropane(1,2,3-)	v				91/7	258-A1/59; A2	261-A8	264-A9			iris	
1319-773	Tricresol											iris	
	Tricyclazole					91/7							
58138-082	Tridiphane											iris	
	Triethanolamine dodecylbenzenesulfonate		122-AD-T5										
126-681	Triethyl phosphorothioate(o,o,o-)		122-AD-T5				258-A2		264-A9				
121-448	Triethylamine										hap	iris	
143-226	Triethylene glycol monobutyl ether											iris	
112-505	Triethylene glycol monoethyl ether											iris	
112-356	Triethylene glycol monomethyl ether											iris	
126-681	Triethylphosphorothioate(o,o,o-)	sv						261-A8					
420-462	Trifluoroethane(1,1,1-)											iris	
75-467	Trifluoromethane											iris	
1582-098	Trifluralin			136.3-T1D/70; 136.3-T2		91/7					hap	iris	
	Trihalomethanes, total				141.30								
	Trimethylamine		122-AD-T5										

SAMPLING, ANALYSIS, AND MONITORING METHODS: A GUIDE TO EPA REQUIREMENTS

CAS RN	Alphabetic Order	type[1]	122[2]	136[3]	141[4]	EPA-[5]	258[6]	261[7]	264[8]	266[9]	hap[10]	iris[11]	pic[12]
	Trimethylbenzene(1,2,4-)					91/7							
	Trimethylbenzene(1,3,5-)					91/7							
540-841	Trimethylpentane(2,2,4-)										hap	iris	
99-354	Trinitrobenzene(1,3,5-)							261-A8				iris	
	Trinitrobenzene(sym-)	sv						261-A8	264-A9				
118-967	Trinitrotoluene(2,4,6-)											iris	
76-879	Triphenyltin hydroxide											iris	
52-244	Tris(1-aziridinyl)phosphine sulfide	v						261-A8					
126-727	Tris(2,3-dibromopropyl) phosphate	sv						261-A8					
72-571	Trypan blue	sv						261-A8					
	Turbidity, NTU:			136.3-T1B/73; 136.3-T2									
66-751	Uracil mustard				141.22			261-A8					
	Uranium		122-AD-T5										
7440-611	Uranium, natural											iris	
	Uranium, soluble salts											iris	
	Vanadic acid, ammonium salt	metal						261-A8					
(Total)	Vanadium		122-AD-T5	136.3-T1B/74; 136.3-T2			258-A1/14; A2		264-A9				
7440-622	Vanadium	metal										iris	
1314-621	Vanadium pentoxide							261-A8		266-A4		iris	
1929-777	Vernam											iris	
	Vernolate					91/7							
50471-448	Vinclozolin		122-AD-T5									iris	
108-054	Vinyl acetate						258-A1/60; A2		264-A9		hap	iris	
593-602	Vinyl bromide										hap	iris	
75-014	Vinyl chloride; Chloroethene	v	122-AD-T2/31V	136.3-T1C/97; 136.3-T2	141.61.a	91/7	258-A1/61; A2	261-A8	264-A9	266-A5	hap	iris	
75-354	Vinylidene chloride										hap*		

SECTION 2. CHEMICAL CROSS-REFERENCE FOR SAMPLING, ANALYSIS, MONITORING AND RISK ASSESSMENT - ALPHABETICAL ORDER

CAS RN	Alphabetic Order	type[1]	122[2]	136[3]	141[4]	EPA.[5]	258[6]	261[7]	264[8]	266[9]	hap[10]	iris[11]	pic[12]
81-812	Warfarin	sv						261-A8		266-A4		iris	
7723-140	White phosphorus											iris	
1330-207	Xylene		122-AD-T5				258-A1/62; A2		264-A9				
108-383	Xylene(m-)					91/7					hap	iris	
95-476	Xylene(o-)					91/7					hap	iris	
106-423	Xylene(p-)					91/7					hap	iris	
1330-207	Xylene (total)				141.61.a					266-A4	hap	iris	
	Xylenol		122-AD-T5										
(Total)	Zinc		122-AD-T3	136.3-T1B/75; 136.3-T2			258-A1/15; A2		264-A9				
7440-666	Zinc and compounds											iris	
557-211	Zinc cyanide	cyanide						261-A8		266-A4		iris	
1314-847	Zinc phosphide	solid						261-A8		266-A4		iris	
12122-677	Zineb		122-AD-T5									iris	
	Zirconium		122-AD-T5										

Notes:

1 type: Compound type including:
- sv: semivolatile
- v: volatile

2 122: 40CFR122 including:
- 122-AD-T1 (40CFR122-Appendix D-Table 1): testing requirements for organic toxic pollutants by industrial category for existing dischargers
- 122-AD-T2/x (40CFR122-Appendix D-Table 2/Chemical Number x): organic toxic pollutants in each of four fractions in analysis by gas chromatography/mass spectroscopy (GC/MS)
- 122-AD-T3 (40CFR122-Appendix D-Table 3): other toxic pollutants (metals and cyanide) and total phenols
- 122-AD-T4 (40CFR122-Appendix D-Table 4): conventional and nonconventional pollutants required to be tested by existing dischargers if expected to be present
- 122-AD-T5 (40CFR122-Appendix D-Table 5): toxic pollutants and hazardous substances required to be identified by existing dischargers if expected to be present

3 136: 40CFR136 including:
- 136.3-T1A/x (40CFR136.3-Table 1A/Chemical Number x): List of approved biological test procedures
- 136.3-T1B/x (40CFR136.3-Table 1B/Chemical Number x): List of approved inorganic test procedures
- 136.3-T1C/x (40CFR136.3-Table 1C/Chemical Number x): List of approved test procedures for non-pesticide organic compounds
- 136.3-T1D/x (40CFR136.3-Table 1D/Chemical Number x): List of approved test procedures for pesticides
- 136.3-T1E/x (40CFR136.3-Table 1E/Chemical Number x): List of approved radiological test procedures
- 136.3-T2 (40CFR136.3-Table 2): required containers, preservation techniques, and holding times

4 141: 40CFR141 including:
- 141.23.a: Detection limits for inorganic contaminants
- 141.23.k: Inorganic analysis methods

2 - 57

SAMPLING, ANALYSIS, AND MONITORING METHODS: A GUIDE TO EPA REQUIREMENTS

- 141.24.f: Organic chemicals other than total trihalomethanes, sampling and analytical requirements
- 141.61.a: Organic contaminants and MCL limits
- 141.61.c: Synthetic organic contaminants and MCL limits
- 141.89: Analytical methods for lead, copper, pH, conductivity, calcium, alkalinity, orthophosphate, silica, and temperature

[5] EPA-: (EPA-91/7), "Methods for the Determination of Organic Compounds in Drinking Water," EPA600-4-88-039, July 1991.

[6] 258: 40CFR258 including:
- 40CFR258 Appendix I: Constituents for detection monitoring
- 40CFR258 Appendix II: Hazardous inorganic and organic constituents

[7] 261: 40CFR261 Appendix VIII: Hazardous constituents

[8] 264: 40CFR264 Appendix IX: Ground-water monitoring list

[9] 266: 40CFR266 including:
- 40CFR266 Appendix IV: Reference air concentrations
- 40CFR266 Appendix V: Risk specific doses

[10] hap: Hazardous Air Pollutants listed in Section 112 of the Clean Air Act
hap*: hap compounds that EPA has developed standards for implementation (40CFR61.01)
hap**: hap compounds that are not listed in CAA Section 112, but EPA is regulating tem under the hazardous air pollutant regulations (40CFR61.01)

[11] iris: Integrated Risk Information System available from EPA's National Risk Management Research Laboratory in Cincinnati, Ohio

[12] pic: Products of incomplete combustion found in stack effluents (40CFR266 Appendix VIII)

Section 3

SOLID WASTE PROGRAMS

3.1. SUMMARY OF SOLID WASTE SAMPLING, ANALYSIS AND MONITORING METHODS 2
    TABLE 3.1A. TEST METHODS--ALPHABETICAL ORDER BY TEST PARAMETER 2
    TABLE 3.1B. TEST METHODS--NUMBER ORDER BY METHOD 11

3.2. MUNICIPAL WASTE (40CFR258) 20
    TABLE 3.2. CONSTITUENTS FOR DETECTION MONITORING[1] (40CFR258 APPENDIX I) 20
    TABLE 3.3. ANALYSIS METHODS FOR MUNICIPAL WASTE CONSTITUENTS[1] 23

3.3. HAZARDOUS WASTE 32
    TABLE 3.4. ANALYSIS METHODS FOR HAZARDOUS WASTE CONSTITUENTS 32

3.4. GROUND-WATER 53
    TABLE 3.5. GROUND-WATER MONITORING LIST (40CFR264 APPENDIX IX) 53

3.5. WASTE ANALYSIS 61

3.6. REFERENCES (40CFR260.11.a) 61

3.1. SUMMARY OF SOLID WASTE SAMPLING, ANALYSIS AND MONITORING METHODS

U.S. EPA publishes the Test Methods for Evaluating Solid Waste, Physical/Chemical Methods (SW846) which provides test procedures and guidance for use in conducting the evaluations and measurements needed to comply with the Resource Conservation and Recovery Act (RCRA), Public Law 94-580. These methods are approved by EPA for obtaining data to satisfy the requirements of 40 CFR Parts 122 through 270. This EPA manual presents state-of-the-art in routine analytical testing for the RCRA program. It contains procedures for field and laboratory quality control, sampling, determining hazardous constituents in wastes, determining the hazardous characteristics of wastes (toxicity, ignitability, reactivity, and corrosivity), and for determining the physical properties of wastes. It also contains guidance on how to select appropriate methods. The methods for sampling, analysis, and monitoring are listed in Tables 3.1A and 3.1B.

	TABLE 3.1A. TEST METHODS--ALPHABETICAL ORDER BY TEST PARAMETER	
Method	Test Parameter	Description
1110	Characteristics • Corrosivity	Corrosivity toward steel
1310	Characteristics • Toxicity	Extraction procedure (EP) toxicity test method and structural integrity test
1010	Characteristics • Ignitability	Pensky-martens closed-cup method for determining ignitability
1020	Characteristics • Ignitability	Setaflash closed-cup method for determining ignitability
1312	Characteristics • Toxicity	Synthetic precipitation leaching procedure
1311	Characteristics • Toxicity	Toxicity characteristic leaching procedure (TCLP)
3010	Metallic analytes • Sample preparation methods	Acid digestion of aqueous samples and extracts for total metals for analysis by flame atomic absorption spectroscopy or inductively coupled plasma spectroscopy
3005	Metallic analytes • Sample preparation methods	Acid digestion of waters for total recoverable or dissolved metals for analysis by flame atomic absorption Spectroscopy or inductively coupled plasma spectroscopy
3050	Metallic analytes • Sample preparation methods	Acid digestion of sediments, sludges, and soils
3020	Metallic analytes • Sample preparation methods	Acid digestion of aqueous samples and extracts for total metals for analysis by graphite furnace atomic absorption spectroscopy
5041	Metallic analytes • Sample preparation methods	Analysis of sorbent cartridges from volatile organic sampling train (VOST) wide-bore capillary gas chromatography/mass spectrometry technique
3541	Metallic analytes • Sample preparation methods	Automated soxhlet extraction
5050	Metallic analytes • Sample preparation methods	Bomb combustion method for solid waste
5110	Metallic analytes • Sample preparation methods	Determination of organic phase vapor pressure in waste samples
5100	Metallic analytes • Sample preparation methods	Determination of the volatile organic content of waste samples
3040	Metallic analytes • Sample preparation methods	Dissolution procedure for oils, greases, or waxes

Method	Test Parameter	Description
— TABLE 3.1A. TEST METHODS--ALPHABETICAL ORDER BY TEST PARAMETER		
3051	Metallic analytes • Sample preparation methods	Microwave assisted acid digestion of sludges, soils, and oils
3015	Metallic analytes • Sample preparation methods	Microwave assisted acid digestion of aqueous samples and extracts
7020	Metals	Aluminum (AA, direct aspiration)
7040	Metals	Antimony (AA, direct aspiration)
7041	Metals	Antimony (AA, furnace technique)
7062	Metals	Antimony and arsenic (AA, gaseous borohydride)
7060	Metals	Arsenic (AA, furnace technique)
7061	Metals	Arsenic (AA, gaseous hydride)
7000	Metals	Atomic absorption methods
7080	Metals	Barium (AA, direct aspiration)
7081	Metals	Barium (AA, furnace technique)
7090	Metals	Beryllium (AA, direct aspiration)
7091	Metals	Beryllium (AA, furnace technique)
7130	Metals	Cadmium (AA, direct aspiration)
7131	Metals	Cadmium (AA, furnace technique)
7140	Metals	Calcium (AA, direct aspiration)
7190	Metals	Chromium (AA, direct aspiration)
7191	Metals	Chromium (AA, furnace technique)
7197	Metals	Chromium, hexavalent (chelation/extraction)
7196	Metals	Chromium, hexavalent (colorimetric)
7195	Metals	Chromium, hexavalent (coprecipitation)
7198	Metals	Chromium, hexavalent (differential pulse polarography)
7200	Metals	Cobalt (AA, direct aspiration)
7201	Metals	Cobalt (AA, furnace technique)
7210	Metals	Copper (AA, direct aspiration)
7211	Metals	Copper (AA, furnace technique)
6020	Metals	Inductively coupled plasma mass spectrometry
6010	Metals	Inductively coupled plasma atomic emission spectroscopy
7380	Metals	Iron (AA, direct aspiration)
7381	Metals	Iron (AA, furnace technique)
7420	Metals	Lead (AA, direct aspiration)
7421	Metals	Lead (AA, furnace technique)

Method	Test Parameter	Description
	TABLE 3.1A. TEST METHODS--ALPHABETICAL ORDER BY TEST PARAMETER	
7430	Metals	Lithium (AA, direct aspiration)
7450	Metals	Magnesium (AA, direct aspiration)
7460	Metals	Manganese (AA, direct aspiration)
7461	Metals	Manganese (AA, furnace technique)
7470	Metals	Mercury in liquid waste (manual cold-vapor technique)
7471	Metals	Mercury in solid or semisolid waste (manual cold-vapor technique)
7480	Metals	Molybdenum (AA, direct aspiration)
7481	Metals	Molybdenum (AA, furnace technique)
7520	Metals	Nickel (AA, direct aspiration)
7550	Metals	Osmium (AA, direct aspiration)
7610	Metals	Potassium (AA, direct aspiration)
7740	Metals	Selenium (AA, furnace technique)
7742	Metals	Selenium (AA, gaseous borohydride)
7741	Metals	Selenium (AA, gaseous hydride)
7760	Metals	Silver (AA, direct aspiration)
7761	Metals	Silver (AA, furnace technique)
7770	Metals	Sodium (AA, direct aspiration)
7780	Metals	Strontium (AA, direct aspiration)
7840	Metals	Thallium (AA, direct aspiration)
7841	Metals	Thallium (AA, furnace technique)
7870	Metals	Tin (AA, direct aspiration)
7910	Metals	Vanadium (AA, direct aspiration)
7911	Metals	Vanadium (AA, furnace technique)
7950	Metals	Zinc (AA, direct aspiration)
7951	Metals	Zinc (AA, furnace technique)
9030	Miscellaneous test methods	Acid-soluble and acid-insoluble sulfides
9251	Miscellaneous test methods	Chloride (colorimetric, automated ferricyanide aaii)
9250	Miscellaneous test methods	Chloride (colorimetric, automated ferricyanide aai)
9252	Miscellaneous test methods	Chloride (titrimetric, mercuric nitrate)
9253	Miscellaneous test methods	Chloride (titrimetric, silver nitrate)
9011	Miscellaneous test methods	Cyanide extraction procedure for solids and oils
9010A	Miscellaneous test methods	Cyanide extraction procedure for solids and oils
9311	Miscellaneous test methods	Determination of gross alpha activity in drinking water by coprecipitation

Method	Test Parameter	Description
	TABLE 3.1A. TEST METHODS--ALPHABETICAL ORDER BY TEST PARAMETER	
9031	Miscellaneous test methods	Extractable sulfides
9056	Miscellaneous test methods	Ion chromatography method
9096	Miscellaneous test methods	Liquid release test procedure
9312	Miscellaneous test methods	Method for gross alpha in solid samples
9200	Miscellaneous test methods	Nitrate
9071	Miscellaneous test methods	Oil & grease extraction method for sludge samples
9066	Miscellaneous test methods	Phenolics (colorimetric, automated 4-AAP with distillation)
9065	Miscellaneous test methods	Phenolics (spectrophotometric, manual 4-AAP with distillation)
9067	Miscellaneous test methods	Phenolics (spectrophotometric, MBTH with distillation)
9021	Miscellaneous test methods	Purgeable organic halides (POX)
9320	Miscellaneous test methods	Radium-228
9035	Miscellaneous test methods	Sulfate (colorimetric, automated, chloranilate)
9036	Miscellaneous test methods	Sulfate (colorimetric, automated, methylthymol blue, AA II)
9038	Miscellaneous test methods	Sulfate (turbidimetric)
9075	Miscellaneous test methods	Test method for total chlorine in used oil by x-ray fluorescence spectrometry (xrf)
9076	Miscellaneous test methods	Test method for total chlorine in new and used petroleum products by oxidative combustion and microcoulometry
9077	Miscellaneous test methods	Test methods for total chlorine in new and used petroleum products (field test kit methods)
9010	Miscellaneous test methods	Total and amenable cyanide (colorimetric, manual)
9012	Miscellaneous test methods	Total and amenable cyanide (colorimetric, automated uv)
9132	Miscellaneous test methods	Total coliformmembrane filter technique
9131	Miscellaneous test methods	Total coliformmultiple tube fermentation technique
9060	Miscellaneous test methods	Total organic carbon
9020	Miscellaneous test methods	Total organic halides (TOX)
9022	Miscellaneous test methods	Total organic halides (TOX) by neutron activation analysis
9073	Miscellaneous test methods	Total recoverable hydrocarbons by infrared spectroscopy
9070	Miscellaneous test methods	Total recoverable oil & grease (gravimetric, separatory funnel extraction)
8011	Organic analytes • Gas chromatographic methods	1,2-dibromoethane and 1,2-dibromo-3-chloropropane by gas chromatography by gas chromatography
3650	Organic analytes • Sample preparation methods - Cleanup	Acid-base partition cleanup
8030	Organic analytes • Gas chromatographic methods	Acrolein, acrylonitrile, acetonitrile by gas chromatography

Method	Test Parameter	Description
\multicolumn{3}{c}{TABLE 3.1A. TEST METHODS--ALPHABETICAL ORDER BY TEST PARAMETER}		
8316	Organic analytes • High performance liquid chromatographic methods	Acrylamide, acrylonitrile and acrolein by high performance liquid chromatography (HPLC)
8032	Organic analytes • Gas chromatographic methods	Acrylamide by gas chromatography
8031	Organic analytes • Gas chromatographic methods	Acrylonitrile by gas chromatography
8145	Organic analytes • Gas chromatographic methods	Alkyl phosphates by gas chromatography (possible incorporation into method 8141)
3610	Organic analytes • Sample preparation methods - Cleanup	Alumina column cleanup
3611	Organic analytes • Sample preparation methods - Cleanup	Alumina column cleanup and separation of petroleum wastes
5041	Organic analytes • Sample preparation methods - Extractions and preparations	Analysis of sorbent cartridges from volatile organic sampling train (VOST) wide-bore capillary gas chromatography/mass spectrometry technique
5040	Organic analytes • Sample preparation methods - Extractions and preparations	Analysis of sorbent cartridges from volatile organic sampling train (VOST) gas chromatograph/mass Spectrometry technique
8020	Organic analytes • Gas chromatographic methods	Aromatic volatile organics by gas chromatography
3541	Organic analytes • Sample preparation methods - Extractions and preparations	Automated soxhlet extraction
8150	Organic analytes • Gas chromatographic methods	Chlorinated herbicides by gas chromatography
8151	Organic analytes • Gas chromatographic methods	Chlorinated herbicides by gas chromatography capillary column technique
8121	Organic analytes • Gas chromatographic methods	Chlorinated hydrocarbons by gas chromatography capillary column technique
8120	Organic analytes • Gas chromatographic methods	Chlorinated hydrocarbons by gas chromatography
3600	Organic analytes • Sample preparation methods - Cleanup	Cleanup
3520B	Organic analytes • Sample preparation methods - Extractions and preparations	Continuous liquid-liquid extraction
5110	Organic analytes • Sample preparation methods - Extractions and preparations	Determination of organic phase vapor pressure in waste samples

Method	Test Parameter	Description
	TABLE 3.1A. TEST METHODS--ALPHABETICAL ORDER BY TEST PARAMETER	
5100	Organic analytes • Sample preparation methods - Extractions and preparations	Determination of the volatile organic content of waste samples
8340	Organic analytes • High performance liquid chromatographic methods	Diquat and paraquat by high performance liquid chromatography (HPLC)
8045	Organic analytes • Gas chromatographic methods	Endothall by derivatization and gas chromatography
3620	Organic analytes • Sample preparation methods - Cleanup	Florisil column cleanup
8315	Organic analytes • High performance liquid chromatographic methods	Formaldehyde by high performance liquid chromatography
8000	Organic analytes • Gas chromatographic methods	Gas chromatography
3640	Organic analytes • Sample preparation methods - Cleanup	Gel-permeation chromatography (gpc) cleanup
8110	Organic analytes • Gas chromatographic methods	Haloethers by gas chromatography
8021	Organic analytes • Gas chromatographic methods	Halogenated and aromatic volatiles by gas chromatography using electrolytic conductivity and photoionization detectors in series capillary column technique
8010	Organic analytes • Gas chromatographic methods	Halogenated volatile organics
3810	Organic analytes • Miscellaneous screening methods	Headspace
3820	Organic analytes • Miscellaneous screening methods	Hexadecane extraction and screening of purgeable organics
8318	Organic analytes • High performance liquid chromatographic methods	N-methyl carbamates by high performance liquid chromatography (HPLC)
8333	Organic analytes • High performance liquid chromatographic methods	Nitro compounds by high performance liquid chromatography (HPLC) (possible incorporation into method 8330)
8090	Organic analytes • Gas chromatographic methods	Nitroaromatics and cyclic ketones by gas chromatography
8330	Organic analytes • High performance liquid chromatographic methods	Nitroaromatics and nitramines by high performance liquid chromatography(HPLC)
8332	Organic analytes • High performance liquid chromatographic methods	Nitroglycerine by high performance liquid chromatography (HPLC)

	TABLE 3.1A. TEST METHODS--ALPHABETICAL ORDER BY TEST PARAMETER	
Method	Test Parameter	Description
8070	Organic analytes • Gas chromatographic methods	Nitrosamines by gas chromatography
8015	Organic analytes • Gas chromatographic methods	Nonhalogenated volatile organics
3500	Organic analytes • Sample preparation methods - Extractions and preparations	Organic extraction and sample preparation
8080	Organic analytes • Gas chromatographic methods	Organochlorine pesticides and polychlorinated biphenyls by gas chromatography
8081	Organic analytes • Gas chromatographic methods	Organochlorine pesticides and polychlorinated biphenyls by gas chromatography capillary column technique
8141	Organic analytes • Gas chromatographic methods	Organophosphorus compounds by gas chromatography capillary column technique
8355	Organic analytes • High performance liquid chromatographic methods	Organophosphorus compounds by high performance liquid chromatography with thermospray/mass Spectrometry (HPLC/TSP/MS) detection
8140	Organic analytes • Gas chromatographic methods	Organophosphorus pesticides by gas chromatography
8040	Organic analytes • Gas chromatographic methods	Phenols by gas chromatography
8061	Organic analytes • Gas chromatographic methods	Phthalate esters by gas chromatography capillary technique
8060	Organic analytes • Gas chromatographic methods	Phthalate esters by gas chromatography
8085	Organic analytes • Gas chromatographic methods	Polychlorinated biphenyls by perchlorination, derivatization and gas chromatography
8290	Organic analytes • Gas chromatographic/mass spectroscopic methods	Polychlorinated dibenzodioxins (pcdds) and polychlorinated dibenzofurans (pcdfs) by high resolution gas chromatography/High Resolution Mass Spectrometry (HRGC/HRMS)
8310	Organic analytes • High performance liquid chromatographic methods	Polynuclear aromatic hydrocarbons
8100	Organic analytes • Gas chromatographic methods	Polynuclear aromatic hydrocarbons by gas chromatography
5030	Organic analytes • Sample preparation methods - Extractions and preparations	Purge-and-trap
8321	Organic analytes • High performance liquid chromatographic methods	Reverse phase high performance liquid chromatography with thermospray/mass spectrometry (HPLC/tsp/ms) detection
8275	Organic analytes • Gas chromatographic/mass spectroscopic methods	Semivolatile organic compounds by thermal chromatography/mass spectrometry (TC/MS) screening technique

Method	Test Parameter	Description
\multicolumn{3}{c}{TABLE 3.1A. TEST METHODS--ALPHABETICAL ORDER BY TEST PARAMETER}		
8250	Organic analytes • Gas chromatographic/mass spectroscopic methods	Semivolatile organic compounds by gas chromatography/mass spectrometry (GC/MS) packed column Technique
8275	Organic analytes • Miscellaneous screening methods	Semivolatile organic compounds by thermal chromatography/mass spectrometry (TC/MS) screening technique
8270	Organic analytes • Gas chromatographic/mass spectroscopic methods	Semivolatile organic compounds by gas chromatography/mass spectrometry (GC/MS) capillary technique
8410	Organic analytes • Fourier transform infrared methods	Semivolatile organics by gas chromatography/fourier transform infrared spectroscopy (GC/FTIR) capillary Column technique
3510B	Organic analytes • Sample preparation methods - Extractions and preparations	Separatory funnel liquid-liquid extraction
3630	Organic analytes • Sample preparation methods - Cleanup	Silica gel cleanup
3660	Organic analytes • Sample preparation methods - Cleanup	Sulfur cleanup
3665	Organic analytes • Sample preparation methods - Cleanup	Sulfuric acid/permanganate cleanup
8331	Organic analytes • High performance liquid chromatographic methods	Tetrazene by high performance liquid chromatography (HPLC)
8280	Organic analytes • Gas chromatographic/mass spectroscopic methods	The analysis of polychlorinated dibenzo-p-dioxins and polychlorinated dibenzofurans. Appendix a signal-to-noise determination methods. Appendix b recommended safety and handling procedures for pcdds/pcdfs
8350	Organic analytes • High performance liquid chromatographic methods	Tris-2,3-dibromopropyl phosphate by high performance liquid chromatography with thermospray/mass spectrometry (HPLC/TSP/MS) detection
8415	Organic analytes • Fourier transform infrared methods	Tris-2,3-dibromopropyl phosphate by gas chromatography/fourier transform infrared spectroscopy (GC/FTIR)
3550	Organic analytes • Sample preparation methods - Extractions and preparations	Ultrasonic extraction
8260	Organic analytes • Gas chromatographic/mass spectroscopic methods	Volatile organic compounds by gas chromatography/mass spectrometry (GC/MS) capillary technique
8240	Organic analytes • Gas chromatographic/mass spectroscopic methods	Volatile organic compounds by gas chromatography/mass spectrometry (GC/MS) packed column technique

Method	Test Parameter	Description

TABLE 3.1A. TEST METHODS--ALPHABETICAL ORDER BY TEST PARAMETER

Method	Test Parameter	Description
3580	Organic analytes • Sample preparation methods - Extractions and preparations	Waste dilution
5031	Organic analytes • Sample preparation methods - Extractions and preparations	Water soluble volatile organic compounds by azeotropic distillation
9315	Properties	Alpha-emitting radium isotopes
9081	Properties	Cation-exchange capacity of soils (sodium acetate)
9080	Properties	Cation-exchange capacity of soils (ammonium acetate)
9090	Properties	Compatibility test for wastes and membrane liners
9311	Properties	Determination of gross alpha activity in drinking water by coprecipitation
1330	Properties	Extraction procedure for oily wastes
9310	Properties	Gross alpha & gross beta
9096	Properties	Liquid release test (LRT) procedure
9312	Properties	Method for gross alpha in solid samples
1320	Properties	Multiple extraction procedure
9095	Properties	Paint filter liquids test
9040	Properties	pH electrometric measurement
9041	Properties	pH paper method
9100	Properties	Saturated hydraulic conductivity, saturated leachate conductivity, and intrinsic permeability
9045	Properties	Solid and waste pH
9050	Properties	Specific conductance
0010	Sampling methods	Modified method 5 sampling train. Appendix A preparation of XAD-2 sorbent resin appendix b total chromatographable organic material analysis
0020	Sampling methods	Source assessment sampling system (SASS)
0030	Sampling methods	Volatile organic sampling train (VOST)

TABLE 3.1B. TEST METHODS--NUMBER ORDER BY METHOD		
Method	Test Parameter	Description
0010	Sampling methods	Modified method 5 sampling train. Appendix A preparation of XAD-2 sorbent resin appendix b total chromatographable organic material analysis
0020	Sampling methods	Source assessment sampling system (SASS)
0030	Sampling methods	Volatile organic sampling train (VOST)
1010	Characteristics • Ignitability	Pensky-martens closed-cup method for determining ignitability
1020	Characteristics • Ignitability	Setaflash closed-cup method for determining ignitability
1110	Characteristics • Corrosivity	Corrosivity toward steel
1310	Characteristics • Toxicity	Extraction procedure (EP) toxicity test method and structural integrity test
1311	Characteristics • Toxicity	Toxicity characteristic leaching procedure (TCLP)
1312	Characteristics • Toxicity	Synthetic precipitation leaching procedure
1320	Properties	Multiple extraction procedure
1330	Properties	Extraction procedure for oily wastes
3005	Metallic analytes • Sample preparation methods	Acid digestion of waters for total recoverable or dissolved metals for analysis by flame atomic absorption Spectroscopy or inductively coupled plasma spectroscopy
3010	Metallic analytes • Sample preparation methods	Acid digestion of aqueous samples and extracts for total metals for analysis by flame atomic absorption spectroscopy or inductively coupled plasma spectroscopy
3015	Metallic analytes • Sample preparation methods	Microwave assisted acid digestion of aqueous samples and extracts
3020	Metallic analytes • Sample preparation methods	Acid digestion of aqueous samples and extracts for total metals for analysis by graphite furnace atomic absorption spectroscopy
3040	Metallic analytes • Sample preparation methods	Dissolution procedure for oils, greases, or waxes
3050	Metallic analytes • Sample preparation methods	Acid digestion of sediments, sludges, and soils
3051	Metallic analytes • Sample preparation methods	Microwave assisted acid digestion of sludges, soils, and oils
3500	Organic analytes • Sample preparation methods - Extractions and preparations	Organic extraction and sample preparation
3510B	Organic analytes • Sample preparation methods - Extractions and preparations	Separatory funnel liquid-liquid extraction

	TABLE 3.1B. TEST METHODS--NUMBER ORDER BY METHOD	
Method	Test Parameter	Description
3520B	Organic analytes • Sample preparation methods - Extractions and preparations	Continuous liquid-liquid extraction
3541	Organic analytes • Sample preparation methods - Extractions and preparations	Automated soxhlet extraction
3541	Metallic analytes • Sample preparation methods	Automated soxhlet extraction
3550	Organic analytes • Sample preparation methods - Extractions and preparations	Ultrasonic extraction
3580	Organic analytes • Sample preparation methods - Extractions and preparations	Waste dilution
3600	Organic analytes • Sample preparation methods - Cleanup	Cleanup
3610	Organic analytes • Sample preparation methods - Cleanup	Alumina column cleanup
3611	Organic analytes • Sample preparation methods - Cleanup	Alumina column cleanup and separation of petroleum wastes
3620	Organic analytes • Sample preparation methods - Cleanup	Florisil column cleanup
3630	Organic analytes • Sample preparation methods - Cleanup	Silica gel cleanup
3640	Organic analytes • Sample preparation methods - Cleanup	Gel-permeation chromatography (gpc) cleanup
3650	Organic analytes • Sample preparation methods - Cleanup	Acid-base partition cleanup
3660	Organic analytes • Sample preparation methods - Cleanup	Sulfur cleanup
3665	Organic analytes • Sample preparation methods - Cleanup	Sulfuric acid/permanganate cleanup
3810	Organic analytes • Miscellaneous screening methods	Headspace
3820	Organic analytes • Miscellaneous screening methods	Hexadecane extraction and screening of purgeable organics

	TABLE 3.1B. TEST METHODS--NUMBER ORDER BY METHOD	
Method	Test Parameter	Description
5030	Organic analytes • Sample preparation methods - Extractions and preparations	Purge-and-trap
5031	Organic analytes • Sample preparation methods - Extractions and preparations	Water soluble volatile organic compounds by azeotropic distillation
5040	Organic analytes • Sample preparation methods - Extractions and preparations	Analysis of sorbent cartridges from volatile organic sampling train (VOST) gas chromatograph/mass Spectrometry technique
5041	Organic analytes • Sample preparation methods - Extractions and preparations	Analysis of sorbent cartridges from volatile organic sampling train (VOST) wide-bore capillary gas chromatography/mass spectrometry technique
5041	Metallic analytes • Sample preparation methods	Analysis of sorbent cartridges from volatile organic sampling train (VOST) wide-bore capillary gas chromatography/mass spectrometry technique
5050	Metallic analytes • Sample preparation methods	Bomb combustion method for solid waste
5100	Metallic analytes • Sample preparation methods	Determination of the volatile organic content of waste samples
5100	Organic analytes • Sample preparation methods - Extractions and preparations	Determination of the volatile organic content of waste samples
5110	Metallic analytes • Sample preparation methods	Determination of organic phase vapor pressure in waste samples
5110	Organic analytes • Sample preparation methods - Extractions and preparations	Determination of organic phase vapor pressure in waste samples
6010	Metals	Inductively coupled plasma atomic emission spectroscopy
6020	Metals	Inductively coupled plasma mass spectrometry
7000	Metals	Atomic absorption methods
7020	Metals	Aluminum (AA, direct aspiration)
7040	Metals	Antimony (AA, direct aspiration)
7041	Metals	Antimony (AA, furnace technique)
7060	Metals	Arsenic (AA, furnace technique)
7061	Metals	Arsenic (AA, gaseous hydride)
7062	Metals	Antimony and arsenic (AA, gaseous borohydride)
7080	Metals	Barium (AA, direct aspiration)
7081	Metals	Barium (AA, furnace technique)
7090	Metals	Beryllium (AA, direct aspiration)
7091	Metals	Beryllium (AA, furnace technique)

	TABLE 3.1B. TEST METHODS--NUMBER ORDER BY METHOD	
Method	Test Parameter	Description
7130	Metals	Cadmium (AA, direct aspiration)
7131	Metals	Cadmium (AA, furnace technique)
7140	Metals	Calcium (AA, direct aspiration)
7190	Metals	Chromium (AA, direct aspiration)
7191	Metals	Chromium (AA, furnace technique)
7195	Metals	Chromium, hexavalent (coprecipitation)
7196	Metals	Chromium, hexavalent (colorimetric)
7197	Metals	Chromium, hexavalent (chelation/extraction)
7198	Metals	Chromium, hexavalent (differential pulse polarography)
7200	Metals	Cobalt (AA, direct aspiration)
7201	Metals	Cobalt (AA, furnace technique)
7210	Metals	Copper (AA, direct aspiration)
7211	Metals	Copper (AA, furnace technique)
7380	Metals	Iron (AA, direct aspiration)
7381	Metals	Iron (AA, furnace technique)
7420	Metals	Lead (AA, direct aspiration)
7421	Metals	Lead (AA, furnace technique)
7430	Metals	Lithium (AA, direct aspiration)
7450	Metals	Magnesium (AA, direct aspiration)
7460	Metals	Manganese (AA, direct aspiration)
7461	Metals	Manganese (AA, furnace technique)
7470	Metals	Mercury in liquid waste (manual cold-vapor technique)
7471	Metals	Mercury in solid or semisolid waste (manual cold-vapor technique)
7480	Metals	Molybdenum (AA, direct aspiration)
7481	Metals	Molybdenum (AA, furnace technique)
7520	Metals	Nickel (AA, direct aspiration)
7550	Metals	Osmium (AA, direct aspiration)
7610	Metals	Potassium (AA, direct aspiration)
7740	Metals	Selenium (AA, furnace technique)
7741	Metals	Selenium (AA, gaseous hydride)
7742	Metals	Selenium (AA, gaseous borohydride)
7760	Metals	Silver (AA, direct aspiration)
7761	Metals	Silver (AA, furnace technique)

Method	Test Parameter	Description
\multicolumn{3}{c}{TABLE 3.1B. TEST METHODS--NUMBER ORDER BY METHOD}
7770	Metals	Sodium (AA, direct aspiration)
7780	Metals	Strontium (AA, direct aspiration)
7840	Metals	Thallium (AA, direct aspiration)
7841	Metals	Thallium (AA, furnace technique)
7870	Metals	Tin (AA, direct aspiration)
7910	Metals	Vanadium (AA, direct aspiration)
7911	Metals	Vanadium (AA, furnace technique)
7950	Metals	Zinc (AA, direct aspiration)
7951	Metals	Zinc (AA, furnace technique)
8000	Organic analytes • Gas chromatographic methods	Gas chromatography
8010	Organic analytes • Gas chromatographic methods	Halogenated volatile organics
8011	Organic analytes • Gas chromatographic methods	1,2-dibromoethane and 1,2-dibromo-3-chloropropane by gas chromatography by gas chromatography
8015	Organic analytes • Gas chromatographic methods	Nonhalogenated volatile organics
8020	Organic analytes • Gas chromatographic methods	Aromatic volatile organics by gas chromatography
8021	Organic analytes • Gas chromatographic methods	Halogenated and aromatic volatiles by gas chromatography using electrolytic conductivity and photoionization detectors in series capillary column technique
8030	Organic analytes • Gas chromatographic methods	Acrolein, acrylonitrile, acetonitrile by gas chromatography
8031	Organic analytes • Gas chromatographic methods	Acrylonitrile by gas chromatography
8032	Organic analytes • Gas chromatographic methods	Acrylamide by gas chromatography
8040	Organic analytes • Gas chromatographic methods	Phenols by gas chromatography
8045	Organic analytes • Gas chromatographic methods	Endothall by derivatization and gas chromatography
8060	Organic analytes • Gas chromatographic methods	Phthalate esters by gas chromatography
8061	Organic analytes • Gas chromatographic methods	Phthalate esters by gas chromatography capillary technique
8070	Organic analytes • Gas chromatographic methods	Nitrosamines by gas chromatography

	TABLE 3.1B. TEST METHODS--NUMBER ORDER BY METHOD	
Method	Test Parameter	Description
8080	Organic analytes • Gas chromatographic methods	Organochlorine pesticides and polychlorinated biphenyls by gas chromatography
8081	Organic analytes • Gas chromatographic methods	Organochlorine pesticides and polychlorinated biphenyls by gas chromatography capillary column technique
8085	Organic analytes • Gas chromatographic methods	Polychlorinated biphenyls by perchlorination, derivatization and gas chromatography
8090	Organic analytes • Gas chromatographic methods	Nitroaromatics and cyclic ketones by gas chromatography
8100	Organic analytes • Gas chromatographic methods	Polynuclear aromatic hydrocarbons by gas chromatography
8110	Organic analytes • Gas chromatographic methods	Haloethers by gas chromatography
8120	Organic analytes • Gas chromatographic methods	Chlorinated hydrocarbons by gas chromatography
8121	Organic analytes • Gas chromatographic methods	Chlorinated hydrocarbons by gas chromatography capillary column technique
8140	Organic analytes • Gas chromatographic methods	Organophosphorus pesticides by gas chromatography
8141	Organic analytes • Gas chromatographic methods	Organophosphorus compounds by gas chromatography capillary column technique
8145	Organic analytes • Gas chromatographic methods	Alkyl phosphates by gas chromatography (possible incorporation into method 8141)
8150	Organic analytes • Gas chromatographic methods	Chlorinated herbicides by gas chromatography
8151	Organic analytes • Gas chromatographic methods	Chlorinated herbicides by gas chromatography capillary column technique
8240	Organic analytes • Gas chromatographic/mass spectroscopic methods	Volatile organic compounds by gas chromatography/mass spectrometry (GC/MS) packed column technique
8250	Organic analytes • Gas chromatographic/mass spectroscopic methods	Semivolatile organic compounds by gas chromatography/mass spectrometry (GC/MS) packed column Technique
8260	Organic analytes • Gas chromatographic/mass spectroscopic methods	Volatile organic compounds by gas chromatography/mass spectrometry (GC/MS) capillary technique
8270	Organic analytes • Gas chromatographic/mass spectroscopic methods	Semivolatile organic compounds by gas chromatography/mass spectrometry (GC/MS) capillary technique
8275	Organic analytes • Miscellaneous screening methods	Semivolatile organic compounds by thermal chromatography/mass spectrometry (TC/MS) screening technique
8275	Organic analytes • Gas chromatographic/mass spectroscopic methods	Semivolatile organic compounds by thermal chromatography/mass spectrometry (TC/MS) screening technique

	TABLE 3.1B. TEST METHODS--NUMBER ORDER BY METHOD	
Method	Test Parameter	Description
8280	Organic analytes • Gas chromatographic/mass spectroscopic methods	The analysis of polychlorinated dibenzo-p-dioxins and polychlorinated dibenzofurans. Appendix a signal-to-noise determination methods. Appendix b recommended safety and handling procedures for pcdds/pcdfs
8290	Organic analytes • Gas chromatographic/mass spectroscopic methods	Polychlorinated dibenzodioxins (pcdds) and polychlorinated dibenzofurans (pcdfs) by high resolution gas chromatography/High Resolution Mass Spectrometry (HRGC/HRMS)
8310	Organic analytes • High performance liquid chromatographic methods	Polynuclear aromatic hydrocarbons
8315	Organic analytes • High performance liquid chromatographic methods	Formaldehyde by high performance liquid chromatography
8316	Organic analytes • High performance liquid chromatographic methods	Acrylamde, acrylonitrile and acrolein by high performance liquid chromatography (HPLC)
8318	Organic analytes • High performance liquid chromatographic methods	N-methyl carbamates by high performance liquid chromatography (HPLC)
8321	Organic analytes • High performance liquid chromatographic methods	Reverse phase high performance liquid chromatography with thermospray/mass spectrometry (HPLC/tsp/ms) detection
8330	Organic analytes • High performance liquid chromatographic methods	Nitroaromatics and nitramines by high performance liquid chromatography(HPLC)
8331	Organic analytes • High performance liquid chromatographic methods	Tetrazene by high performance liquid chromatography (HPLC)
8332	Organic analytes • High performance liquid chromatographic methods	Nitroglycerine by high performance liquid chromatography (HPLC)
8333	Organic analytes • High performance liquid chromatographic methods	Nitro compounds by high performance liquid chromatography (HPLC) (possible incorporation into method 8330)
8340	Organic analytes • High performance liquid chromatographic methods	Diquat and paraquat by high performance liquid chromatography (HPLC)
8350	Organic analytes • High performance liquid chromatographic methods	Tris-2,3-dibromopropyl phosphate by high performance liquid chromatography with thermospray/mass spectrometry (HPLC/TSP/MS) detection
8355	Organic analytes • High performance liquid chromatographic methods	Organophosphorus compounds by high performance liquid chromatography with thermospray/mass Spectrometry (HPLC/TSP/MS) detection
8410	Organic analytes • Fourier transform infrared methods	Semivolatile organics by gas chromatography/fourier transform infrared spectroscopy (GC/FTIR) capillary Column technique

Method	Test Parameter	Description
\multicolumn{3}{c}{TABLE 3.1B. TEST METHODS--NUMBER ORDER BY METHOD}		
8415	Organic analytes • Fourier transform infrared methods	Tris-2,3-dibromopropyl phosphate by gas chromatography/fourier transform infrared spectroscopy (GC/FTIR)
9010	Miscellaneous test methods	Total and amenable cyanide (colorimetric, manual)
9010A	Miscellaneous test methods	Cyanide extraction procedure for solids and oils
9011	Miscellaneous test methods	Cyanide extraction procedure for solids and oils
9012	Miscellaneous test methods	Total and amenable cyanide (colorimetric, automated uv)
9020	Miscellaneous test methods	Total organic halides (TOX)
9021	Miscellaneous test methods	Purgeable organic halides (POX)
9022	Miscellaneous test methods	Total organic halides (TOX) by neutron activation analysis
9030	Miscellaneous test methods	Acid-soluble and acid-insoluble sulfides
9031	Miscellaneous test methods	Extractable sulfides
9035	Miscellaneous test methods	Sulfate (colorimetric, automated, chloranilate)
9036	Miscellaneous test methods	Sulfate (colorimetric, automated, methylthymol blue, AA II)
9038	Miscellaneous test methods	Sulfate (turbidimetric)
9040	Properties	pH electrometric measurement
9041	Properties	pH paper method
9045	Properties	Solid and waste pH
9050	Properties	Specific conductance
9056	Miscellaneous test methods	Ion chromatography method
9060	Miscellaneous test methods	Total organic carbon
9065	Miscellaneous test methods	Phenolics (spectrophotometric, manual 4-AAP with distillation)
9066	Miscellaneous test methods	Phenolics (colorimetric, automated 4-AAP with distillation)
9067	Miscellaneous test methods	Phenolics (spectrophotometric, MBTH with distillation)
9070	Miscellaneous test methods	Total recoverable oil & grease (gravimetric, separatory funnel extraction)
9071	Miscellaneous test methods	Oil & grease extraction method for sludge samples
9073	Miscellaneous test methods	Total recoverable hydrocarbons by infrared spectroscopy
9075	Miscellaneous test methods	Test method for total chlorine in used oil by x-ray fluorescence spectrometry (xrf)
9076	Miscellaneous test methods	Test method for total chlorine in new and used petroleum products by oxidative combustion and microcoulometry
9077	Miscellaneous test methods	Test methods for total chlorine in new and used petroleum products (field test kit methods)
9080	Properties	Cation-exchange capacity of soils (ammonium acetate)
9081	Properties	Cation-exchange capacity of soils (sodium acetate)

TABLE 3.1B. TEST METHODS--NUMBER ORDER BY METHOD		
Method	Test Parameter	Description
9090	Properties	Compatibility test for wastes and membrane liners
9095	Properties	Paint filter liquids test
9096	Properties	Liquid release test (LRT) procedure
9096	Miscellaneous test methods	Liquid release test procedure
9100	Properties	Saturated hydraulic conductivity, saturated leachate conductivity, and intrinsic permeability
9131	Miscellaneous test methods	Total coliformmultiple tube fermentation technique
9132	Miscellaneous test methods	Total coliformmembrane filter technique
9200	Miscellaneous test methods	Nitrate
9250	Miscellaneous test methods	Chloride (colorimetric, automated ferricyanide aai)
9251	Miscellaneous test methods	Chloride (colorimetric, automated ferricyanide aaii)
9252	Miscellaneous test methods	Chloride (titrimetric, mercuric nitrate)
9253	Miscellaneous test methods	Chloride (titrimetric, silver nitrate)
9310	Properties	Gross alpha & gross beta
9311	Miscellaneous test methods	Determination of gross alpha activity in drinking water by coprecipitation
9311	Properties	Determination of gross alpha activity in drinking water by coprecipitation
9312	Miscellaneous test methods	Method for gross alpha in solid samples
9312	Properties	Method for gross alpha in solid samples
9315	Properties	Alpha-emitting radium isotopes
9320	Miscellaneous test methods	Radium-228

3.2. MUNICIPAL WASTE (40CFR258)

- Ground-water monitoring systems (40CFR258.51)
- Ground-water sampling and analysis requirements (40CFR258.53)
- Detection monitoring program (40CFR258.54)
- Assessment monitoring program (40CFR258.55)

TABLE 3.2. CONSTITUENTS FOR DETECTION MONITORING[1] (40CFR258 APPENDIX I)			
CAS RN[3] order		Common name[2]--Alphabetical order	
Inorganic Constituents			
(Total)[3]	(1)[4] Antimony		
(Total)	(2) Arsenic		
(Total)	(3) Barium		
(Total)	(4) Beryllium		
(Total)	(5) Cadmium		
(Total)	(6) Chromium		
(Total)	(7) Cobalt		
(Total)	(8) Copper		
(Total)	(9) Lead		
(Total)	(10) Nickel		
(Total)	(11) Selenium		
(Total)	(12) Silver		
(Total)	(13) Thallium		
(Total)	(14) Vanadium		
(Total)	(15) Zinc		
Organic Constituents			
CAS RN order		Common name--Alphabetical order	
56-235	(23) Carbon tetrachloride	67-641	(16) Acetone
67-641	(16) Acetone	107-131	(17) Acrylonitrile
67-663	(26) Chloroform; Trichloromethane	71-432	(18) Benzene
71-432	(18) Benzene	74-975	(19) Bromochloromethane
71-556	(55) 1,1,1-Trichloroethane; Methylchloroform	75-274	(20) Bromodichloromethane
74-975	(19) Bromochloromethane	75-252	(21) Bromoform; Tribromomethane
74-839	(43) Methyl bromide; Bromomethane	56-235	(23) Carbon tetrachloride
74-873	(44) Methyl chloride; Chloromethane	108-907	(24) Chlorobenzene
74-953	(45) Methylene bromide; Dibromomethane	75-003	(25) Chloroethane; Ethyl chloride
74-884	(48) Methyl iodide; Iodomethane	67-663	(26) Chloroform; Trichloromethane

TABLE 3.2. CONSTITUENTS FOR DETECTION MONITORING[1] (40CFR258 APPENDIX I)			
CAS RN[3] order		Common name[2]--Alphabetical order	
75-274	(20) Bromodichloromethane	124-481	(27) Dibromochloromethane; Chlorodibromomethane
75-252	(21) Bromoform; Tribromomethane	96-128	(28) 1,2-Dibromo-3-chloropropane; DBCP
75-003	(25) Chloroethane; Ethyl chloride	106-934	(29) 1,2-Dibromoethane; Ethylene dibromide; EDB
75-343	(33) 1,1-Dichloroethane; Ethylidene chloride	95-501	(30) o-Dichlorobenzene; 1,2-Dichlorobenzene
75-354	(35) 1,1-Dichloroethylene; 1,1-Dichloroethene; Vinylidene chloride	106-467	(31) p-Dichlorobenzene; 1,4-Dichlorobenzene
75-092	(46) Methylene chloride; Dichloromethane	110-576	(32) trans-1,4-Dichloro-2-butene
75-694	(58) Trichlorofluoromethane; CFC-11	75-343	(33) 1,1-Dichloroethane; Ethylidene chloride
75-014	(61) Vinyl chloride	107-062	(34) 1,2-Dichloroethane; Ethylene dichloride
78-875	(38) 1,2-Dichloropropane; Propylene dichloride	75-354	(35) 1,1-Dichloroethylene; 1,1-Dichloroethene; Vinylidene chloride
78-933	(47) Methyl ethyl ketone; MEK; 2-Butanone	156-592	(36) cis-1,2-Dichloroethylene; cis-1,2-Dichloroethene
79-345	(52) 1,1,2,2-Tetrachloroethane	156-605	(37) trans-1,2-Dichloroethylene; trans-1,2-Dichloroethene
79-005	(56) 1,1,2-Trichloroethane	78-875	(38) 1,2-Dichloropropane; Propylene dichloride
79-016	(57) Trichloroethylene; Trichloroethene	10061-015	(39) cis-1,3-Dichloropropene
95-501	(30) o-Dichlorobenzene; 1,2-Dichlorobenzene	10061-026	(40) trans-1,3-Dichloropropene
96-128	(28) 1,2-Dibromo-3-chloropropane; DBCP	100-414	(41) Ethylbenzene
96-184	(59) 1,2,3-Trichloropropane	591-786	(42) 2-Hexanone; Methyl butyl ketone
100-414	(41) Ethylbenzene	74-839	(43) Methyl bromide; Bromomethane
100-425	(50) Styrene	74-873	(44) Methyl chloride; Chloromethane
106-934	(29) 1,2-Dibromoethane; Ethylene dibromide; EDB	74-953	(45) Methylene bromide; Dibromomethane
106-467	(31) p-Dichlorobenzene; 1,4-Dichlorobenzene	75-092	(46) Methylene chloride; Dichloromethane
107-131	(17) Acrylonitrile	78-933	(47) Methyl ethyl ketone; MEK; 2-Butanone
107-062	(34) 1,2-Dichloroethane; Ethylene dichloride	74-884	(48) Methyl iodide; Iodomethane
108-907	(24) Chlorobenzene	108-101	(49) 4-Methyl-2-pentanone; Methyl isobutyl ketone
108-101	(49) 4-Methyl-2-pentanone; Methyl isobutyl ketone	100-425	(50) Styrene
108-883	(54) Toluene	630-206	(51) 1,1,1,2-Tetrachloroethane
108-054	(60) Vinyl acetate	79-345	(52) 1,1,2,2-Tetrachloroethane
110-576	(32) trans-1,4-Dichloro-2-butene	127-184	(53) Tetrachloroethylene; Tetrachloroethene; Perchloroethylene
124-481	(27) Dibromochloromethane; Chlorodibromomethane	108-883	(54) Toluene

SAMPLING, ANALYSIS, AND MONITORING METHODS: A GUIDE TO EPA REQUIREMENTS

TABLE 3.2. CONSTITUENTS FOR DETECTION MONITORING[1] (40CFR258 APPENDIX I)			
CAS RN[3] order		Common name[2]--Alphabetical order	
127-184	(53) Tetrachloroethylene; Tetrachloroethene; Perchloroethylene	71-556	(55) 1,1,1-Trichloroethane; Methylchloroform
156-592	(36) cis-1,2-Dichloroethylene; cis-1,2-Dichloroethene	79-005	(56) 1,1,2-Trichloroethane
156-605	(37) trans-1,2-Dichloroethylene; trans-1,2-Dichloroethene	79-016	(57) Trichloroethylene; Trichloroethene
591-786	(42) 2-Hexanone; Methyl butyl ketone	75-694	(58) Trichlorofluoromethane; CFC-11
630-206	(51) 1,1,1,2-Tetrachloroethane	96-184	(59) 1,2,3-Trichloropropane
1330-207	(62) Xylenes	108-054	(60) Vinyl acetate
10061-015	(39) cis-1,3-Dichloropropene	75-014	(61) Vinyl chloride
10061-026	(40) trans-1,3-Dichloropropene	1330-207	(62) Xylenes

[1] This list contains 47 volatile organics for which possible analytical procedures provided in EPA Report SW-846 "Test Methods for Evaluating Solid Waste," third edition, November 1986, as revised December 1987, includes Method 8260; and 15 metals for which SW-846 provides either Method 6010 or a method from the 7000 series of methods.

This table was provided in a two major column format for easy cross reference. The left hand side column under CAS RN order is for a reader to use CAS number to find the compound he needs and the right hand side column is to use a chemical name to locate where the compound is.

[2] Common names are those widely used in government regulations, scientific publications, and commerce; synonyms exist for many chemicals.

[3] Chemical Abstracts Service registry number. Where "Total" is entered, all species in the ground water that contain this element are included, i.e., (Total) means the total amount of the inorganic compound from all species in the sample.

[4] The numerical number in front of each compound was given by EPA in 40CFR258 Appendix I.

TABLE 3.3. ANALYSIS METHODS FOR MUNICIPAL WASTE CONSTITUENTS[1] (40CFR258-APPENDIX II)				
CAS RN[3] order		Common name[2]--alphabetical order		Methods[5]
(Total)[3]	Antimony	83-329	Acenaphthene	8100, 8270
(Total)	Arsenic	208-968	Acenaphthylene	8100, 8270
(Total)	Barium	67-641	Acetone	8260
(Total)	Beryllium	75-058	Acetonitrile; Methyl cyanide	8015
(Total)	Cadmium	98-862	Acetophenone	8270
(Total)	Chromium	53-963	Acetylaminofluorene(2-); 2-AAF	8270
(Total)	Cobalt	107-028	Acrolein	8030, 8260
(Total)	Copper	107-131	Acrylonitrile	8030, 8260
(Total)	Lead	309-002	Aldrin	8080, 8270
(Total)	Mercury	107-051	Allyl chloride	8010, 8260
(Total)	Nickel	319-846	alpha-BHC	8080, 8270
(Total)	Selenium	92-671	Aminobiphenyl(4-)	8270
(Total)	Silver	120-127	Anthracene	8100, 8270
(Total)	Thallium	(Total)	Antimony	6010, 7040, 7041
(Total)	Tin	(Total)	Arsenic	6010, 7060, 7061
(Total)	Vanadium	(Total)	Barium	6010, 7080
(Total)	Zinc	71-432	Benzene	8020, 8021, 8260
See note 8	Chlordane	56-553	Benzo(a)anthracene; Benzanthracene	8100, 8270
See note 9	Polychlorinated biphenyls; PCBs; Aroclors	50-328	Benzo(a)pyrene	8100, 8270
See note 10	Toxaphene	205-992	Benzo(b)fluoranthene	8100, 8270
See note 11	Xylene (total)	191-242	Benzo(ghi)perylene	8100, 8270
50-293	DDT(4,4-)	207-089	Benzo(k)fluoranthene	8100, 8270
50-328	Benzo(a)pyrene	100-516	Benzyl alcohol	8270
51-285	Dinitrophenol;(2,4-)	(Total)	Beryllium	6010, 7090, 7091
52-857	Famphur	319-857	beta-BHC	8080, 8270
53-703	Dibenz(a,h)anthracene	58-899	BHC(gamma-); Lindane	8080, 8270
53-963	Acetylaminofluorene(2-); 2-AAF	111-911	Bis(2-chloroethoxy)methane	8110, 8270
55-185	Nitrosodiethylamine(n-)	111-444	Bis(2-chloroethyl) ether;	8110, 8270
56-235	Carbon tetrachloride	117-817	Bis(2-ethylhexyl) phthalate	8060

TABLE 3.3. ANALYSIS METHODS FOR MUNICIPAL WASTE CONSTITUENTS[1] (40CFR258-APPENDIX II)				
CAS RN[3] order		Common name[2]--alphabetical order		Methods[5]
56-382	Parathion	108-601	Bis-(2-chloro-1-methylethyl) ether	8110, 8270
56-495	Methylcholanthrene(3-)	74-975	Bromochloromethane; Chlorobromomethane	8021, 8260
56-553	Benzo(a)anthracene; Benzanthracene	75-274	Bromodichloromethane; Dibromochloromethane	8010, 8021, 8260
57-125	Cyanide	75-252	Bromoform; Tribromomethane	8010, 8021, 8260
57-976	Dimethylbenz(a)anthracene(7,12-)	101-553	Bromophenyl phenyl ethe(4-)	8110, 8270
58-899	BHC(gamma-); Lindane	85-687	Butyl benzyl phthalate; Benzyl butyl phthalate	8060, 8270
58-902	Tetrachlorophenol(2,3,4,6-)	(Total)	Cadmium	6010, 7130, 7131
59-507	Chloro-m-cresol(p-); 4-Chloro-3-methylphenol	75-150	Carbon disulfide	8260
60-117	Dimethylamino-azobenzene(p-)	56-235	Carbon tetrachloride	8010, 8021, 8260
60-515	Dimethoate	See note 8	Chlordane	8080, 8270
60-571	Dieldrin	59-507	Chloro-m-cresol(p-); 4-Chloro-3-methylphenol	8040, 8270
62-442	Phenacetin	106-478	Chloroaniline(p-)	8270
62-500	Ethyl methanesulfonate	108-907	Chlorobenzene	8010, 8020, 8021, 8260
62-759	Nitrosodimethylamine(n-)	510-156	Chlorobenzilate	8270
66-273	Methyl methanesulfonate	75-003	Chloroethane; Ethyl chloride	8010, 8021, 8260
67-641	Acetone	67-663	Chloroform; Trichloromethane	8010, 8021, 8260
67-663	Chloroform; Trichloromethane	91-587	Chloronaphthalene(2-)	8120, 8270
67-721	Hexachloroethane	95-578	Chlorophenol(2-)	8040, 8270
71-432	Benzene	7005-723	Chlorophenyl phenyl ether(4-)	8110, 8270
71-556	Trichloroethane(1,1,1-); Methylchloroform	126-998	Chloroprene	8010, 8260
72-208	Endrin	(Total)	Chromium	6010, 7190, 7191
72-435	Methoxychlor	218-019	Chrysene	8100, 8270
72-548	DDD(4,4-)	(Total)	Cobalt	6010, 7200, 7201

TABLE 3.3. ANALYSIS METHODS FOR MUNICIPAL WASTE CONSTITUENTS[1]
(40CFR258-APPENDIX II)

CAS RN[3] order		Common name[2]--alphabetical order		Methods[5]
72-559	DDE(4,4-)	(Total)	Copper	6010, 7210, 7211
74-839	Methyl bromide; Bromomethane	108-394	Cresol(m-); 3-methylphenol	8270
74-873	Methyl chloride; Chloromethane	95-487	Cresol(o-); 2-methylphenol	8270
74-884	Methyl iodide; Iodomethane	106-445	Cresol(p-); 4-methylphenol	8270
74-953	Methylene bromide; Dibromomethane	57-125	Cyanide	9010
74-975	Bromochloromethane; Chlorobromomethane	94-757	D(2,4-); 2,4-Dichlorophenoxyaceticacid	8150
75-003	Chloroethane; Ethyl chloride	72-548	DDD(4,4-)	8080, 8270
75-014	Vinyl chloride; Chloroethene	72-559	DDE(4,4-)	8080, 8270
75-058	Acetonitrile; Methyl cyanide	50-293	DDT(4,4-)	8080, 8270
75-092	Methylene chloride; Dichloromethane	319-868	delta-BHC	8080, 8270
75-150	Carbon disulfide	84-742	Di-n-butyl phthalate	8060, 8270
75-252	Bromoform; Tribromomethane	117-840	Di-n-octyl phthalate	8060, 8270
75-274	Bromodichloromethane; Dibromochloromethane	2303-164	Diallate	8270
75-343	Dichloroethane(1,1-); Ethyldidene chloride	132-649	Dibenzofuran	8270
75-354	Dichloroethylene(1,1-); 1,1-Dichloroethene;, Vinylidene chloride	53-703	Dibenz(a,h)anthracene	8100, 8270
75-694	Trichlorofluoromethane; CFC-11	96-128	Dibromo-3-chloropropane(1,2-); DBCP	8011, 8021, 8260
75-718	Dichlorodifluoromethane; CFC-12	124-481	Dibromochloromethane; Chlorodibromomethane	8010, 8021, 8260
76-448	Heptachlor	106-934	Dibromoethane(1,2-); Ethylene dribromide; EDB	8011, 8021, 8260
77-474	Hexachlorocyclopentadiene	110-576	Dichloro-2-butene(trans-1,4-)	8260
78-591	Isophorone	541-731	Dichlorobenzene(m-); 1,3-Dichlorobenzene	8010, 8020, 8021, 8120, 8260, 8270
78-831	Isobutyl alcohol	95-501	Dichlorobenzene(o-); 1,2-Dichlorobenzene	8010, 8020, 8021, 8120, 8260, 8270
78-875	Dichloropropane(1,2-); Propylene dichloride	106-467	Dichlorobenzene(p-); 1,4-Dichlorobenzene	8010, 8020, 8021, 8120, 8260, 8270
78-933	Methyl ethyl ketone; MEK; 2-Butanone	91-941	Dichlorobenzidine(3,3-)	8270
79-005	Trichloroethane(1,1,2-)	75-718	Dichlorodifluoromethane; CFC-12	8021, 8260

TABLE 3.3. ANALYSIS METHODS FOR MUNICIPAL WASTE CONSTITUENTS[1] (40CFR258-APPENDIX II)				
CAS RN[3] order		Common name[2]--alphabetical order		Methods[5]
79-016	Trichloroethylene; Trichloroethene	75-343	Dichloroethane(1,1-); Ethyldidene chloride	8010, 8021, 8260
79-345	Tetrachloroethane(1,1,2,2-)	107-062	Dichloroethane(1,2-); Ethylene dichloride	8010, 8021, 8260
80-626	Methyl methacrylate	75-354	Dichloroethylene(1,1-); 1,1-Dichloroethene;, Vinylidene chloride	8010, 8021, 8260
82-688	Pentachloronitrobenzene	156-592	Dichloroethylene(cis-1,2-); cis-1,2-Dichloroethene	8021, 8260
83-329	Acenaphthene	156-605	Dichloroethylene(trans-1,2-); trans-1,2-Dichloroethene	8010, 8021, 8260
84-662	Diethyl phthalate	120-832	Dichlorophenol(2,4-)	8040, 8270
84-742	Di-n-butyl phthalate	87-650	Dichlorophenol(2,6-)	8270
85-018	Phenanthrene	78-875	Dichloropropane(1,2-); Propylene dichloride	8010, 8021, 8260
85-687	Butyl benzyl phthalate; Benzyl butyl phthalate	142-289	Dichloropropane(1,3-); Trimethylene dichloride	8021, 8260
86-306	Nitrosodiphenylamine(n-)	594-207	Dichloropropane(2,2-); Isopropylidene chloride	8021, 8260
86-737	Fluorene	563-586	Dichloropropene(1,1-)	8021, 8260
87-650	Dichlorophenol(2,6-)	10061-015	Dichloropropene(cis-1,3-)	8010, 8260
87-683	Hexachlorobutadiene	10061-026	Dichloropropene(trans-1,3-)	8010, 8260
87-865	Pentachlorophenol	60-571	Dieldrin	8080, 8270
88-062	Trichlorophenol(2,4,6-)	84-662	Diethyl phthalate	8060, 8270
88-744	Nitroaniline(o-); 2-Nitroaniline	297-972	Diethyl(o,o-); o-2-pyrazinyl phosphorothioate; Thionazin	8141, 8270
88-755	Nitrophenol(o-); 2-Nitrophenol	60-515	Dimethoate	8141, 8270
88-857	Dinoseb; DNBP; 2-sec-Butyl-4,6-dinitrophenol	131-113	Dimethyl phthalate	8060, 8270
91-203	Naphthalene	60-117	Dimethylamino-azobenzene(p-)	8270
91-576	Methylnaphthalene(2-)	57-976	Dimethylbenz(a)anthracene(7,12-)	8270
91-587	Chloronaphthalene(2-)	119-937	Dimethylbenzidine(3,3-)	8270
91-598	Naphthylamine(2-)	105-679	Dimethylphenol(2,4-); m-Xylenol	8040, 8270
91-805	Methapyrilene	534-521	Dinitro-o-cresol 4,6-Dinitro-2-methylphenol(4,6-)	8040, 8270
91-941	Dichlorobenzidine(3,3-)	99-650	Dinitrobenzene(m-)	8270
92-671	Aminobiphenyl(4-)	51-285	Dinitrophenol;(2,4-)	8040, 8270
93-721	Silvex; 2,4,5-TP	121-142	Dinitrotoluene(2,4-)	8090, 8270

TABLE 3.3. ANALYSIS METHODS FOR MUNICIPAL WASTE CONSTITUENTS[1] (40CFR258-APPENDIX II)				
CAS RN[3] order		Common name[2]--alphabetical order		Methods[5]
93-765	T(2,4,5-); 2,4,5-Trichlorophenoxyacetic acid	606-202	Dinitrotoluene(2,6-)	8090, 8270
94-597	Safrole	88-857	Dinoseb; DNBP; 2-sec-Butyl-4,6-dinitrophenol	8150, 8270
94-757	D(2,4-); 2,4-Dichlorophenoxyaceticacid	122-394	Diphenylamine	8270
95-487	Cresol(o-); 2-methylphenol	298-044	Disulfoton	8140, 8141, 8270
95-501	Dichlorobenzene(o-); 1,2-Dichlorobenzene	959-988	Endosulfan I	8080, 8270
95-534	Toluidine(o-)	33213-659	Endosulfan II	8080, 8270
95-578	Chlorophenol(2-)	1031-078	Endosulfan sulfate	8080, 8270
95-943	Tetrachlorobenzene(1,2,4,5-)	72-208	Endrin	8080, 8270
95-954	Trichlorophenol(2,4,5-)	7421-934	Endrin aldehyde	8080, 8270
96-128	Dibromo-3-chloropropane(1,2-); DBCP	97-632	Ethyl methacrylate	8015, 8260, 8270
96-184	Trichloropropane(1,2,3-)	62-500	Ethyl methanesulfonate	8270
97-632	Ethyl methacrylate	100-414	Ethylbenzene	8020, 8221, 8260
98-862	Acetophenone	52-857	Famphur	8270
98-953	Nitrobenzene	206-440	Fluoranthene	8100, 8270
99-092	Nitroaniline(m-); 3-Nitroanile	86-737	Fluorene	8100, 8270
99-354	sym-Trinitrobenzene	76-448	Heptachlor	8080, 8270
99-558	Nitro-o-toluidine(5-)	1024-573	Heptachlor epoxide	8080, 8270
99-650	Dinitrobenzene(m-)	118-741	Hexachlorobenzene	8120, 8270
100-016	Nitroaniline(p-); 4-Nitroaniline	87-683	Hexachlorobutadiene	8021, 8120, 8260, 8270
100-027	Nitrophenol(p-); 4-Nitrophenol	77-474	Hexachlorocyclopentadiene	8120, 8270
100-414	Ethylbenzene	67-721	Hexachloroethane	8120, 8260, 8270
100-425	Styrene	1888-717	Hexachloropropene	8270
100-516	Benzyl alcohol	591-786	Hexanone(2-); Methyl butyl ketone	8260
100-754	Nitrosopiperidine(n-)	193-395	Indeno(1,2,3-cd)pyrene	8100, 8270
101-553	Bromophenyl phenyl ethe(4-)	78-831	Isobutyl alcohol	8015, 8240
105-679	Dimethylphenol(2,4-); m-Xylenol	465-736	Isodrin	8270, 8260
106-445	Cresol(p-); 4-methylphenol	78-591	Isophorone	8090, 8270

TABLE 3.3. ANALYSIS METHODS FOR MUNICIPAL WASTE CONSTITUENTS[1] (40CFR258-APPENDIX II)				
CAS RN[3] order		Common name[2]--alphabetical order		Methods[5]
106-467	Dichlorobenzene(p-); 1,4-Dichlorobenzene	120-581	Isosafrole	8270
106-478	Chloroaniline(p-)	143-500	Kepone	8270
106-503	Phenylenediamine(p-)	(Total)	Lead	6010, 7420, 7421
106-934	Dibromoethane(1,2-); Ethylene dribromide; EDB	(Total)	Mercury	7470
107-028	Acrolein	126-987	Methacrylonitrile	8015, 8260
107-051	Allyl chloride	91-805	Methapyrilene	8270
107-062	Dichloroethane(1,2-); Ethylene dichloride	72-435	Methoxychlor	8080, 8270
107-120	Propionitrile; Ethyl cyanide	74-839	Methyl bromide; Bromomethane	8010, 8021
107-131	Acrylonitrile	74-873	Methyl chloride; Chloromethane	8010, 8021
108-054	Vinyl acetate	78-933	Methyl ethyl ketone; MEK; 2-Butanone	8015, 8260
108-101	Methyl-2-pentanone(4-); Methyl isobutyl ketone	74-884	Methyl iodide; Iodomethane	8010, 8260
108-394	Cresol(m-); 3-methylphenol	80-626	Methyl methacrylate	8015, 8260
108-601	Bis-(2-chloro-1-methylethyl) ether	66-273	Methyl methanesulfonate	8270
108-883	Toluene	298-000	Methyl parathion; Parathion methyl	8140, 8141, 8270
108-907	Chlorobenzene	108-101	Methyl-2-pentanone(4-); Methyl isobutyl ketone	8015, 8260
108-952	Phenol	56-495	Methylcholanthrene(3-)	8270
110-576	Dichloro-2-butene(trans-1,4-)	74-953	Methylene bromide; Dibromomethane	8010, 8021, 8260
111-444	Bis(2-chloroethyl) ether;	75-092	Methylene chloride; Dichloromethane	8010, 8021, 8260
111-911	Bis(2-chloroethoxy)methane	91-576	Methylnaphthalene(2-)	8270
117-817	Bis(2-ethylhexyl) phthalate	91-203	Naphthalene	8021, 8100, 8260, 8270
117-840	Di-n-octyl phthalate	130-154	Naphthoquinone(1,4-)	8270
118-741	Hexachlorobenzene	134-327	Naphthylamine(1-)	8270
119-937	Dimethylbenzidine(3,3-)	91-598	Naphthylamine(2-)	8270
120-127	Anthracene	(Total)	Nickel	6010, 7520
120-581	Isosafrole	99-558	Nitro-o-toluidine(5-)	8270
120-821	Trichlorobenzene(1,2,4-)	99-092	Nitroaniline(m-); 3-Nitroanile	8270
120-832	Dichlorophenol(2,4-)	88-744	Nitroaniline(o-); 2-Nitroaniline	8270

TABLE 3.3. ANALYSIS METHODS FOR MUNICIPAL WASTE CONSTITUENTS[1]
(40CFR258-APPENDIX II)

CAS RN[3] order		Common name[2]--alphabetical order		Methods[5]
121-142	Dinitrotoluene(2,4-)	100-016	Nitroaniline(p-); 4-Nitroaniline	8270
122-394	Diphenylamine	98-953	Nitrobenzene	8090, 8270
124-481	Dibromochloromethane; Chlorodibromomethane	88-755	Nitrophenol(o-); 2-Nitrophenol	8040, 8270
126-681	Triethyl phosphorothioate(o,o,o-)	100-027	Nitrophenol(p-); 4-Nitrophenol	8040, 8270
126-987	Methacrylonitrile	924-163	Nitrosodi-n-butylamine(n-)	8270
126-998	Chloroprene	55-185	Nitrosodiethylamine(n-)	8270
127-184	Tetrachloroethylene; Tetrachloroethene; Perchloroethylene	62-759	Nitrosodimethylamine(n-)	8070
129-000	Pyrene	86-306	Nitrosodiphenylamine(n-)	8070
130-154	Naphthoquinone(1,4-)	621-647	Nitrosodipropylamine(n-); n-Nitroso-N-dipropylamine; Di-n-propylnitrosamine	8070
131-113	Dimethyl phthalate	10595-956	Nitrosomethylethalamine(n-)	8270
132-649	Dibenzofuran	100-754	Nitrosopiperidine(n-)	8270
134-327	Naphthylamine(1-)	930-552	Nitrosopyrrolidine(n-)	8270
142-289	Dichloropropane(1,3-); Trimethylene dichloride	56-382	Parathion	8141, 8270
143-500	Kepone	608-935	Pentachlorobenzene	8270
156-592	Dichloroethylene(cis-1,2-); cis-1,2-Dichloroethene	82-688	Pentachloronitrobenzene	8270
156-605	Dichloroethylene(trans-1,2-); trans-1,2-Dichloroethene	87-865	Pentachlorophenol	8040, 8270
191-242	Benzo(ghi)perylene	62-442	Phenacetin	8270
193-395	Indeno(1,2,3-cd)pyrene	85-018	Phenanthrene	8100, 8270
205-992	Benzo(b)fluoranthene	108-952	Phenol	8040
206-440	Fluoranthene	106-503	Phenylenediamine(p-)	8270
207-089	Benzo(k)fluoranthene	298-022	Phorate	8140, 8141, 8270
208-968	Acenaphthylene	See note 9	Polychlorinated biphenyls; PCBs; Aroclors	8080, 8270
218-019	Chrysene	23950-585	Pronamide	8270
297-972	Diethyl(o,o-); o-2-pyrazinyl phosphorothioate; Thionazin	107-120	Propionitrile; Ethyl cyanide	8015, 8260
298-000	Methyl parathion; Parathion methyl	129-000	Pyrene	8100, 8270
298-022	Phorate	94-597	Safrole	8270
298-044	Disulfoton	(Total)	Selenium	6010, 7740, 7741

TABLE 3.3. ANALYSIS METHODS FOR MUNICIPAL WASTE CONSTITUENTS[1] (40CFR258-APPENDIX II)				
CAS RN[3] order		Common name[2]--alphabetical order		Methods[5]
309-002	Aldrin	(Total)	Silver	6010, 7760, 7761
319-846	alpha-BHC	93-721	Silvex; 2,4,5-TP	8150
319-857	beta-BHC	100-425	Styrene	8020, 8021, 8260
319-868	delta-BHC	18496-258	Sulfide	9030
465-736	Isodrin	99-354	sym-Trinitrobenzene	8270
510-156	Chlorobenzilate	93-765	T(2,4,5-); 2,4,5-Trichlorophenoxyacetic acid	8150
534-521	Dinitro-o-cresol 4,6-Dinitro-2-methylphenol(4,6-)	95-943	Tetrachlorobenzene(1,2,4,5-)	8270
541-731	Dichlorobenzene(m-); 1,3-Dichlorobenzene	79-345	Tetrachloroethane(1,1,2,2-)	8010, 8021, 8260
563-586	Dichloropropene(1,1-)	630-206	Tetrachloroethane(1,1,1,2-)	8010, 8021, 8260
591-786	Hexanone(2-); Methyl butyl ketone	127-184	Tetrachloroethylene; Tetrachloroethene; Perchloroethylene	8010, 8021, 8260
594-207	Dichloropropane(2,2-); Isopropylidene chloride	58-902	Tetrachlorophenol(2,3,4,6-)	8270
606-202	Dinitrotoluene(2,6-)	(Total)	Thallium	6010, 7840, 7841
608-935	Pentachlorobenzene	(Total)	Tin	6010
621-647	Nitrosodipropylamine(n-); n-Nitroso-N-dipropylamine; Di-n-propylnitrosamine	108-883	Toluene	8020, 8021, 8260
630-206	Tetrachloroethane(1,1,1,2-)	95-534	Toluidine(o-)	8270
924-163	Nitrosodi-n-butylamine(n-)	See note 10	Toxaphene	8080
930-552	Nitrosopyrrolidine(n-)	120-821	Trichlorobenzene(1,2,4-)	8021, 8120, 8260, 8270
959-988	Endosulfan I	71-556	Trichloroethane(1,1,1-); Methylchloroform	8010, 8021, 8260
1024-573	Heptachlor epoxide	79-005	Trichloroethane(1,1,2-)	8010, 8260
1031-078	Endosulfan sulfate	79-016	Trichloroethylene; Trichloroethene	8010, 8021, 8260
1888-717	Hexachloropropene	75-694	Trichlorofluoromethane; CFC-11	8010, 8021, 8260
2303-164	Diallate	95-954	Trichlorophenol(2,4,5-)	8270
7005-723	Chlorophenyl phenyl ether(4-)	88-062	Trichlorophenol(2,4,6-)	8040, 8270
7421-934	Endrin aldehyde	96-184	Trichloropropane(1,2,3-)	8010, 8021, 8260

CAS RN[3] order		Common name[2]--alphabetical order		Methods[5]
10061-015	Dichloropropene(cis-1,3-)	126-681	Triethyl phosphorothioate(o,o,o-)	8270
10061-026	Dichloropropene(trans-1,3-)	(Total)	Vanadium	6010, 7910, 7911
10595-956	Nitrosomethylethalamine(n-)	108-054	Vinyl acetate	8260
18496-258	Sulfide	75-014	Vinyl chloride; Chloroethene	8010, 8021, 8260
23950-585	Pronamide	See note 11	Xylene (total)	8020, 8021, 8260
33213-659	Endosulfan II	(Total)	Zinc	6010, 7950

TABLE 3.3. ANALYSIS METHODS FOR MUNICIPAL WASTE CONSTITUENTS[1]
(40CFR258-APPENDIX II)

Note:

[1] The regulatory requirements pertain only to the list of substances; the right hand columns (Methods and PQL) are given for informational purposes only. See also footnotes 5 and 6.

This table was provided in a two major column format for easy cross reference. The left hand side column under CAS RN order is for a reader to use CAS number to find the compound he needs and the right hand side column is to use a chemical name to locate where the compound is.

[2] Common names are those widely used in government regulations, scientific publications, and commerce; synonyms exist for many chemicals.

[3] Chemical Abstracts Service registry number. Where "Total" is entered, all species in the ground water that contain this element are included, i.e., (Total) means the total amount of the inorganic compound from all species in the sample.

[4] Reserved.

[5] Suggested Methods refer to analytical procedure numbers used in EPA Report SW-846 "Test Methods for Evaluating Solid Waste", third edition, November 1986, as revised, December 1987. Analytical details can be found in SW-846 and in documentation on file at the agency. CAUTION: The methods listed are representative SW-846 procedures and may not always be the most suitable method(s) for monitoring an analyte under the regulations.

[6] Reserved.

[7] This substance is often called Bis(2-chloroisopropyl) ether, the name Chemical Abstracts Service applies to its noncommercial isomer, Propane, 2,2"-oxybis(2-chloro- (CAS RN 39638-329).

[8] Chlordane: This entry includes alpha-chlordane (CAS RN 5103-719), beta-chlordane (CAS RN 5103-742), gamma-chlordane (CAS RN 5566-347), and constituents of chlordane (CAS RN 57-749 and CAS RN 12789-036). PQL shown is for technical chlordane. PQLs of specific isomers are about 20 μg/L by method 8270.

[9] Polychlorinated biphenyls (CAS RN 1336-363); this category contains congener chemicals, including constituents of Aroclor 1016 (CAS RN 12674-112), Aroclor 1221 (CAS RN 11104-282), Aroclor 1232 (CAS RN 11141-165), Aroclor 1242 (CAS RN 53469-219), Aroclor 1248 (CAS RN 12672-296), Aroclor 1254 (CAS RN 11097-691), and Aroclor 1260 (CAS RN 11096-825). The PQL shown is an average value for PCB congeners.

[10] Toxaphene: This entry includes congener chemicals contained in technical toxaphene (CAS RN 8001-352), i.e., chlorinated camphene.

[11] Xylene (total): This entry includes o-xylene (CAS RN 96-476), m-xylene (CAS RN 108-383), p-xylene (CAS RN 106-423), and unspecified xylenes (dimethylbenzenes) (CAS RN 1330-207). PQLs for method 8021 are 0.2 for o-xylene and 0.1 for m- or p-xylene. The PQL for m-xylene is 2.0 μg/L by method 8020 or 8260.

3.3. HAZARDOUS WASTE

EPA published a guidance document (EPA-89/6) which has been used to assist permit writers in reviewing the measurement aspects of incineration permit applications and trial burn activities. The guidance deals specially with commonly required measurement parameters and measurement methods for process monitoring, sampling and analysis aspects of trial burns and subsequent operation of the incinerator, and QA/QC associated with these activities. Key information of test methods in the document is provided in Table 3.4. below. Because the compounds listed in Appendix VIII to 40CFR261 are the main elements of hazardous waste, the test methods are provided in accordance with Appendix VIII compounds.

TABLE 3.4. ANALYSIS METHODS FOR HAZARDOUS WASTE CONSTITUENTS[1] (40CFR261 APPENDIX VIII)						
CAS RN order		Alphabetic Order		(A)	(B)	(C)
50-000	Formaldehyde	75-058	Acetonitrile	8030, 8240	GC, GC/MS	A101
50-077	Mitomycin C	129-066	Acetonylbenzyl-4-hydroxycoumarin(3-alpha-), see warfarin	8250	HPLC, GC/MS	A122
50-180	Cyclophosphamide	98-862	Acetophenone	8250	HPLC, GC/MS	A121
50-293	DDT	75-365	Acetyl chloride		GC/MS	A144
50-328	Benzo(a)pyrene	591-082	Acetyl-2-thiourea(-1)	8250	HPLC, GC/MS	A123
50-555	Reserpine	53-963	Acetylaminofluorene(2-)	8250	GC/MS	A121
51-285	Dinitrophenol(2,4-)	107-028	Acrolein	8030, 8240	GC, GC/MS	A101
51-434	Epinephrine	79-061	Acrylamide	8015, 8240	GC, GC/MS	A101
51-525	Propylthiouracil	107-131	Acrylonitrile	8030, 8240	GC, GC/MS	A101
51-752	Nitrogen mustard	1402-682	Aflatoxins	8250	HPLC, GC/MS	A145
51-796	Ethyl carbamate (urethane)	116-063	Aldicarb			
52-244	Tris(1-aziridinyl)phosphine sulfide	309-002	Aldrin	8080, 8250	GC, GC/MS	A121
52-857	Famphur	107-186	Allyl alcohol	8240	GC/MS	A134
53-703	Dibenz(ah)anthracene	107-186	Allyl chloride			
53-963	Acetylaminofluorene(2-)	20859-738	Aluminum phosphide		GC/FPD	A253
54-115	Nicotine	92-671	Aminobiphenyl(4-)	8250	GC/MS	A121
55-185	Nitrosodiethylamine(n-)	2763-964	Aminomethyl-3-isoxazolol(5-)	8250	GC/MS	A121
55-630	Nitroglycerin	504-245	Aminopyridine(4-)			
55-914	Diisopropylfluorophosphate (DFP)	61-825	Amitrole	8250	GC/MS	A121
56-042	Methylthiouracil	7803-556	Ammonium vanadate			
56-235	Carbon tetrachloride	62-533	Aniline	8250	GC/MS	A121

TABLE 3.4. ANALYSIS METHODS FOR HAZARDOUS WASTE CONSTITUENTS[1]
(40CFR261 APPENDIX VIII)

CAS RN order			Alphabetic Order		(A)	(B)	(C)
56-382	Parathion	7440-360	Antimony and compounds, N.O.S.		7040, 7041	AAS	A221
56-495	Methylcholanthrene(3-)	140-578	Aramite		8250	GC/MS	A121
56-531	Diethylstilbesterol		Arcolein		8030, 8240	GC, GC/MS	A101
56-553	Benz(a)anthracene	7778-394	Arsenic acid $HAsO_4$		7060, 7061	AAS	A222
57-147	Dimethylhydrazine(1,1-)	7440-382	Arsenic and compounds, N.O.S.		7060, 7061	AAS	A222
57-249	Strychnine	1303-282	Arsenic pentoxide As_2O_5		7060, 7061	AAS	A222
57-749	Chlordane (alpha and gamma isomers)	1327-533	Arsenic trioxide As_2O_3		7060, 7061	AAS	A222
57-976	Dimethylbenz(a)anthracene(7,12-)	492-808	Auramine		8250	GC/MS	A121
58-899	Lindane	115-026	Azaserine			HPLC	A123
58-902	Tetrachlorophenol(2,3,4,6-)	7440-393	Barium and compounds, N.O.S.		7080, 7081	ICAP, AAS	A223
59-507	Chloro-m-cresol(p-)	542-621	Barium cyanide		7080	ICAP, AAS	A223
59-892	Nitrosomorpholine(n-)	98-873	Benzal chloride				
60-117	Dimethylaminoazobenzene(p-)	71-432	Benzene		8020, 8240	GC, GC/MS	A101
60-344	Methyl hydrazine		Benzene, 2-amino-1-methyl				
60-515	Dimethoate		Benzene, 4-amino-1-methyl				
60-571	Dieldrin	98-873	Benzene, dichloromethyl-		8120, 8250	GC, GC/MS	A121
61-825	Amitrole	98-055	Benzenearsonic acid		7060, 7061	AAS	A222
62-384	Phenylmercury acetate	108-985	Benzenethiol		8250	GC/MS	A121
62-442	Phenacetin	92-875	Benzidine		8250	GC/MS	A121
62-500	Ethyl methanesulfonate	106-514	Benzoquinone(p-)		8250	GC/MS	A121
62-533	Aniline	98-077	Benzotrichloride		8120, 8250	GC, GC/MS	A121
62-555	Thioacetamide	50-328	Benzo(a)pyrene		8100, 8250, 8310	GC, GC/MS	A121
62-566	Thiourea	205-992	Benzo(b)fluoranthene		8100, 8250, 8310	GC, GC/MS	A121

CAS RN order		Alphabetic Order		(A)	(B)	(C)
TABLE 3.4. ANALYSIS METHODS FOR HAZARDOUS WASTE CONSTITUENTS[1] (40CFR261 APPENDIX VIII)						
62-748	Fluoroacetic acid, sodium salt	205-823	Benzo(j)fluoranthene	8100, 8250, 8310	GC, GC/MS	A121
62-759	Nitrosodimethylamine(n-)	207-089	Benzo(k)fluoranthene			
64-186	Formic acid	100-447	Benzyl chloride	8010, 8120, 8250	GC, GC/MS	A121
66-273	Methyl methanesulfonate	56-553	Benz(a)anthracene			
66-751	Uracil mustard	225-514	Benz(c)acridine	8250	GC/MS	A121
67-663	Chloroform		Benz(c)anthracene	8100, 8250, 8310	GC, GC/MS	A121
67-721	Hexachloroethane	7440-417	Beryllium and compounds, N.O.S.		GC, GC/MS	A224
70-257	MNNG (Guanidine, n-methyl-n'-nitro-n-nitroso-)	111-911	Bis(2-chloroethoxy)methane	8010, 8240, 8250	GC, GC/MS	A121
70-304	Hexachlorophene	494-031	Bis(2-chloroethyl)-2-naphthylamine(n,n-)		GC/MS	A121
71-432	Benzene	111-444	Bis(2-chloroethyl)ether	8010, 8240, 8250	GC, GC/MS	A121
71-556	Methyl chloroform	108-601	Bis(2-chloroisopropyl)ether	8010, 8240, 8250	GC, GC/MS	A121
72-208	Endrin	117-817	Bis(2-ethylhexyl)phthalate	8060, 8250	GC, GC/MS	A121
72-435	Methoxychlor	542-881	Bis(chloromethyl)ether	8010, 8250	GC, GC/MS	A121
72-548	DDD	598-312	Bromoacetone		GC/MS	A101
72-559	DDE	75-252	Bromoform			
72-571	Trypan blue	74-839	Bromomethane	8010, 8240	GC, GC/MS	A101
74-839	Bromomethane	101-553	Bromophenyl phenyl ether(4-)	8250	GC/MS	A121
74-839	Methyl bromide	357-573	Brucine	8250	GC/FID, HPLC	A148
74-873	Chloromethane		Butanone peroxide(2-)	8250	GC/MS	A121
74-873	Methyl chloride	85-687	Butyl benzyl phthalate			
74-884	Methyl iodide		Butyl-4,6-dinitrophenol(2-sec-)	8040, 8250	GC, GC/MS	A121

TABLE 3.4. ANALYSIS METHODS FOR HAZARDOUS WASTE CONSTITUENTS[1]
(40CFR261 APPENDIX VIII)

CAS RN order		Alphabetic Order		(A)	(B)	(C)
74-908	Hydrogen cyanide		Butylbenzyl phthalate	8060, 8250	GC, GC/MS	A121
74-931	Thiomethanol	75-605	Cacodylic acid			
74-953	Dibromomethane	7440-439	Cadmium and compounds, N.O.S.	7130, 7131	ICAP, AAS	A225
74-953	Methylene bromide	13765-190	Calcium chromate H_2CrO_4	7190, 7191	ICAP, AAS	A226
75-014	Vinyl chloride	592-018	Calcium cyanide $Ca(CN)_2$	9010	ICAP, T, C	A252
75-058	Acetonitrile	75-150	Carbon disulfide	8015, 8240	GC/MS	A101
75-092	Methylene chloride	353-504	Carbon oxyfluoride		GC/MS	A101
75-150	Carbon disulfide	56-235	Carbon tetrachloride			
75-218	Ethylene oxide	75-876	Chloral	8010, 8240	GC, GC/MS, HPLC	A131
75-252	Bromoform	305-033	Chlorambucil		HPLC	A122
75-343	Ethylidene dichloride	57-749	Chlordane (alpha and gamma isomers)	8080, 8250	GC, GC/MS	A121
75-354	Dichloroethylene(1,1-)		Chlorinated benzenes, N.O.S.	8010, 8240	GC, GC/MS	A101
75-365	Acetyl chloride		Chlorinated ethane, N.O.S.	8010, 8240	GC, GC/MS	A101
75-445	Phosgene		Chlorinated fluorocarbons, N.O.S.		GC/MS	A101
75-558	Propylenimine(1,2-)		Chlorinated naphthalene, N.O.S.	8120, 8250	GC, GC/MS	A121
75-605	Cacodylic acid		Chlorinated phenol, N.O.S.	8040, 8250	GC, GC/MS	A121
75-694	Trichloromonofluoromethane	106-898	Chloro-2,3-epoxypropane(1-)		GC/MS	A101
75-707	Trichloromethanithiol		Chloro-1,3-butadiene(2-)			
75-718	Dichlorodifluoromethane	59-507	Chloro-m-cresol(p-)	8040, 8250	GC, GC/MS	A121
75-865	Methyllactonitrile(2-)	107-200	Chloroacetaldehyde	8010, 8240	GC, GC/MS, HPLC	A131
75-876	Chloral		Chloroalkyl ethers, N.O.S.		GC/MS	A101
76-017	Pentachloroethane	106-478	Chloroaniline(p-)		GC/MS	A121
76-448	Heptachlor	108-907	Chlorobenzene	8020, 8240	GC, GC/MS	A101

TABLE 3.4. ANALYSIS METHODS FOR HAZARDOUS WASTE CONSTITUENTS[1] (40CFR261 APPENDIX VIII)						
CAS RN order		Alphabetic Order		(A)	(B)	(C)
77-474	Hexachlorocyclopentadiene	510-156	Chlorobenzilate		GC/MS	A121
77-781	Dimethyl sulfate	110-758	Chloroethyl vinyl ether(2-)	8010, 8240	GC, GC/MS	A101
78-002	Tetraethyl lead	67-663	Chloroform	8010, 8240	GC, GC/MS	A101
78-831	Isobutyl alcohol	74-873	Chloromethane	8010, 8240	GC, GC/MS	A101
78-875	Propylene dichloride	107-302	Chloromethyl methyl ether	8010	GC, GC/MS	A101
78-933	Methyl ethyl ketone (MEK)	91-587	Chloronaphthalene(2-)	8120, 8250	GC, GC/MS	A121
79-005	Trichloroethane(1,1,2-)	91-587	Chloronaphthalene(beta-)			
79-016	Trichloroethylene	95-578	Chlorophenol(2-)	8040, 8250	GC, GC/MS, HPLC	A121
79-061	Acrylamide	95-578	Chlorophenol(o-)			
79-196	Thiosemicarbazide		Chlorophenyl((1-)o-)thiourea		HPLC	A123
79-221	Methylchlorocarbonate	5344-821	Chlorophenyl(1-(o-))			
79-345	Tetrachloroethane(1,1,2,2-)		Chlorophenyl thiourea(1-)			
79-447	Dimethylcarbamoyl chloride	126-998	Chloroprene			
79-469	Nitropropane(2-)		Chloropropene(3-)			
80-626	Methyl methacrylate	542-767	Chloropropionitrile(3-)	8250	GC/MS	A121
81-072	Saccharin	7440-473	Chromium and compounds, N.O.S.	7190, 7191, 7195, 7196, 7197	ICAP, AAS	A226
81-812	Warfarin	218-019	Chrysene	8100, 8250, 8310	GC, GC/MS	A121
82-688	Pentachloronitrobenzene	6358-538	Citrus red No. 2		HPLC	A149
84-662	Diethyl phthalate	8007-452	Coal tar creosote		GC/MS	A121
84-742	Dibutyl phthalate	544-923	Copper cyanide	9010	T, C	A252
85-449	Phthalic anhydride	8001-589	Creosote	8100, 8250	GC, GC/MS	A121
85-687	Butyl benzyl phthalate	1319-773	Cresol; Cresylic acid	8040, 8250	GC, GC/MS	A121
86-884	Naphthylthiourea(alpha-)	4170-303	Crotonaldehyde		HPLC	A123

TABLE 3.4. ANALYSIS METHODS FOR HAZARDOUS WASTE CONSTITUENTS[1] (40CFR261 APPENDIX VIII)						
CAS RN order			Alphabetic Order	(A)	(B)	(C)
87-650	Dichlorophenol(2,6-)		Cyanides (soluble salts and complexes), N.O.S.	9010	T, C	A252
87-663	Hexachlorobutadiene	460-195	Cyanogen	9010	GC/TCD, T, C	A138
87-865	Pentachlorophenol	506-683	Cyanogen bromide	9010	GC/TCD, T, C	A138
88-062	Trichlorophenol(2,4,6-)	506-774	Cyanogen chloride	9010	GC/TCD, T, C	A138
88-857	Dinoseb	14901-087	Cycasin			A150
91-203	Naphthalane	131-895	Cyclohexyl-4,6-dinitrophenol(2-)	8040, 8250	GC, GC/MS	A121
91-587	Chloronaphthalene(2-)	50-180	Cyclophosphamide			
91-587	Chloronaphthalene(beta-)	94-757	D(2,4-)			
91-598	Naphthylamine(beta-)		D(2,4-), salts, esters			
91-805	Methapyrilene	20830-813	Daunomycin		HPLC	A122
91-941	Dichlorobenzidine(3,3'-)	72-548	DDD	8080, 8250	GC, GC/MS	A121
92-671	Aminobiphenyl(4-)	72-559	DDE	8080, 8250	GC, GC/MS	A121
92-875	Benzidine	50-293	DDT	8080, 8250	GC, GC/MS	A121
93-721	Silvex (2,4,5-TP)		Di-n-butyl phthalate	8060, 8250	GC, GC/MS	A121
93-765	T(2,4,5-)	117-840	Di-n-octyl phthalate	8060, 8250	GC, GC/MS	A121
94-586	Dihydrosafrole	621-647	Di-n-propylnitrosamine		GC/MS	A121
94-597	Safrole	2303-164	Diallate		GC/MS	A121
94-757	D(2,4-)	192-654	Dibenzo(ae)pyrene	8100	GC, GC/MS	A121
95-501	Dichlorobenzene(o-)	189-640	Dibenzo(ah)pyrene	8100	GC, GC/MS	A121
95-534	Toluidine(o-)	189-559	Dibenzo(ai)pyrene	8100	GC, GC/MS	A121
95-578	Chlorophenol(2-)	194-592	Dibenzo(cg)carbazole(7h-)	8100	GC, GC/MS	A121
95-578	Chlorophenol(o-)	226-368	Dibenz(ah)acridine	8100	GC, GC/MS	A121

TABLE 3.4. ANALYSIS METHODS FOR HAZARDOUS WASTE CONSTITUENTS[1] (40CFR261 APPENDIX VIII)							
CAS RN order		Alphabetic Order		(A)	(B)	(C)	
95-807	Toluene-2,4-diamine	53-703	Dibenz(ah)anthracene	8100, 8310, 8250	GC, GC/MS	A121	
95-943	Tetrachlorobenzene(1,2,4,5-)	224-420	Dibenz(aj)acridine	8100	GC, GC/MS	A121	
95-954	Trichlorophenol(2,4,5-)	96-128	Dibromo-3-chloropropane(1,2-)	8010, 8240	GC, GC/MS	A101	
96-128	Dibromo-3-chloropropane(1,2-)	106-934	Dibromoethane(1,2-)	8010, 8240	GC, GC/MS	A101	
96-184	Trichloropropane(1,2,3-)	74-953	Dibromomethane	8010, 8240	GC, GC/MS	A101	
96-457	Ethylenethiourea	84-742	Dibutyl phthalate				
97-632	Ethyl methacrylate	764-410	Dichloro-2-butene(1,4-)	8010, 8240	GC, GC/MS	A101	
98-055	Benzenearsonic acid		Dichlorobenzene (meta, ortho and para isomers)	8010, 8120	GC, GC/MS	A101	
98-077	Benzotrichloride	25321-226	dichlorobenzene, N.O.S.	8010, 8120	GC, GC/MS	A101	
98-862	Acetophenone	541-731	Dichlorobenzene(m-)				
98-873	Benzal chloride	95-501	Dichlorobenzene(o-)				
98-873	Benzene, dichloromethyl-	106-467	Dichlorobenzene(p-)				
98-953	Nitrobenzene	91-941	Dichlorobenzidine(3,3'-)	8250	GC/MS	A121	
99-354	Trinitrobenzene(1,3,5-)	75-718	Dichlorodifluoromethane	8010	GC, GC/MS	A101	
99-558	Nitro-o-toluidine(5-)		Dichloroethane(1,1-)	8010, 8240	GC, GC/MS	A101	
100-016	Nitroaniline(p-)		Dichloroethane(1,2-)	8010, 8240	GC, GC/MS	A101	
100-027	Nitrophenol(p-)		Dichloroethene(trans-1,2-)	8010, 8240	GC, GC/MS	A101	
100-447	Benzyl chloride	111-444	Dichloroethyl ether				
100-754	Nitrosopiperidine(n-)	25323-302	Dichloroethylene, N.O.S.	8010	GC, GC/MS	A101	
101-144	Methylene(4,4'-)bis(2-chloroaniline)	75-354	Dichloroethylene(1,1-)	8010	GC, GC/MS	A101	
101-553	Bromophenyl phenyl ether(4-)	156-605	Dichloroethylene(1,2-)				
103-855	Phenylthiourea	108-601	Dichloroisopropyl ether				
105-679	Dimethylphenol(2,4-)		Dichloromethane	8010, 8240	GC, GC/MS	A101	

TABLE 3.4. ANALYSIS METHODS FOR HAZARDOUS WASTE CONSTITUENTS[1]
(40CFR261 APPENDIX VIII)

CAS RN order		Alphabetic Order		(A)	(B)	(C)
106-467	Dichlorobenzene(p-)	111-911	Dichloromethoxy ethane			
106-478	Chloroaniline(p-)	542-881	Dichloromethyl ether			
106-490	Toluidine(p-)	120-832	Dichlorophenol(2,4-)	8040, 8250	GC, GC/MS	A121
106-514	Benzoquinone(p-)	87-650	Dichlorophenol(2,6-)	8040, 8250	GC, GC/MS	A121
106-898	Chloro-2,3-epoxypropane(1-)	696-286	Dichlorophenylarsine	7060, 7061	AAS	A222
106-898	Epichlorohydrin	26638-197	Dichloropropane, N.O.S.	8010, 8240	GC, GC/MS	A101
106-934	Dibromoethane(1,2-)		Dichloropropane(1,2-)	8010, 8240	GC, GC/MS	A101
106-934	Ethylene dibromide	542-756	Dichloropropane(1,3-), N.O.S.			
107-028	Acrolein	26545-733	Dichloropropanol, N.O.S.	8120, 8250	GC, GC/MS	A121
107-062	Ethylene dichloride	26952-238	Dichloropropene, N.O.S.	8240	GC/MS	A101
107-108	Propylamine(n-)		Dichloropropene(1,3-)	8240	GC/MS	A101
107-120	Ethyl cyanide		Dichlrophenoxyacetic(2,4-) acid	8150, 8250	GC, GC/MS	A122
107-131	Acrylonitrile	60-571	Dieldrin	8080	GC, GC/MS	A121
107-186	Allyl alcohol	1464-535	Diepoxybutane(1,2:3,4-)		GC/MS	A121
107-186	Allyl chloride	84-662	Diethyl phthalate	8060, 8250	GC, GC/MS	A121
107-197	Propargyl alcohol	297-972	Diethyl o-pyrazinyl phosphorothioate(o,o-)	8250	GC/MS	A121
107-200	Chloroacetaldehyde		Diethyl s-methyl ester of phosphorodithioic acid(o,o-)	8250	GC/MS	A121
107-302	Chloromethyl methyl ether	692-422	Diethylarsine	7060, 7061	AAS	A222
107-493	Tetraethyl pyrophosphate	123-911	Diethyleneoxide(1,4-), see dioxane(1,4-)			
108-316	Maleic anhydride	1615-801	Diethylhydrazine(n,n'-)		GC/MS	A121
108-463	Resorcinol		Diethylphosphoric acid, o-p-nitrophenyl ester	8250	GC/MS	A121
108-601	Bis(2-chloroisopropyl)ether	56-531	Diethylstilbesterol		HPLC	A123
108-601	Dichloroisopropyl ether	94-586	Dihydrosafrole		GC/MS	A121
108-883	Toluene	55-914	Diisopropyfluorophosphate (DFP)		GC/MS	A121

TABLE 3.4. ANALYSIS METHODS FOR HAZARDOUS WASTE CONSTITUENTS[1]
(40CFR261 APPENDIX VIII)

CAS RN order		Alphabetic Order		(A)	(B)	(C)
108-907	Chlorobenzene	60-515	Dimethoate	8140	GC, GC/MS	A121
108-952	Phenol	119-904	Dimethoxybenzidine(3,3'-)		GC/MS	A121
108-985	Benzenethiol	131-113	Dimethyl phthalate	8060, 8250	GC, GC/MS	A121
108-985	Thiophenol	77-781	Dimethyl sulfate	8250	GC/MS	A121
109-068	Picoline(2-)		Dimethyl(3,3-)-1-(methylthio)-2-butanone, o-((methylamino)carbonyl)oxime, see thiofanox			
109-773	Malononitrile	57-976	Dimethylbenz(a)anthracene(7,12-)		GC/MS	A121
110-758	Chloroethyl vinyl ether(2-)	122-098	Dimethylphenethylamine(alpha, alpha-)		GC/MS	A121
110-805	Ethylene glycol monoethyl ether	60-117	Dimethylaminoazobenzene(p-)		GC/MS	A121
110-861	Pyridine	119-937	Dimethylbenzidine(3,3'-)		GC/MS	A121
111-444	Bis(2-chloroethyl)ether	79-447	Dimethylcarbamoyl chloride			A144
111-444	Dichloroethyl ether	57-147	Dimethylhydrazine(1,1-)		GC/MS	A121
111-546	Ethylenebisdithiocarbamic acid	540-738	Dimethylhydrazine(1,2-)		GC/MS	A121
111-911	Bis(2-chloroethoxy)methane	105-679	Dimethylphenol(2,4-)	8040, 8250	GC, GC/MS	A121
111-911	Dichloromethoxy ethane	534-521	Dinitro-o-cresol(4,6-)	8040, 8250	GC, GC/MS	A121
115-026	Azaserine		Dinitro-o-cresol salts(4,6-)	8040, 8250	GC, GC/MS	A121
115-297	Endosulfan	25154-545	Dinitrobenzene, N.O.S.	8090, 8250	GC, GC/MS	A121
116-063	Aldicarb	51-285	Dinitrophenol(2,4-)	8040, 8250	GC, GC/MS	A121
117-817	Bis(2-ethylhexyl)phthalate	121-142	Dinitrotoluene(2,4-)	8040, 8250	GC, GC/MS	A121
117-840	Di-n-octyl phthalate	606-202	Dinitrotoluene(2,6-)	8040, 8250	GC, GC/MS	A121
118-741	Hexachlorobenzene	88-857	Dinoseb			
119-904	Dimethoxybenzidine(3,3'-)		Dioxane(1,4-)		GC/MS	A101
119-937	Dimethylbenzidine(3,3'-)	122-394	Diphenylamine		GC/MS	A121
120-581	Isosafrole	122-667	Diphenylhydrazine(1,2-)		GC/MS	A121
120-821	Trichlorobenzene(1,2,4-)	298-044	Disulfoton	8140	GC, GC/MS	A121

TABLE 3.4. ANALYSIS METHODS FOR HAZARDOUS WASTE CONSTITUENTS[1] (40CFR261 APPENDIX VIII)							
CAS RN order		Alphabetic Order		(A)	(B)	(C)
120-832	Dichlorophenol(2,4-)	541-537	Dithiobiuret			
121-142	Dinitrotoluene(2,4-)		Dithiobiuret(2,4-)		GC/MS	A121
122-098	Dimethylphenethylamine(alpha, alpha-)	115-297	Endosulfan	8080, 8250	GC, GC/MS	A121
122-394	Diphenylamine	145-733	Endothall			
122-667	Diphenylhydrazine(1,2-)	72-208	Endrin	8080, 8250	GC, GC/MS	A121
123-331	Maleic hydrazide		Endrin metabolites	8080, 8250	GC, GC/MS	A121
123-637	Paraldehyde	106-898	Epichlorohydrin			
123-911	Diethyleneoxide(1,4-), see dioxane(1,4-)	51-434	Epinephrine			
126-681	Triethylphosphorothioate(o,o,o-)	51-796	Ethyl carbamate (urethane)		GC/MS	A121
126-727	Tris(2,3-dibromopropyl) phosphate	107-120	Ethyl cyanide	9010	T, C	A252
126-852	Nitrogen mustard n-oxide	97-632	Ethyl methacrylate		GC/MS	A121
126-987	Methacrylonitrile	62-500	Ethyl methanesulfonate		GC/MS	A121
126-998	Chloroprene	106-934	Ethylene dibromide			
127-184	Tetrachloroethylene	107-062	Ethylene dichloride			
129-066	Acetonylbenzyl-4-hydroxycoumarin(3-alpha-), see warfarin	110-805	Ethylene glycol monoethyl ether			
130-154	Naphthoquinone(1,4-)	75-218	Ethylene oxide		GC/FID	A156
131-113	Dimethyl phthalate	111-546	Ethylenebisdithiocarbamic acid			
131-895	Cyclohexyl-4,6-dinitrophenol(2-)		Ethylenebisdithiocarbamic acid, salts and esters.			
134-327	Naphthylamine(alpha-)	151-564	Ethyleneimine		GC/MS	A121
137-268	Thiram	96-457	Ethylenethiourea		HPLC	A123
140-578	Aramite	75-343	Ethylidene dichloride			
143-339	Sodium cyanide	52-857	Famphur	8140	GC, GC/MS	A121
143-500	Kepone	206-440	Fluoranthene	8100, 8250, 8310	GC, GC/MS	A121
145-733	Endothall	7782-414	Fluorine			A137
148-823	Melphalan	640-197	Fluoroacetamide			
151-508	Potassium cyanide		Fluoroacetamide(2-)		GC/FID	A157

TABLE 3.4. ANALYSIS METHODS FOR HAZARDOUS WASTE CONSTITUENTS[1] (40CFR261 APPENDIX VIII)						
CAS RN order		Alphabetic Order		(A)	(B)	(C)
151-564	Ethyleneimine	62-748	Fluoroacetic acid, sodium salt		GC/MS	A121
152-169	Octamethylpyro-phosphoramide	50-000	Formaldehyde	8015, 8240	GC, GC/MS, HPLC	A131
156-605	Dichloroethylene(1,2-)	64-186	Formic acid		GC/MS	A101
189-559	Dibenzo(ai)pyrene		Formic acid		GC/MS	A101
189-640	Dibenzo(ah)pyrene	765-344	Glycidylaldehyde		GC/MS, HPLC	A131
192-654	Dibenzo(ae)pyrene		Halomethanes, N.O.S.		GC/MS	A101
193-395	Indeno(1,2,3-cd)pyrene	76-448	Heptachlor	8080, 8250	GC, GC/MS	A121
194-592	Dibenzo(cg)carbazole(7h-)	1024-573	Heptachlor epoxide			
205-823	Benzo(j)fluoranthene		Heptachlor epoxide (alpha, beta, and gamma isomers)	8080, 8250	GC, GC/MS	A121
205-992	Benzo(b)fluoranthene		Heptachlorodibenzo-p-dioxins			
206-440	Fluoranthene		Heptachlorodibenzofurans			
207-089	Benzo(k)fluoranthene		Hexachloro-1,4,4a,5,8,8a-hexahydro-1,4:5,8-endo, endo-dimethanonaphthalene(1,2,3,4,10,10-)		GC/MS	A121
218-019	Chrysene	118-741	Hexachlorobenzene	8120, 8250	GC, GC/MS	A121
224-420	Dibenz(aj)acridine	87-663	Hexachlorobutadiene	8120, 8250	GC, GC/MS	A121
225-514	Benz(c)acridine		Hexachlorocyclohexane(all isomers)	8120	GC, GC/MS	A121
226-368	Dibenz(ah)acridine	77-474	Hexachlorocyclopentadiene	8120, 8250	GC, GC/MS	A121
297-972	Diethyl o-pyrazinyl phosphorothioate(o,o-)		Hexachlorodibenzo-p-dioxins	8280	GC/MS	
298-000	Methyl parathion		Hexachlorodibenzofurans	8280	GC/MS	
298-022	Phorate	67-721	Hexachloroethane	8010, 8240	GC, GC/MS	A101
298-044	Disulfoton	70-304	Hexachlorophene		GC/MS	A121
301-042	Lead acetate	1888-717	Hexachloropropene		GC/MS	A101
302-012	Hydrazine	757-584	Hexaethyl tetraphosphate		GC/MS	A121
303-341	Lasiocarpine	302-012	Hydrazine		GC/MS	A101
305-033	Chlorambucil		Hydrocyanic acid		GC/TCD	A141

TABLE 3.4. ANALYSIS METHODS FOR HAZARDOUS WASTE CONSTITUENTS[1]
(40CFR261 APPENDIX VIII)

CAS RN order		Alphabetic Order		(A)	(B)	(C)
309-002	Aldrin		Hydrofluoric acid		IC	A251
353-504	Carbon oxyfluoride	74-908	Hydrogen cyanide			
357-573	Brucine	7664-393	Hydrogen fluoride			
460-195	Cyanogen	7783-064	Hydrogen sulfide		GC/TCD	A141
465-736	Isodrin		Hydroxydimethylarsine oxide	7060, 7061	AAS	A222
492-808	Auramine	193-395	Indeno(1,2,3-cd)pyrene	8100, 8250, 8310	GC, GC/MS	A121
494-031	Bis(2-chloroethyl)-2-naphthylamine(n,n-)		Iodomethane		GC/MS	A101
496-720	Toluene-3,4-diamine		Iron dextran (complex)			
504-245	Aminopyridine(4-)	78-831	Isobutyl alcohol		GC/MS	A134
505-602	Mustard gas		Isocyanic acid, methyl ester		GC/MS	A101
506-616	Potassium silver cyanide	465-736	Isodrin			
506-649	Silver cyanide	120-581	Isosafrole		GC/MS	A121
506-683	Cyanogen bromide	143-500	Kepone	8080	GC, GC/MS	A121
506-774	Cyanogen chloride	303-341	Lasiocarpine			A160
509-148	Tetranitromethane	301-042	Lead acetate	7420, 7421	ICAP, AAS	A227
510-156	Chlorobenzilate	7439-921	Lead and compounds, N.O.S.	7420, 7421	ICAP, AAS	A227
534-521	Dinitro-o-cresol(4,6-)	7446-277	Lead phosphate	7420, 7421	ICAP, AAS	A227
540-738	Dimethylhydrazine(1,2-)	1335-326	Lead subacetate	7420, 7421	ICAP, AAS	A227
541-537	Dithiobiuret	58-899	Lindane			
541-731	Dichlorobenzene(m-)	108-316	Maleic anhydride	8250	GC/MS	A121
542-621	Barium cyanide	123-331	Maleic hydrazide		GC/MS	A121
542-756	Dichloropropane(1,3-), N.O.S.	109-773	Malononitrile		GC/MS	A121
542-767	Chloropropionitrile(3-)	148-823	Melphalan		HPLC	A122
542-881	Bis(chloromethyl)ether	7439-976	Mercury and compounds, N.O.S.	7470, 7471	Cv/AAS	A228
542-881	Dichloromethyl ether	628-864	Mercury fulminate	7470, 7471	Cv/AAS	A228
544-923	Copper cyanide	126-987	Methacrylonitrile		GC/MS	A121

TABLE 3.4. ANALYSIS METHODS FOR HAZARDOUS WASTE CONSTITUENTS[1] (40CFR261 APPENDIX VIII)						
CAS RN order		Alphabetic Order		(A)	(B)	(C)
557-197	Nickel cyanide		Methanethiol		GC/MS	A101
557-211	Zinc cyanide	91-805	Methapyrilenc		GC/MS	A121
563-688	Thallium(l) acetate	16752-775	Methomyl	8250	GC, GC/MS, HPLC	A122
591-082	Acetyl-2-thiourea(-1)	72-435	Methoxychlor	8080	GC, GC/MS	A121
592-018	Calcium cyanide Ca(CN)$_2$	74-839	Methyl bromide			
598-312	Bromoacetone	74-873	Methyl chloride			
606-202	Dinitrotoluene(2,6-)	71-556	Methyl chloroform			
608-935	Pentachlorobenzene	78-933	Methyl ethyl ketone (MEK)	8015, 8240	GC, GC/MS	A101
615-532	Nitroso-n-methylurethane(n-)	1338-234	Methyl ethyl ketone peroxide			
621-647	Di-n-propylnitrosamine	60-344	Methyl hydrazine		GC/MS	A101
624-839	Methyl isocyanate	74-884	Methyl iodide			
628-864	Mercury fulminate	624-839	Methyl isocyanate			
630-104	Selenourea	80-626	Methyl methacrylate		GC/MS	A121
630-206	Tetrachloroethane(1,1,1,2-)	66-273	Methyl methanesulfonate		GC/MS	A121
636-215	Toluidine hydrochloride(o-)	298-000	Methyl parathion	8140	GC, GC/MS	A121
640-197	Fluoroacetamide		Methyl-2-(methylthio)propionaldehyde-o-(methylcarbonyl)oxime(2-)		GC/FPD	A183
684-935	Nitroso-n-methylurea(n-)		Methyl-n'-nitro-n-nitroso-guanidine(n-)		GC/MS	A121
692-422	Diethylarsine		Methylaziridine(2-)		GC/MS	A121
696-286	Dichlorophenylarsine	79-221	Methylchlorocarbonate			
757-584	Hexaethyl tetraphosphate	56-495	Methylcholanthrene(3-)	8100	GC, GC/MS	A121
759-739	Nitroso-n-ethylurea(n-)	74-953	Methylene bromide			
764-410	Dichloro-2-butene(1,4-)	75-092	Methylene chloride			
765-344	Glycidylaldehyde	101-144	Methylene(4,4'-)bis(2-chloroaniline)		GC/MS	A121
823-405	Toluene-2,6-diamine	75-865	Methyllactonitrile(2-)		GC/MS	A121
924-163	Nitrosodi-n-butylamine(n-)	56-042	Methylthiouracil		GC/MS	A121
930-552	Nitrosopyrrolidine(n-)	50-077	Mitomycin C		GC/MS	A122

TABLE 3.4. ANALYSIS METHODS FOR HAZARDOUS WASTE CONSTITUENTS[1] (40CFR261 APPENDIX VIII)					
CAS RN order		Alphabetic Order		(A)	(B)	(C)
1024-573	Heptachlor epoxide	70-257	MNNG (Guanidine, n-methyl-n'-nitro-n-nitroso-)			
1116-547	Nitrosodiethanolamine(n-)	505-602	Mustard gas		GC/FPD	A139
1120-714	Propane(1,3-) sultone	91-203	Naphthalane	8100, 8250, 8310	GC, GC/MS	A121
1303-282	Arsenic pentoxide As$_2$O$_5$	130-154	Naphthoquinone(1,4-)	8090, 8250	GC, GC/MS	A121
1314-325	Thallic oxide		Naphthyl-2-thiourea(1-)		HPLC	A123
1314-621	Vanadium pentoxide		Naphthylamine(1-)		GC/MS	A121
1314-847	Zinc phosphide		Naphthylamine(2-)		GC/MS	A121
1319-773	Cresol; Cresylic acid	134-327	Naphthylamine(alpha-)			
1327-533	Arsenic trioxide As$_2$O$_3$	91-598	Naphthylamine(beta-)			
1335-326	Lead subacetate	86-884	Naphthylthiourea(alpha-)			
1338-234	Methyl ethyl ketone peroxide	7440-020	Nickel and compounds, N.O.S.	7520, 7521	ICAP, AAS	A229
1402-682	Aflatoxins	13463-393	Nickel carbonyl	7520, 7521	ICAP, AAS	A229
1464-535	Diepoxybutane(1,2:3,4-)	557-197	Nickel cyanide	7520, 7521	ICAP, AAS	A229
1615-801	Diethylhydrazine(n,n'-) ˜	54-115	Nicotine		GC/MS	A121
1746-016	Tetrachlorodibenzo-p-dioxin(2,3,7,8-) (TCDD)		Nicotine salts		GC/MS	A121
1888-717	Hexachloropropene	10102-439	Nitric oxide		GC/TCD	A141
2303-164	Diallate	99-558	Nitro-o-toluidine(5-)		HPLC	A122
2763-964	Aminomethyl-3-isoxazolol(5-)	100-016	Nitroaniline(p-)		GC/MS	A121
3689-245	Tetraethyldithiopyro-phosphate	98-953	Nitrobenzene	8090, 8250	GC, GC/MS	A121
4170-303	Crotonaldehyde	10102-440	Nitrogen dioxide		GC/TCD	A141
4549-400	Nitrosomethylvinylamine(n-)	51-752	Nitrogen mustard		GC/FPD	A139
5344-821	Chlorophenyl(1-(o-))		Nitrogen mustard, hydrochloride salt		GC/FPD	A139
6358-538	Citrus red No. 2	126-852	Nitrogen mustard n-oxide		GC/FPD	A139
6533-739	Thallium(l) carbonate		Nitrogen mustard, n-oxide, hydrochloride salt		GC/FPD	A139
7439-921	Lead and compounds, N.O.S.	55-630	Nitroglycerin		GC/MS	A121

TABLE 3.4. ANALYSIS METHODS FOR HAZARDOUS WASTE CONSTITUENTS[1]
(40CFR261 APPENDIX VIII)

CAS RN order		Alphabetic Order		(A)	(B)	(C)
7439-976	Mercury and compounds, N.O.S.		Nitrophenol(4-)	8040, 8240	GC, GC/MS	A121
7440-020	Nickel and compounds, N.O.S.	100-027	Nitrophenol(p-)			
7440-224	Silver and compounds, N.O.S.	79-469	Nitropropane(2-)			
7440-280	Thallium and compounds, N.O.S.		Nitroquinoline(4-)-1-oxide		GC/MS, HPLC	
7440-360	Antimony and compounds, N.O.S.	35576-911	Nitrosamines, N.O.S.	8250	GC/MS	A121
7440-382	Arsenic and compounds, N.O.S.	759-739	Nitroso-n-ethylurea(n-)	8250	GC/MS	A121
7440-393	Barium and compounds, N.O.S.	615-532	Nitroso-n-methylurethane(n-)	8250	GC/MS	A121
7440-417	Beryllium and compounds, N.O.S.	684-935	Nitroso-n-methylurea(n-)	8250	GC/MS	A121
7440-439	Cadmium and compounds, N.O.S.	924-163	Nitrosodi-n-butylamine(n-)	8250	GC/MS	A121
7440-473	Chromium and compounds, N.O.S.	1116-547	Nitrosodiethanolamine(n-)	8250	GC/MS	A121
7446-186	Thallium(l) sulfate	55-185	Nitrosodiethylamine(n-)	8250	GC/MS	A121
7446-277	Lead phosphate	62-759	Nitrosodimethylamine(n-)	8250	GC/MS	A121
7488-564	Selenium sulfide	10595-956	Nitrosomethylethylamine(n-)	8250	GC/MS	A121
7664-393	Hydrogen fluoride	4549-400	Nitrosomethylvinylamine(n-)	8250	GC/MS	A121
7778-394	Arsenic acid $HAsO_4$	59-892	Nitrosomorpholine(n-)	8250	GC/MS	A121
7782-414	Fluorine	16543-558	Nitrosonornicotine(n-)	8250	GC/MS	A121
7782-492	Selenium and compounds, N.O.S.	100-754	Nitrosopiperidine(n-)	8250	GC/MS	A121
7783-008	Selenium dioxide	930-552	Nitrosopyrrolidine(n-)	8250	GC/MS	A121
7783-064	Hydrogen sulfide	13256-229	Nitrososarcosine(n-)	8250	GC/MS	A121
7791-120	Thallium(l) chloride	152-169	Octamethylpyro-phosphoramide		GC/MS	A121
7803-512	Phosphine	20816-120	Osmium tetroxide		ICAP, AAS	A230
7803-556	Ammonium vanadate		Oxabicyclo(2.2.1)heptane-2,3-dicarboxylic acid		GC/MS	A133
8001-352	Toxaphene	123-637	Paraldehyde	8015, 8240	GC, GC/MS, HPLC	A131
8001-589	Creosote	56-382	Parathion	8140	GC, GC/MS	A121
8007-452	Coal tar creosote	608-935	Pentachlorobenzene		GC/MS	A121
10102-439	Nitric oxide		Pentachlorobenzofurans	8280	GC/MS	
10102-440	Nitrogen dioxide		Pentachlorodibenzo-p-dioxins	8280	GC/MS	
10102-451	Thallium(l) nitrate	76-017	Pentachloroethane		GC/MS	A121

TABLE 3.4. ANALYSIS METHODS FOR HAZARDOUS WASTE CONSTITUENTS[1]
(40CFR261 APPENDIX VIII)

CAS RN order		Alphabetic Order		(A)	(B)	(C)
10595-956	Nitrosomethylethylamine(n-)	82-688	Pentachloronitrobenzene		GC/MS	A121
12039-520	Thallium selenite	87-865	Pentachlorophenol	8040, 8250	GC, GC/MS	A121
13256-229	Nitrososarcosine(n-)	62-442	Phenacetin		HPLC	A174
13463-393	Nickel carbonyl	108-952	Phenol	8040, 8250	GC, GC/MS	A121
13765-190	Calcium chromate H_2CrO_4	25265-763	Phenylenediamine		GC/MS	A121
14901-087	Cycasin	62-384	Phenylmercury acetate	7470, 7471	Cv/AAS	A228
16543-558	Nitrosonornicotine(n-)	103-855	Phenylthiourea			
16752-775	Methomyl	298-022	Phorate	8140	GC, GC/MS	A121
18883-664	Streptozotocin	75-445	Phosgene		GC/TCD	A138
20816-120	Osmium tetroxide	7803-512	Phosphine		GC/FPD	A136
20830-813	Daunomycin		Phosphorodithioic acid, o,o-diethyl s-((ethylthio)methyl) ester, see phorate			
20859-738	Aluminum phosphide		Phosphorothioic acid, o,o-dimethyl o-(p-(dimethylamino)sulfonyl)phenyl ester, see famphur			
23950-585	Pronamide		Phthalic acid ester, N.O.S.	8060	GC, GC/MS	A121
25154-545	Dinitrobenzene, N.O.S.	85-449	Phthalic anhydride	8090, 8250	GC, GC/MS	A121
25265-763	Phenylenediamine	109-068	Picoline(2-)	8090, 8250	GC, GC/MS	A121
25321-226	dichlorobenzene, N.O.S.		Polychlorinated biphenyls, N.O.S.	8090, 8250	GC, GC/MS	A121
25322-207	Tetrachloroethane, N.O.S.	151-508	Potassium cyanide	9010	T, C	A252
25323-302	Dichloroethylene, N.O.S.	506-616	Potassium silver cyanide	7760, 7761	ICAP, AAS	A232
25376-458	Toluenediamine	23950-585	Pronamide		GC/MS	A121
25735-299	Trichloropropane, N.O.S.	1120-714	Propane(1,3-) sultone		GC/MS	A121
26471-625	Toluene diisocyanate	107-197	Propargyl alcohol			
26545-733	Dichloropropanol, N.O.S.	107-108	Propylamine(n-)		GC/MS	A121
26638-197	Dichloropropane, N.O.S.	78-875	Propylene dichloride			
26952-238	Dichloropropene, N.O.S.	75-558	Propylenimine(1,2-)			

TABLE 3.4. ANALYSIS METHODS FOR HAZARDOUS WASTE CONSTITUENTS[1] (40CFR261 APPENDIX VIII)						
CAS RN order		Alphabetic Order		(A)	(B)	(C)
35576-911	Nitrosamines, N.O.S.	51-525	Propylthiouracil		GC/MS	A121
39196-184	Thiofanox		Propyn-1-ol(2-)		GC/FID, GC/MS	A134
	Arcolein	110-861	Pyridine	8090, 8250	GC, GC/MS	A121
	Benzene, 2-amino-1-methyl	50-555	Reserpine		HPLC	A122
	Benzene, 4-amino-1-methyl	108-463	Resorcinol		GC/FID, GC/MS	A134
	Benz(c)anthracene	81-072	Saccharin		GC/MS	A121
	Butanone peroxide(2-)		Saccharin salts		GC/MS	A121
	Butyl-4,6-dinitrophenol(2-sec-)	94-597	Safrole		GC/MS	A121
	Butylbenzyl phthalate	7782-492	Selenium and compounds, N.O.S.	7740, 7741	AAS	A231
	Chlorinated benzenes, N.O.S.	7783-008	Selenium dioxide			
	Chlorinated ethane, N.O.S.	7488-564	Selenium sulfide	7740, 7741	AAS	A231
	Chlorinated fluorocarbons, N.O.S.	630-104	Selenourea	7740, 7741	AAS	A231
	Chlorinated naphthalene, N.O.S.	7440-224	Silver and compounds, N.O.S.	7760, 7761	ICAP, AAS	A232
	Chlorinated phenol, N.O.S.	506-649	Silver cyanide	7760, 7761	ICAP, AAS	A232
	Chloro-1,3-butadiene(2-)	93-721	Silvex (2,4,5-TP)	8150, 8250	GC, GC/MS	A122
	Chloroalkyl ethers, N.O.S.	143-339	Sodium cyanide	9010	T, C	A252
	Chlorophenyl thiourea(1-)	18883-664	Streptozotocin		HPLC	A122
	Chlorophenyl((1-)o-)thiourea		Strontium sulfide		ICAP, AAS	A233
	Chloropropene(3-)	57-249	Strychnine		HPLC	A180
	Cyanides (soluble salts and complexes), N.O.S.		Strychnine salts		HPLC	A180
	D(2,4-), salts, esters	93-765	T(2,4,5-)	8150, 8250	GC, GC/MS	A122
	Di-n-butyl phthalate	95-943	Tetrachlorobenzene(1,2,4,5-)	8120, 8250	GC, GC/MS	A121
	Dichlorobenzene (meta, ortho and para isomers)	1746-016	Tetrachlorodibenzo-p-dioxin(2,3,7,8-) (TCDD)	8280	GC/MS	A121
	Dichloroethane(1,1-)		Tetrachlorodibenzofurans	8280	GC/MS	

TABLE 3.4. ANALYSIS METHODS FOR HAZARDOUS WASTE CONSTITUENTS[1] (40CFR261 APPENDIX VIII)					
CAS RN order		Alphabetic Order	(A)	(B)	(C)
Dichloroethane(1,2-)	25322-207	Tetrachloroethane, N.O.S.	8010, 8240	GC, GC/MS	A101
Dichloroethene(trans-1,2-)	79-345	Tetrachloroethane(1,1,2,2-)	8010, 8240	GC, GC/MS	A101
Dichloromethane	630-206	Tetrachloroethane(1,1,1,2-)	8010, 8240	GC, GC/MS	A101
Dichloropropane(1,2-)	127-184	Tetrachloroethylene	8010, 8240	GC, GC/MS	A101
Dichloropropene(1,3-)	58-902	Tetrachlorophenol(2,3,4,6-)	8040, 8250	GC, GC/MS	A121
Dichlrophenoxyacetic(2,4-) acid	78-002	Tetraethyl lead	7420, 7421	ICAP, AAS	A227
Diethyl s-methyl ester of phosphorodithioic acid(o,o-)	107-493	Tetraethyl pyrophosphate		GC/MS	A121
Diethylphosphoric acid, o-p-nitrophenyl ester	3689-245	Tetraethyldithiopyro-phosphate		GC/MS	A121
Dimethyl(3,3-)-1-(methylthio)-2-butanone, o-((methylamino)carbonyl)oxime, see thiofanox	509-148	Tetranitromethane		GC/MS	A101
Dinitro-o-cresol salts(4,6-)	1314-325	Thallic oxide		ICAP, AAS	A234
Dioxane(1,4-)	7440-280	Thallium and compounds, N.O.S.		ICAP, AAS	A234
Dithiobiuret(2,4-)	12039-520	Thallium selenite		ICAP, AAS	A234
Endrin metabolites	563-688	Thallium(l) acetate		ICAP, AAS	A234
Ethylenebisdithiocarbamic acid, salts and esters.	6533-739	Thallium(l) carbonate		ICAP, AAS	A234
Fluoroacetamide(2-)	7791-120	Thallium(l) chloride		ICAP, AAS	A234
Formic acid	10102-451	Thallium(l) nitrate		ICAP, AAS	A234
Halomethanes, N.O.S.	7446-186	Thallium(l) sulfate		ICAP, AAS	A234
Heptachlor epoxide (alpha, beta, and gamma isomers)	62-555	Thioacetamide		HPLC	A123
Heptachlorodibenzo-p-dioxins	39196-184	Thiofanox		GC/FPD	A183
Heptachlorodibenzofurans	74-931	Thiomethanol			

TABLE 3.4. ANALYSIS METHODS FOR HAZARDOUS WASTE CONSTITUENTS[1] (40CFR261 APPENDIX VIII)					
CAS RN order		Alphabetic Order	(A)	(B)	(C)
Hexachloro-1,4,4a,5,8,8a-hexahydro-1,4:5,8-endo, endo-dimethanonaphthalene(1,2,3,4,10,10-)	108-985	Thiophenol			
Hexachlorocyclohexane(all isomers)	79-196	Thiosemicarbazide			
Hexachlorodibenzo-p-dioxins	62-566	Thiourea		HPLC	A123
Hexachlorodibenzofurans	137-268	Thiram		HPLC	A122
Hydrocyanic acid	108-883	Toluene	8020, 8240	GC, GC/MS	A101
Hydrofluoric acid	26471-625	Toluene diisocyanate			
Hydroxydimethylarsine oxide	95-807	Toluene-2,4-diamine	8250	GC/MS	A121
Iodomethane	823-405	Toluene-2,6-diamine	8250	GC/MS	A121
Iron dextran (complex)	496-720	Toluene-3,4-diamine	8250	GC/MS	A121
Isocyanic acid, methyl ester	25376-458	Toluenediamine			
Methanethiol		Toluenediamine, N.O.S.	8250	GC/MS	A121
Methyl-2-(methylthio)propionaldehyde-o-(methylcarbonyl)oxime(2-)	95-534	Toluidine(o-)			
Methyl-n'-nitro-n-nitroso-guanidine(n-)	636-215	Toluidine hydrochloride(o-)			
Methylaziridine(2-)	106-490	Toluidine(p-)			
Naphthyl-2-thiourea(1-)		Tolylene diisocyanate	8250	GC/MS	A121
Naphthylamine(1-)	8001-352	Toxaphene	8080, 8250	GC, GC/MS	A121
Naphthylamine(2-)		Tribromomethane	8010, 8240	GC, GC/MS	A121
Nicotine salts	120-821	Trichlorobenzene(1,2,4-)	8080, 8250	GC, GC/MS	A121
Nitrogen mustard, n-oxide, hydrochloride salt		Trichloroethane(1,1,1-)	8010, 8240	GC, GC/MS	A101
Nitrogen mustard, hydrochloride salt	79-005	Trichloroethane(1,1,2-)	8010, 8240	GC, GC/MS	A101
Nitrophenol(4-)	79-016	Trichloroethylene	8010, 8240	GC, GC/MS	A101
Nitroquinoline(4-)-1-oxide	75-707	Trichloromethanithiol		GC/MS	A121
Oxabicyclo(2.2.1)heptane-2,3-dicarboxylic acid	75-694	Trichloromonofluoromethane	8010, 8240	GC, GC/MS	A101

TABLE 3.4. ANALYSIS METHODS FOR HAZARDOUS WASTE CONSTITUENTS[1]
(40CFR261 APPENDIX VIII)

CAS RN order		Alphabetic Order	(A)	(B)	(C)
Pentachlorobenzofurans	95-954	Trichlorophenol(2,4,5-)	8040, 8250	GC, GC/MS	A121
Pentachlorodibenzo-p-dioxins	88-062	Trichlorophenol(2,4,6-)	8040, 8250	GC, GC/MS	A121
Phosphorodithioic acid, o,o-diethyl s-((ethylthio)methyl) ester, see phorate		Trichlorophenoxyacetic acid(2,4,5-), see T(2,4,5-)			
Phosphorothioic acid, o,o-dimethyl o-(p-(dimethylamino)sulfonyl)phenyl ester, see famphur		Trichlorophenoxypropionic(2,4,5-) acid = TP(2,4,5-), see silvex			
Phthalic acid ester, N.O.S.	25735-299	Trichloropropane, N.O.S.	8010, 8240	GC, GC/MS	A101
Polychlorinated biphenyls, N.O.S.	96-184	Trichloropropane(1,2,3-)	8010, 8240	GC, GC/MS	A101
Propyn-1-ol(2-)	126-681	Triethylphosphorothioate(o,o,o-)		GC/MS	A121
Saccharin salts	99-354	Trinitrobenzene(1,3,5-)			
Strontium sulfide		Trinitrobenzene(sym-)		GC/MS	A121
Strychnine salts	52-244	Tris(1-aziridinyl)phosphine sulfide		GC/FPD	A190
Tetrachlorodibenzofurans	126-727	Tris(2,3-dibromopropyl) phosphate		GC/MS	A121
Toluenediamine, N.O.S.	72-571	Trypan blue		HPLC	A123
Tolylene diisocyanate	66-751	Uracil mustard			
Tribromomethane		Vanadic acid, ammonium salt		ICAP, AAS	A235
Trichloroethane(1,1,1-)	1314-621	Vanadium pentoxide		ICAP, AAS	A235
Trichlorophenoxyacetic acid(2,4,5-), see T(2,4,5-)	75-014	Vinyl chloride	8010, 8240	GC, GC/MS	A101
Trichlorophenoxypropionic(2,4,5-) acid = TP(2,4,5-), see silvex	81-812	Warfarin	8250	HPLC, GC/MS	A122
Trinitrobenzene(sym-)	557-211	Zinc cyanide	9010	T, C	A252
Vanadic acid, ammonium salt	1314-847	Zinc phosphide		GC/FPD	A253

TABLE 3.4. ANALYSIS METHODS FOR HAZARDOUS WASTE CONSTITUENTS[1]
(40CFR261 APPENDIX VIII)

CAS RN order	Alphabetic Order	(A)	(B)	(C)

[1] This table was provided in a two major column format for easy cross reference. The left hand side column under CAS RN order is for a reader to use CAS number to find the compound he needs and the right hand side column is to use a chemical name to locate where the compound is.

Column (A): (SW-846), "Test Methods for Evaluating Solid Waste, Volumes 1a - 1c: Laboratory Manual, Physical/Chemical Methods, and Volume II: Field Manual, Physical/Chemical Methods", Third Edition, Office of Solid Waste, US Environmental Protection Agency, Document Control No. 955-001-00000-1, November 1986.

Column (B): (EPA-89/6), "Hazardous Waste Incineration Measurement Guidance Manual, Volume III of the Hazardous Waste Incineration Guidance Series," Prepared by MRI, EPA625-6-89-021, June 1989. See acronym section at the end of book.

Column (C): (EPA-84/2), "Sampling and Analysis Methods for Hazardous Waste Combustion, Appendix C," Prepared by ADL, EPA600-8-84-002, PB84-155845, February 1984.

3.4. GROUND-WATER

- Ground-water protection standard (40CFR264.92)
- General ground-water monitoring requirements (40CFR264.97)
- Detection monitoring program (40CFR264.98 and Table 3.5 below)
- Ground-water monitoring system (interim status) (40CFR265.91)
- Sampling and analysis (interim status) (40CFR265.92)

TABLE 3.5. GROUND-WATER MONITORING LIST (40CFR264 APPENDIX IX)[1]				
CAS RN[3] order		Common name[2]--alphabetical order		Methods[5]
(Total)[3]	Antimony	83-329	Acenaphthene	8100, 8270
(Total)	Arsenic	208-968	Acenaphthylene	8100, 8270
(Total)	Barium	67-641	Acetone	8240
(Total)	Beryllium	75-058	Acetonitrile; Methyl cyanide	8015
(Total)	Cadmium	98-862	Acetophenone	8270
(Total)	Chromium	53-963	Acetylaminofluorene(2-); 2-AAF	8270
(Total)	Cobalt	107-028	Acrolein	8030, 8240
(Total)	Copper	107-131	Acrylonitrile	8030, 8240
(Total)	Lead	309-002	Aldrin	8080, 8270
(Total)	Mercury	107-051	Allyl chloride	8010, 8240
(Total)	Nickel	92-671	Aminobiphenyl(4-)	8270
(Total)	Selenium	62-533	Aniline	8270
(Total)	Silver	120-127	Anthracene	8100, 8270
(Total)	Thallium	(Total)	Antimony	6010, 7040, 7041
(Total)	Tin	140-578	Aramite	8270
(Total)	Vanadium	(Total)	Arsenic	6010, 7060, 7061
(Total)	Zinc	(Total)	Barium	6010, 7080
See note 7	Polychlorinated biphenyls; PCBs	71-432	Benzene	8020, 8240
See note 8	Polychlorinated dibenzo-p-dioxins; PCDDs	56-553	Benzo(a)anthracene; Benzanthracene	8100, 8270
See note 9	Polychlorinated dibenzofurans; PCDFs	50-328	Benzo(a)pyrene	8100, 8270
50-293	DDT(4,4'-)	205-992	Benzo(b)fluoranthene	8100, 8270
50-328	Benzo(a)pyrene	191-242	Benzo(ghi)perylene	8100, 8270
51-285	Dinitrophenol(2,4-)	207-089	Benzo(k)fluoranthene	8100, 8270
52-857	Famphur	100-516	Benzyl alcohol	8270
53-703	Dibenz(a,h)anthracene	(Total)	Beryllium	6010, 7090, 7091
53-963	Acetylaminofluorene(2-); 2-AAF	319-857	beta-BHC	8080, 8250

TABLE 3.5. GROUND-WATER MONITORING LIST (40CFR264 APPENDIX IX)[1]				
CAS RN[3] order		Common name[2]--alphabetical order		Methods[5]
55-185	Nitrosodiethylamine(n-)	319-846	BHC(alpha-)	8080, 8250
56-235	Carbon tetrachloride	319-868	BHC(delta-)	8080, 8250
56-382	Parathion	58-899	BHC(gamma-); Lindane	8080, 8250
56-495	Methylcholanthrene(3-)	108-601	Bis(2-chloro-1-methylethyl) ether; 2,2'-Di-chlorodiisopropyl ether	8010, 8270
56-553	Benzo(a)anthracene; Benzanthracene	111-911	Bis(2-chloroethoxy) methane	8270
56-575	Nitroquinoline 1-oxide(4-)	111-444	Bis(2-chloroethyl) ether	8270
57-125	Cyanide	117-817	Bis(2-ethylhexyl) phthalate	8060, 8270
57-749	Chlordane	75-274	Bromodichloromethane	8010, 8240
57-976	Dimethylbenz(a) anthracene(7,12-)	75-252	Bromoform; Tribromomethane	8010, 8240
58-899	BHC(gamma-); Lindane	101-553	Bromophenyl phenyl ether(4-)	8270
58-902	Tetrachlorophenol(2,3,4,6-)	85-687	Butyl benzyl phthalate; Benzyl butyl phthalate	8060, 8270
59-507	Chloro-m-cresol(p-)	(Total)	Cadmium	6010, 7130, 7131
59-892	Nitrosomorpholine(n-)	75-150	Carbon disulfide	8240
60-117	Dimethylamino azobenzene(p-)	56-235	Carbon tetrachloride	8010, 8240
60-515	Dimethoate	57-749	Chlordane	8080, 8250
60-571	Dieldrin	59-507	Chloro-m-cresol(p-)	8040, 8270
62-442	Phenacetin	106-478	Chloroaniline(p-)	8270
62-500	Ethyl methanesulfonate	108-907	Chlorobenzene	8010, 8020, 8240
62-533	Aniline	510-156	Chlorobenzilate	8270
62-759	Nitrosodimethylamine(n-)	75-003	Chloroethane; Ethyl chloride	8010, 8240
66-273	Methyl methanesulfonate	67-663	Chloroform	8010, 8240
67-641	Acetone	91-587	Chloronaphthalene(2-)	8120, 8270
67-663	Chloroform	95-578	Chlorophenol(2-)	8040, 8270
67-721	Hexachloroethane	7005-723	Chlorophenyl phenyl ether(4-)	8270
70-304	Hexachlorophene	126-998	Chloroprene	8010, 8240
71-432	Benzene	(Total)	Chromium	6010, 7190, 7191
71-556	Trichloroethane(1,1,1-); Methylchloroform	218-019	Chrysene	8100, 8270
72-208	Endrin	(Total)	Cobalt	6010, 7200, 7201
72-435	Methoxychlor	(Total)	Copper	6010, 7210

TABLE 3.5. GROUND-WATER MONITORING LIST (40CFR264 APPENDIX IX)[1]				Methods[5]
CAS RN[3] order		Common name[2]--alphabetical order		
72-548	DDD(4,4'-)	108-394	Cresol(m-)	8270
72-559	DDE(4,4'-)	95-487	Cresol(o-)	8270
74-839	Methyl bromide; Bromomethane	106-445	Cresol(p-)	8270
74-873	Methyl chloride; Chloromethane	57-125	Cyanide	9010
74-884	Methyl iodide; Iodomethane	94-757	D(2,4-); 2,4-Dichlorophenoxyaceti acid	8150
74-953	Methylene bromide; Dibromomethane	72-548	DDD(4,4'-)	8080, 8270
75-003	Chloroethane; Ethyl chloride	72-559	DDE(4,4'-)	8080, 8270
75-014	Vinyl chloride	50-293	DDT(4,4'-)	8080, 8270
75-058	Acetonitrile; Methyl cyanide	84-742	Di-n-butyl phthalate	8060, 8270
75-092	Methylene chloride; Dichloromethane	117-840	Di-n-octyl phthalate	8060, 8270
75-150	Carbon disulfide	2303-164	Diallate	8270
75-252	Bromoform; Tribromomethane	132-649	Dibenzofuran	8270
75-274	Bromodichloromethane	53-703	Dibenz(a,h)anthracene	8100, 8270
75-343	Dichloroethane(1,1-)	96-128	Dibromo-3-chloropropane(1,2-); DBCP	8010, 8240, 8270
75-354	Dichloroethylene(1,1-); Vinylidene chloride	124-481	Dibromochloromethane; Chlorodibromomethane	8010, 8240
75-694	Trichlorofluoromethane	106-934	Dibromoethane(1,2-); Ethylene dibromide	8010, 8240
75-718	Dichlorodifluoromethane	110-576	Dichloro-2-butene(trans-1,4-)	8240
76-017	Pentachloroethane	541-731	Dichlorobenzene(m-)	8010, 8020, 8120, 8270
76-448	Heptachlor	95-501	Dichlorobenzene(o-)	8010, 8020, 8120, 8270
77-474	Hexachlorocyclopentadiene	106-467	Dichlorobenzene(p-)	8010, 8020, 8120, 8270
78-591	Isophorone	91-941	Dichlorobenzidine(3,3'-)	8270
78-831	Isobutyl alcohol	75-718	Dichlorodifluoromethane	8010, 8240
78-875	Dichloropropane(1,2-)	75-343	Dichloroethane(1,1-)	8010, 8240
78-933	Methyl ethyl ketone; MEK	107-062	Dichloroethane(1,2-); Ethylene dichloride	8010, 8240
79-005	Trichloroethane(1,1,2-)	75-354	Dichloroethylene(1,1-); Vinylidene chloride	8010, 8240
79-016	Trichloroethylene; Trichloroethene	156-605	Dichloroethylene(trans-1,2-)	8010, 8240
79-345	Tetrachloroethane(1,1,2,2-)	120-832	Dichlorophenol(2,4-)	8040, 8270
80-626	Methyl methacrylate	87-650	Dichlorophenol(2,6-)	8270
82-688	Pentachloronitrobenzene	78-875	Dichloropropane(1,2-)	8010, 8240
83-329	Acenaphthene	10061-015	Dichloropropene(cis-1,3-)	8010, 8240

TABLE 3.5. GROUND-WATER MONITORING LIST (40CFR264 APPENDIX IX)[1]				
CAS RN[3] order		Common name[2]--alphabetical order		Methods[5]
84-662	Diethyl phthalate	10061-026	Dichloropropene(trans-1,3-)	8010, 8240
84-742	Di-n-butyl phthalate	60-571	Dieldrin	8080, 8270
85-018	Phenanthrene	297-972	Diethyl o-2-pyrazinyl phosphorothioate(o,o-); Thionazin	8270
85-687	Butyl benzyl phthalate; Benzyl butyl phthalate	84-662	Diethyl phthalate	8060, 8270
86-306	Nitrosodiphenylamine(n-)	60-515	Dimethoate	8270
86-737	Fluorene	131-113	Dimethyl phthalate	8060, 8270
87-650	Dichlorophenol(2,6-)	122-098	Dimethyl-phenethylamine(alpha, alpha-)	8270
87-683	Hexachlorobutadiene	60-117	Dimethylamino azobenzene(p-)	8270
87-865	Pentachlorophenol	119-937	Dimethylbenzidine(3,3'-)	8270
88-062	Trichlorophenol(2,4,6-)	57-976	Dimethylbenz(a) anthracene(7,12-)	8270
88-744	Nitroaniline(o-)	105-679	Dimethylphenol(2,4-)	8040, 8270
88-755	Nitrophenol(o-)	534-521	Dinitro-o-cresol(4,6-)	8040, 8270
88-857	Dinoseb; DNBP; 2-sec-Butyl-4,6-dinitrophenol	99-650	Dinitrobenzene(m-)	8270
91-203	Naphthalene	51-285	Dinitrophenol(2,4-)	8040, 8270
91-576	Methylnaphthalene(2-)	121-142	Dinitrotoluene(2,4-)	8090, 8270
91-587	Chloronaphthalene(2-)	606-202	Dinitrotoluene(2,6-)	8090, 8270
91-598	Naphthylamine(2-)	88-857	Dinoseb; DNBP; 2-sec-Butyl-4,6-dinitrophenol	8150, 8270
91-805	Methapyrilene	123-911	Dioxane(1,4-)	8015
91-941	Dichlorobenzidine(3,3'-)	122-394	Diphenylamine	8270
92-671	Aminobiphenyl(4-)	298-044	Disulfoton	8140, 8270
93-721	Silvex; 2,4,5-TP	959-988	Endosulfan I	8080, 8250
93-765	T(2,4,5-); 2,4,5-Trichlorophenoxyacetic acid	33213-659	Endosulfan II	8080
94-597	Safrole	1031-078	Endosulfan sulfate	8080, 8270
94-757	D(2,4-); 2,4-Dichlorophenoxyaceti acid	72-208	Endrin	8080, 8250
95-487	Cresol(o-)	7421-934	Endrin aldehyde	8080, 8270
95-501	Dichlorobenzene(o-)	97-632	Ethyl methacrylate	8015, 8240, 8270
95-534	Toluidine(o-)	62-500	Ethyl methanesulfonate	8270
95-578	Chlorophenol(2-)	100-414	Ethylbenzene	8020, 8240
95-943	Tetrachlorobenzene(1,2,4,5-)	52-857	Famphur	8270
95-954	Trichlorophenol(2,4,5-)	206-440	Fluoranthene	8100, 8270

TABLE 3.5. GROUND-WATER MONITORING LIST (40CFR264 APPENDIX IX)[1]

CAS RN[3] order		Common name[2]--alphabetical order		Methods[5]
96-128	Dibromo-3-chloropropane(1,2-); DBCP	86-737	Fluorene	8100, 8270
96-184	Trichloropropane(1,2,3-)	76-448	Heptachlor	8080, 8270
97-632	Ethyl methacrylate	1024-573	Heptachlor epoxide	8080, 8270
98-862	Acetophenone	118-741	Hexachlorobenzene	8120, 8270
98-953	Nitrobenzene	87-683	Hexachlorobutadiene	8120, 8270
99-092	Nitroaniline(m-)	77-474	Hexachlorocyclopentadiene	8120, 8270
99-354	sym-Trinitrobenzene	67-721	Hexachloroethane	8120, 8270
99-558	Nitro-o-toluidine(5-)	70-304	Hexachlorophene	8270
99-650	Dinitrobenzene(m-)	1888-717	Hexachloropropene	8270
100-016	Nitroaniline(p-)	591-786	Hexanone(2-)	8240
100-027	Nitrophenol(p-)	193-395	Indeno(1,2,3-cd) pyrene	8100, 8270
100-414	Ethylbenzene	78-831	Isobutyl alcohol	8015
100-425	Styrene	465-736	Isodrin	8270
100-516	Benzyl alcohol	78-591	Isophorone	8090, 8270
100-754	Nitrosopiperidine(n-)	120-581	Isosafrole	8270
101-553	Bromophenyl phenyl ether(4-)	143-500	Kepone	8270
105-679	Dimethylphenol(2,4-)	(Total)	Lead	6010, 7420, 7421
106-445	Cresol(p-)	(Total)	Mercury	7470
106-467	Dichlorobenzene(p-)	126-987	Methacrylonitrile	8015, 8240
106-478	Chloroaniline(p-)	91-805	Methapyrilene	8270
106-503	Phenylenediamine(p-)	72-435	Methoxychlor	8080, 8270
106-934	Dibromoethane(1,2-); Ethylene dibromide	74-839	Methyl bromide; Bromomethane	8010, 8240
107-028	Acrolein	74-873	Methyl chloride; Chloromethane	8010, 8240
107-051	Allyl chloride	78-933	Methyl ethyl ketone; MEK	8015, 8240
107-062	Dichloroethane(1,2-); Ethylene dichloride	74-884	Methyl iodide; Iodomethane	8010, 8240
107-120	Propionitrile; Ethyl cyanide	80-626	Methyl methacrylate	8015, 8240
107-131	Acrylonitrile	66-273	Methyl methanesulfonate	8270
108-054	Vinyl acetate	298-000	Methyl parathion; Parathion methyl	8140, 8270
108-101	Methyl-2-pentanone(4-); Methyl isobutyl ketone	108-101	Methyl-2-pentanone(4-); Methyl isobutyl ketone	8015, 8240
108-394	Cresol(m-)	56-495	Methylcholanthrene(3-)	8270
108-601	Bis(2-chloro-1-methylethyl) ether; 2,2'-Di-chlorodiisopropyl ether	74-953	Methylene bromide; Dibromomethane	8010, 8240

TABLE 3.5. GROUND-WATER MONITORING LIST (40CFR264 APPENDIX IX)[1]				
CAS RN[3] order		Common name[2]--alphabetical order		Methods[5]
108-883	Toluene	75-092	Methylene chloride; Dichloromethane	8010, 8240
108-907	Chlorobenzene	91-576	Methylnaphthalene(2-)	8270
108-952	Phenol	91-203	Naphthalene	8100, 8270
109-068	Picoline(2-)	130-154	Naphthoquinone(1,4-)	8270
110-576	Dichloro-2-butene(trans-1,4-)	134-327	Naphthylamine(1-)	8270
110-861	Pyridine	91-598	Naphthylamine(2-)	8270
111-444	Bis(2-chloroethyl) ether	(Total)	Nickel	6010, 7520
111-911	Bis(2-chloroethoxy) methane	99-558	Nitro-o-toluidine(5-)	8270
117-817	Bis(2-ethylhexyl) phthalate	99-092	Nitroaniline(m-)	8270
117-840	Di-n-octyl phthalate	88-744	Nitroaniline(o-)	8270
118-741	Hexachlorobenzene	100-016	Nitroaniline(p-)	8270
119-937	Dimethylbenzidine(3,3'-)	98-953	Nitrobenzene	8090, 8270
120-127	Anthracene	88-755	Nitrophenol(o-)	8040, 8270
120-581	Isosafrole	100-027	Nitrophenol(p-)	8040, 8270
120-821	Trichlorobenzene(1,2,4-)	56-575	Nitroquinoline 1-oxide(4-)	8270
120-832	Dichlorophenol(2,4-)	924-163	Nitrosodi-n-butylamine(n-)	8270
121-142	Dinitrotoluene(2,4-)	55-185	Nitrosodiethylamine(n-)	8270
122-098	Dimethyl-phenethylamine(alpha, alpha-)	62-759	Nitrosodimethylamine(n-)	8270
122-394	Diphenylamine	86-306	Nitrosodiphenylamine(n-)	8270
123-911	Dioxane(1,4-)	621-647	Nitrosodipropylamine(n-); Di-n-propylnitrosamine	8270
124-481	Dibromochloromethane; Chlorodibromomethane	10595-956	Nitrosomethylethylamine(n-)	8270
126-681	Triethyl phosphorothioate(o,o,o-)	59-892	Nitrosomorpholine(n-)	8270
126-987	Methacrylonitrile	100-754	Nitrosopiperidine(n-)	8270
126-998	Chloroprene	930-552	Nitrosopyrrolidine(n-)	8270
127-184	Tetrachloroethylene; Perchloroethylene; Tetrachloroethene	56-382	Parathion	8270
129-000	Pyrene	608-935	Pentachlorobenzene	8270
130-154	Naphthoquinone(1,4-)	76-017	Pentachloroethane	8240, 8270
131-113	Dimethyl phthalate	82-688	Pentachloronitrobenzene	8270
132-649	Dibenzofuran	87-865	Pentachlorophenol	8040, 8270
134-327	Naphthylamine(1-)	62-442	Phenacetin	8270
140-578	Aramite	85-018	Phenanthrene	8100, 8270

TABLE 3.5. GROUND-WATER MONITORING LIST (40CFR264 APPENDIX IX)[1]				
CAS RN[3] order		Common name[2]--alphabetical order		Methods[5]
143-500	Kepone	108-952	Phenol	8040, 8270
156-605	Dichloroethylene(trans-1,2-)	106-503	Phenylenediamine(p-)	8270
191-242	Benzo(ghi)perylene	298-022	Phorate	8140, 8270
193-395	Indeno(1,2,3-cd) pyrene	109-068	Picoline(2-)	8240, 8270
205-992	Benzo(b)fluoranthene	See note 7	Polychlorinated biphenyls; PCBs	8080, 8250
206-440	Fluoranthene	See note 8	Polychlorinated dibenzo-p-dioxins; PCDDs	8280
207-089	Benzo(k)fluoranthene	See note 9	Polychlorinated dibenzofurans; PCDFs	8280
208-968	Acenaphthylene	23950-585	Pronamide	8270
218-019	Chrysene	107-120	Propionitrile; Ethyl cyanide	8015, 8240
297-972	Diethyl o-2-pyrazinyl phosphorothioate(o,o-); Thionazin	129-000	Pyrene	8100, 8270
298-000	Methyl parathion; Parathion methyl	110-861	Pyridine	8240, 8270
298-022	Phorate	94-597	Safrole	8270
298-044	Disulfoton	(Total)	Selenium	6010, 7740, 7741
309-002	Aldrin	(Total)	Silver	6010, 7760
319-846	BHC(alpha-)	93-721	Silvex; 2,4,5-TP	8150
319-857	beta-BHC	100-425	Styrene	8020, 8240, 9030
319-868	BHC(delta-)	99-354	sym-Trinitrobenzene	8270
465-736	Isodrin	93-765	T(2,4,5-); 2,4,5-Trichlorophenoxyacetic acid	8150
510-156	Chlorobenzilate	1746-016	TCDD(2,3,7,8-); 2,3,7,8-Tetrachlorodibenzo-p-dioxin	8280
534-521	Dinitro-o-cresol(4,6-)	95-943	Tetrachlorobenzene(1,2,4,5-)	8270
541-731	Dichlorobenzene(m-)	79-345	Tetrachloroethane(1,1,2,2-)	8010, 8240
591-786	Hexanone(2-)	630-206	Tetrachloroethane(1,1,1,2-)	8010, 8240
606-202	Dinitrotoluene(2,6-)	127-184	Tetrachloroethylene; Perchloroethylene; Tetrachloroethene	8010, 8240
608-935	Pentachlorobenzene	58-902	Tetrachlorophenol(2,3,4,6-)	8270
621-647	Nitrosodipropylamine(n-); Di-n-propylnitrosamine	3689-245	Tetraethyl dithiopyrophosphate; Sulfotepp	8270
630-206	Tetrachloroethane(1,1,1,2-)	(Total)	Thallium	6010, 7840, 7841
924-163	Nitrosodi-n-butylamine(n-)	(Total)	Tin	7870
930-552	Nitrosopyrrolidine(n-)	108-883	Toluene	8020, 8240
959-988	Endosulfan I	95-534	Toluidine(o-)	8270

TABLE 3.5. GROUND-WATER MONITORING LIST (40CFR264 APPENDIX IX)[1]				
CAS RN[3] order		Common name[2]--alphabetical order		Methods[5]
1024-573	Heptachlor epoxide	8001-352	Toxaphene	8080, 8250
1031-078	Endosulfan sulfate	120-821	Trichlorobenzene(1,2,4-)	8270
1330-207	Xylene (total)	71-556	Trichloroethane(1,1,1-); Methylchloroform	8240
1746-016	TCDD(2,3,7,8-); 2,3,7,8-Tetrachlorodibenzo-p-dioxin	79-005	Trichloroethane(1,1,2-)	8010, 8240
1888-717	Hexachloropropene	79-016	Trichloroethylene; Trichloroethene	8010, 8240
2303-164	Diallate	75-694	Trichlorofluoromethane	8010, 8240
3689-245	Tetraethyl dithiopyrophosphate; Sulfotepp	95-954	Trichlorophenol(2,4,5-)	8270
7005-723	Chlorophenyl phenyl ether(4-)	88-062	Trichlorophenol(2,4,6-)	8040, 8270
7421-934	Endrin aldehyde	96-184	Trichloropropane(1,2,3-)	8010, 8240
8001-352	Toxaphene	126-681	Triethyl phosphorothioate(o,o,o-)	8270
10061-015	Dichloropropene(cis-1,3-)	(Total)	Vanadium	6010, 7910, 7911
10061-026	Dichloropropene(trans-1,3-)	108-054	Vinyl acetate	8240
10595-956	Nitrosomethylethylamine(n-)	75-014	Vinyl chloride	8010, 8240
23950-585	Pronamide	1330-207	Xylene (total)	8020, 8240
33213-659	Endosulfan II	(Total)	Zinc	6010, 7950

[1] The list was provided in a two major column format for easy cross reference. The left hand side column under CAS RN order is for a reader to use CAS number to find the compound he needs and the right hand side column is to use a chemical name to locate where the compound is.

[2] Common names are those widely used in government regulations, scientific publications, and commerce; synonyms exist for many chemicals.

[3] Chemical Abstracts Service registry number. Where "Total" is entered, all species in the ground water that contain this element are included, i.e., (Total) means the total amount of the inorganic compound from all species in the sample.

3.5. WASTE ANALYSIS

- General waste analysis (40CFR264.13)
- Waste analysis (for incinerators) (40CFR264.341)
- General waste analysis (40CFR265.13)
- Waste analysis and trial tests (for tank systems) (40CFR265.200)
- Waste analysis and trial tests (for surface impoundments) (40CFR265.225)
- Waste analysis (for waste piles) (40CFR265.252)
- Waste analysis (for land treatment) (40CFR265.273)
- Waste analysis (for incinerators) (40CFR265.341)
- Waste analysis (for thermal treatment) (40CFR265.375)
- Waste analysis and trial tests (for chemical, physical, and biological treatment) (40CFR265.402)
- Waste analysis and recordkeeping (for land disposal restrictions) (40CFR268.7)

3.6. REFERENCES (40CFR260.11.a)

When used in parts 40CFR260 through 270 of this chapter, the following publications are incorporated by reference:

(ASTM, D3278-78), "ASTM Standard Test Methods for Flash Point of Liquids by Setaflash Closed Tester," ASTM Standard D3278-78, available from American Society for Testing and Materials, 1916 Race Street, Philadelphia, PA 19103.

(ASTM, D93-79 or D93-80, D93-80), "ASTM Standard Test Methods for Flash Point by Pensky-Martens Closed Tester," ASTM Standard D93-79 or D93-80, D93-80 is available from American Society for Testing and Materials, 1916 Race Street, Philadelphia, PA 19103.

(ASTM, D1946-82), "ASTM Standard Method for Analysis of Reformed Gas by Gas Chromatography," ASTM Standard D1946-82, available from American Society for Testing and Materials, 1916 Race Street, Philadelphia, PA 19103.

(ASTM, D2267-88), "ASTM Standard Test Method for Aromatics in Light Naphthas and Aviation Gasolines by Gas Chromatography," ASTM Standard D2267-88, available from American Society for Testing and Materials, 1916 Race Street, Philadelphia, PA 19103.

(ASTM, D2382-83), "ASTM Standard Test Method for Heat of Combustion of Hydrocarbon Fuels by Bomb Calorimeter (High-Precision Method)," ASTM Standard D2382-83, available from American Society for Testing and Materials, 1916 Race Street, Philadelphia, PA 19103.

(ASTM, D2879-86), "ASTM Standard Test Method for Vapor Pressure-Temperature Relationship and Initial Decomposition Temperature of Liquids by Isoteriscope," ASTM Standard D2879-86, available from American Society for Testing and Materials, 1916 Race Street, Philadelphia, PA 19103.

(ASTM, E168-88), "ASTM Standard Practices for General Techniques of Infrared Quantitative Analysis," ASTM Standard E168-88, available from American Society for Testing and Materials, 1916 Race Street, Philadelphia, PA 19103.

(ASTM, E169-87), "ASTM Standard Practices for General Techniques of Ultraviolet-Visible Quantitative Analysis," ASTM Standard E169-87, available from American Society for Testing and Materials, 1916 Race Street, Philadelphia, PA 19103.

(ASTM, E260-85), "ASTM Standard Practice for Packed Column Gas Chromatography," ASTM Standard E260-85, available from American Society for Testing and Materials, 1916 Race Street, Philadelphia, PA 19103.

(ASTM, E926-88), "ASTM Standard Test Methods for Preparing Refuse-Derived Fuel (RDF) Samples for Analyses of Metals," ASTM Standard E926-88, Test Method C-Bomb, Acid Digestion Method, available from American Society for Testing Materials, 1916 Race Street, Philadelphia, PA 19103.

(EPA-81/12), "APTI Course 415: Control of Gaseous Emissions," EPA Publication EPA450-2-81-005, December 1981, available from National Technical Information Service, 5285 Port Royal Road, Springfield, VA 2216.

(NFPA-81), "Flammable and Combustible Liquids Code" (1977 or 1981), available from the National Fire Protection Association (NFPA), 470 Atlantic Avenue, Boston, MA 02210.

(SW-846), "Test Methods for Evaluating Solid Waste, Physical/Chemical Methods," EPA Publication SW-846 (Third Edition (November, 1986), as amended by Updates I (July, 1992), II (September, 1994), IIA (August, 1993), and IIB (January, 1995)). The Third Edition of SW-846 and Updates I, II, IIA, and IIB (document number 955-001-00000-1) are available from the Superintendent of Documents, U.S. Government Printing Office, Washington, DC 20402, (202) 512-1800. Copies may be inspected at the Library, U.S. Environmental Protection Agency, 401 M Street, SW, Washington, DC 20460.

0010 Modified Method 5 Sampling Train
0020 Source Assessment Sampling System (SASS)
0030 Volatile Organic Sampling Train (VOST)
1320 Multiple Extraction Procedure
1330 Extraction Procedure for Oily Wastes
3611 Alumina Column Cleanup and Separation of Petroleum Wastes
5040 Protocol for Analysis of Sorbent Cartridges from Volatile Organic Sampling Train
6010 Inductively Coupled Plasma Atomic Emission Spectroscopy
7090 Beryllium (AA, Direct Aspiration)
7091 Beryllium (AA, Furnace Technique)
7198 Chromium, Hexavalent (Differential Pulse Polarography)
7210 Copper (AA, Direct Aspiration)
7211 Copper (AA, Furnace Technique)
7380 Iron (AA, Direct Aspiration)
7381 Iron (AA, Furnace Technique)
7460 Manganese (AA, Direct Aspiration)
7461 Manganese (AA, Furnace Technique)
7550 Osmium (AA, Direct Aspiration)
7770 Sodium (AA, Direct Aspiration)
7840 Thallium (AA, Direct Aspiration)
7841 Thallium (AA, Furnace Technique)
7910 Vanadium (AA, Direct Aspiration)
7911 Vanadium (AA, Furnace Technique)
7950 Zinc (AA, Direct Aspiration)
7951 Zinc (AA, Furnace Technique)
9022 Total Organic Halides (TOX) by Neutron Activation Analysis
9035 Sulfate (Colorimetric, Automated, Chloranilate)
9036 Sulfate (Colorimetric, Automated, Methylthymol Blue, AA II)
9038 Sulfate (Turbidimetric)
9060 Total Organic Carbon
9065 Phenolics (Spectrophotometric, Manual 4-AAP with Distillation)
9066* Phenolics (Colorimetric, Automated 4-AAP with Distillation)
9067 Phenolics (Spectrophotometric, MBTH with Distillation)
9070 Total Recoverable Oil and Grease (Gravimetric, Separatory Funnel Extraction)
9071 Oil and Grease Extraction Method for Sludge Samples
9080 Cation-Exchange Capacity of Soils (Ammonium Acetate)
9081 Cation-Exchange Capacity of Soils (Sodium Acetate)
9100 Saturated Hydraulic Conductivity, Saturated Leachate Conductivity, and Intrinsic Permeability
9131 Total Coliform: Multiple Tube Fermentation Technique
9132 Total Coliform: Membrane Filter Technique
9200 Nitrate
9250 Chloride (Colorimetric, Automated Ferricyanide AAI)
9251 Chloride (Colorimetric, Automated Ferricyanide AAII)
9252 Chloride (Titrimetric, Mercuric Nitrate)
9310 Gross Alpha and Gross Beta
9315 Alpha-Emitting Radium Isotopes
9320 Radium-228

* When Method 9066 is used it must be preceded by the manual distillation specified in procedure 7.1 of Method 9065. Just prior to distillation in Method 9065, adjust the sulfuric acid-preserved sample to pH 4 with 1 + 9 NaOH. After the manual distillation is completed, the autoanalyzer manifold is simplified by connecting the re-sample line directly to the sampler.

Section 4

DRINKING WATER PROGRAMS

4. I. MONITORING, SAMPLING AND ANALYTICAL REQUIREMENTS IN 40CFR 2
    4.1.1. COLIFORM SAMPLING (40CFR141.21) 2
    4.1.2. TURBIDITY SAMPLING AND ANALYTICAL REQUIREMENTS (40CFR141.22) 2
    4.1.3. INORGANIC CHEMICALS 2
        TABLE 4.1. DETECTION LIMITS FOR INORGANIC CONTAMINANTS
        (40CFR141.23.a) 2
        TABLE 4.2. ANALYTICAL METHODS FOR THE LISTED INORGANIC
        CONTAMINANTS (40CFR141.23.k) 4
    4.1.4. ORGANIC CHEMICALS 6
        TABLE 4.3. ORGANIC CONTAMINANTS AND MCL LIMITS 6
        TABLE 4.4. SYNTHETIC ORGANIC CONTAMINANTS AND MCL LIMITS 7
    4.1.5. ANALYTICAL METHODS FOR RADIOACTIVITY (40CFR141.25 AND TABLE 4.3) .. 9
        TABLE 4.5. MCL STANDARDS AND ANALYTICAL METHODS
        FOR RADIOACTIVITY 9
    4.1.6. TOTAL TRIHALOMETHANES SAMPLING, ANALYTICAL
        AND OTHER REQUIREMENTS (40CFR141.30) 10
    4.1.7. ANALYSIS OF TRIHALOMETHANES (40CFR141, APPENDIX C) 10
    4.1.8. MICROBIOLOGICAL CONTAMINANTS 10

4.2. MONITORING, SAMPLING AND ANALYTICAL METHODS IN EPA PUBLICATIONS 10
    4.2.1. METHODS FOR THE DETERMINATION OF INORGANIC SUBSTANCES
        IN ENVIRONMENTAL SAMPLES 10
    4.2.2. METHODS FOR THE DETERMINATION OF ORGANIC COMPOUNDS
        IN DRINKING WATER 10
        TABLE 4.6. ANALYTE - METHOD CROSS-REFERENCE (EPA-91/7) 12
    4.2.3. METHODS FOR CHEMICAL ANALYSIS OF WATER AND WASTES 25
        TABLE 4.7A. TEST METHODS--ALPHABETICAL ORDER BY PARAMETER 25
        TABLE 4.7B. TEST METHODS--NUMBER ORDER BY METHOD 31

4.3. REFERENCES 37

4.1. MONITORING, SAMPLING AND ANALYTICAL REQUIREMENTS IN 40CFR

4.1.1. COLIFORM SAMPLING (40CFR141.21)

4.1.2. TURBIDITY SAMPLING AND ANALYTICAL REQUIREMENTS (40CFR141.22)

4.1.3. INORGANIC CHEMICALS

- Inorganic chemical sampling and analytical requirements (40CFR141.23 and Tables 4.1. and 4.2. below)
  - Detection limits for inorganic contaminants (40CFR141.23.a)
  - Inorganic analysis methods (40CFR141.23.k)
  - Analytical methods for lead, copper, pH, conductivity, calcium, alkalinity, orthophosphate, silica, and temperature (40CFR141.89)

- Organic chemicals other than total trihalomethanes, sampling and analytical requirements
  - Organic contaminants and MCL limits (40CFR141.61.a and 141.24.f)
  - Synthetic organic contaminants and MCL limits (40CFR141.61.c)

- Maximum contaminant level goals for inorganic contaminants (40CFR141.51)

- Maximum contaminant levels for inorganic contaminants (40CFR141.62)

TABLE 4.1. DETECTION LIMITS FOR INORGANIC CONTAMINANTS (40CFR141.23.a)			
Contaminant	MCL(mg/l)	Methodology	Detection limit (mg/l)
Antimony	0.006	Atomic Absorption; Furnace	0.003
		Atomic Absorption; Platform	0.0008[5]
		ICP-Mass Spectrometry	0.0004
		Hydride-Atomic Absorption	0.001
Asbestos	7 MFL[1]	Transmission Electron Microscopy	0.01 MFL
Barium	2	Atomic Absorption; furnace technique	0.002
		Atomic Absorption; direct spiration	0.1
		Inductively Coupled Plasma	0.002 (0.001)
Beryllium	0.004	Atomic Absorption; Furnace	0.0002
		Atomic Absorbtion; Platform	0.00002[5]
		Inductively Coupled Plasma[2]	0.0003
		ICP-Mass Spectrometry	0.0003
Cadmium	0.005	Atomic Absorption; furnace technique	0.0001
		Inductively Coupled Plasma	0.001
Chromium	0.1	Atomic Absorption; furnace technique	0.001
		Inductively Coupled Plasma	0.007 (0.001)
Cyanide	0.2	Distillation, Spectrophotometric[3]	0.02
		Distillation, Automated,	0.005
		Spectrophotometric[3]	
		Distillation, Selective Electrode[3]	0.05
		Distillation, Amenable,	0.02

TABLE 4.1. DETECTION LIMITS FOR INORGANIC CONTAMINANTS (40CFR141.23.a)			
Contaminant	MCL(mg/l)	Methodology	Detection limit (mg/l)
		Spectrophotometric[4]	
Mercury	0.002	Manual Cold Vapor Technique	0.0002
		Automated Cold Vapor Technique	0.0002
Nickel	0.1	Atomic Absorption; Furnace	0.001
		Atomic Absorbtion; Platform	0.0006[5]
		Inductively Coupled Plasma[2]	0.005
		ICP-Mass Spectrometry	0.0005
Nitrate	10 (as N)	Manual Cadmium Reduction	0.01
		Automated Hydrazine Reduction	0.01
		Automated Cadmium Reduction	0.05
		Ion Selective Electrode	1
		Ion Chromatography	0.01
Nitrite	1 (as N)	Spectrophotometric	0.01
		Automated Cadmium Reduction	0.05
		Manual Cadmium Reduction	0.01
		Ion Chromatography	0.004
Selenium	0.05	Atomic Absorption; furnace	0.002
		Atomic Absorption; gaseous hydride	0.002
Thallium	0.002	Atomic Absorbtion; Furnace	0.001
		Atomic Absorption; Platform	0.0007[5]
		ICP-Mass Spectrometry	0.0003

[1] MFL = million fibers per liter >10 μm.
[2] Using a 2X preconcentration step as noted in Method 200.7. Lower MDLs may be achieved when using a 4X preconcentration.
[3] Screening method for total cyanides.
[4] Measures "free" cyanides.
[5] Lower MDLs are reported using stabilized temperature graphite furnace atomic absorption.

TABLE 4.2. ANALYTICAL METHODS FOR THE LISTED INORGANIC CONTAMINANTS (40CFR141.23.k)		EPA[1,5,12]	ASTM[2]	SM[3]	USGS[4]	Other
Antimony	Atomic Absorption; Platform[6]	[5]2009				
	Atomic Absorption; Furnace[6]	[1]204.2		3113		
	ICP-Mass Spectrometry	[5]2008				
	Hydride-Atomic Absorption		D3697-92			
Asbestos	Transmission Electron n Microscopy	[12]EPA				
Barium	Atomic Absorption; furnace technique[6]	[2]208.2		3120B		
	Atomic Absorption; direct spiration[6]	[2]208.1		3111D		
	Inductively Coupled Plasma[6]	[5]200.7		3120		
Beryllium	Atomic Absorption; Furnace[6]	[1]210.2	D3645-84B	3113		
	Atomic Absorbtion; Platform[6]	[5]200.9				
	Inductively Coupled Plasma[6]	[5]200.7		3120		
	ICP-Mass Spectrometry[6]	[5]200.8				
Cadmium	Atomic Absorption; furnace technique[6]	[1]213.2		3113B		
	Inductively Coupled Plasma[6]	[5]200.7				
Chromium	Atomic Absorption; furnace technique[6]	[1]218.2		2113B		
	Inductively Coupled Plasma[6]	[5]200.7		3120		
Cyanide	Distillation, Spectrophotometric	[1]335.2	D2036-89A	4500-CN-D	1330085	
	Distillation, Automated, Spectrophotometric[3]	[1]335.3		4500-CN-E		
	Distillation, Selective Electrode[3]		D2036-89A	4500-CN-F		
	Distillation, Amenable, Spectrophotometric[4]	[1]335.1	D2036-89B	4500-CN-G		
Mercury	Manual Cold Vapor Technique[9]	[1]245.1	D3223-86	3112B		
	Automated Cold Vapor Technique[9]	[1]245.2				
Nickel	Atomic Absorption; Furnace[6]	[1]249.2		3113		
	Atomic Absorbtion; Platform[6]	[5]200.9				
	Atomic Absorption; direct[6]	[1]249.1		3111B		
	Inductively Coupled Plasma[6]	[5]200.7		3120		
	ICP-Mass Spectrometry[6]	[5]200.8				
Nitrate	Manual Cadmium Reduction	[1]353.3	D3867-90	4500-NO3-E		
	Automated Hydrazine Reduction	[1]353.1				
	Automated Cadmium Reduction	[1]353.2		4500-NO3-F		

TABLE 4.2. ANALYTICAL METHODS FOR THE LISTED INORGANIC CONTAMINANTS (40CFR141.23.k)		EPA[1,5,12]	ASTM[2]	SM[3]	USGS[4]	Other
	Ion Selective Electrode					WeWWG/5880[7]
	Ion Chromatography	[11]300.0				B-1011[8]
Nitrite	Spectrophotometric	[1]354.1		3120B		
	Automated Cadmium Reduction	[1]353.2	D3867-90	4500-NO3-F		
	Manual Cadmium Reduction	[1]353.3	D3867-90	4500-NO3-E		
	Ion Chromatography	[11]300.0				B-1011[8]
Selenium	Hydride-Atomic Absorption[9]		D3859-84A	3114B		
	Atomic Absorption; Furnace[6,10]	[1]272.2	D3859-88	3113B		
Thallium	Atomic Absorption; Furnace[6]	[1]279.2		3113		
	Atomic Absorption; Platform[6]	[5]200.9		4500-CN-C		
	ICP-Mass Spectrometry[6]	[5]200.8				

Notes:

[1] "Methods of Chemical Analysis of Water and Wastes," EPA Environmental Monitoring Systems Laboratory, Cincinnati, OH 45268 March 11983, EPA600-4-79-020, which is available at NTIS, PB84-128677.

[2] Annual Book of ASTM Standards, Vols 11.01 and 11.02, 1991, American Society for Testing and Materials, 1916 Race Street, 1916 Race Street, Philadelphia, PA 19103.

[3] Standard Methods for the Examination of Water and Wastewater, 17th edition, American Public Health Association, American Water Works Association, Water Pollution Control Federation, 1989.

[4] Techniques of Water Resources Investigations of the U.S. Geological Survey, "Methods for the Determination of Inorganic Substances in Water and Fluvial Sediments," Book 5, Chapter A-1, Third Edition, 1989. Available at Superintendent of Documents, U.S. Government Printing Office, Washington, DC 20402.

[5] "Method for the Determination of Metals in Environmental Samples," Available at NTIS, PB 91-231498.

[6] Samples that contain less than 1 NTU (nephelometric turbidity unit) and are properly preserved (conc HNO3 to pH < 2) may be analyzed directly (without digestion) for total metals, otherwise digestion is required. Turbidity must be measured on the preserved samples just prior to the initiation of metal analysis. The total recoverable techniques as defined in the method must be used.

[7] "Orion Guide to Water and Wastewater Analysis." Form WeWWG/5880. p.5, 1985. Orion Research, Inc., Cambridge, MA.

[8] "Waters Test Method for Determination of Nitrite/Nitrate in Water Using Single Column Ion Chromatography," Method B-1011 Millipore Corporation, Waters Chromatography Division, 34 Maple Street, Milford, MA 01757.

[9] For the gaseous hydride determinations of antimony and selenium and for the determination of mercury by the cold vapor techniques, the proper digestion technique as defined in the method must be used to ensure the element is in the proper state for analysis.

[10] Add 2 ml of 30% H_2O_2 and an appropriate concentration of matrix modifier $Ni(NO_2)+6H_2O$ (nickel nitrate) to samples.

[11] "Method 300. Determination of Inorganic Anions in Water by Ion Chromatography." Inorganic Chemistry Branch, Environmental Monitoring Systems Laboratory. August 1991.

[12] "Analytical Method for Determination of Asbestos fibers in Water," EPA600-4-83-043, September 1083, U.S. EPA Environmental Research Laboratory, Athens, GA 30613.

4.1.4. ORGANIC CHEMICALS

- Organic chemicals other than total trihalomethanes, sampling and analytical requirements (40CFR141.24)
- Maximum contaminant level goals for organic contaminants (40CFR141.50)
- Maximum contaminant levels (revised) for organic contaminants (40CFR141.61)

TABLE 4.3. ORGANIC CONTAMINANTS AND MCL LIMITS (40CFR141.61.a)			
	CAS No.	Contaminant	MCL (mg/l)
(1)[1]	75-014	Vinyl chloride	0.002
(2)	71-432	Benzene	0.005
(3)	56-235	Carbon tetrachloride	0.005
(4)	107-062	Dichloroethane(1,2-)	0.005
(5)	79-016	Trichloroethylene	0.005
(6)	106-467	Dichlorobenzene(para-)	0.075
(7)	75-354	Dichloroethylene(1,1-)	0.007
(8)	71-556	Trichloroethane(1,1,1-)	0.2
(9)	156-592	Dichloroethylene(cis-1,2-)	0.07
(10)	78-875	Dichloropropane(1,2-)	0.005
(11)	100-414	Ethylbenzene	0.7
(12)	108-907	Monochlorobenzene	0.1
(13)	95-501	Dichlorobenzene(o-)	0.6
(14)	100-425	Styrene	0.1
(15)	127-184	Tetrachloroethylene	0.005
(16)	108-883	Toluene	1
(17)	156-605	Dichloroethylene(trans-1,2-)	0.1
(18)	1330-207	Xylenes (total)	10
(19)	75-092	Dichloromethane	0.005
(20)	120-821	Trichloro-benzene(1,2,4-)	0.07
(21)	79-005	Trichloro-ethane(1,1,2-)	0.005
[1] The numerical number was designated by EPA in 40CFR141.61.a			

	CAS No.	TABLE 4.4. SYNTHETIC ORGANIC CONTAMINANTS AND MCL LIMITS (40CFR141.61.c) Contaminant	MCL (mg/l)
(1)[1]	15972-608	Alachlor	0.002
(2)	116-063	Aldicarb	0.003
(3)	1646-873	Aldicarb sulfoxide	0.004
(4)	1646-884	Aldicarb sulfone	0.002
(5)	1912-249	Atrazine	0.003
(6)	1563-662	Carbofuran	0.04
(7)	57-749	Chlordane	0.002
(8)	96-128	Dibromochloropropane	0.0002
(9)	94-757	D(2,4-)	0.07
(10)	106-934	Ethylene dibromide	0.00005
(11)	76-448	Heptachlor	0.0004
(12)	1024-573	Heptachlor epoxide	0.0002
(13)	58-899	Lindane	0.0002
(14)	72-435	Methoxychlor	0.04
(15)	1336-363	Polychlorinated biphenyls	0.0005
(16)	87-865	Pentachlorophenol	0.001
(17)	8001-352	Toxaphene	0.003
(18)	93-721	TP(2,4,5-)	0.05
(19)	50-328	Benzo[a]pyrene	0.0002
(20)	75-990	Dalapon	0.2
(21)	103-231	Di(2-ethylhexyl) adipate	0.4
(22)	117-817	Di(2-ethylhexyl) phthalate	0.006
(23)	88-857	Dinoseb	0.007
(24)	85-007	Diquat	0.02
(25)	145-733	Endothall	0.1
(26)	72-208	Endrin	0.002
(27)	1071-536	Glyphosate	0.7
(28)	118-741	Hexachlorobenzene	0.001
(29)	77-474	Hexachlorocyclopentadiene	0.05
(30)	23135-220	Oxamyl; Vydate	0.2
(31)	1918-021	Picloram	0.5
(32)	122-349	Simazine	0.004

TABLE 4.4. SYNTHETIC ORGANIC CONTAMINANTS AND MCL LIMITS (40CFR141.61.c)			
CAS No.		Contaminant	MCL (mg/l)
(33)	1746-016	TCDD(2,3,7,8-); Dioxin	3×10^{-8}

[1] The numerical number was designated by EPA in 40CFR141.61.c

4.1.5. ANALYTICAL METHODS FOR RADIOACTIVITY (40CFR141.25 AND TABLE 4.3.)

TABLE 4.5. MCL STANDARDS AND ANALYTICAL METHODS FOR RADIOACTIVITY				
Radionuclide	MCL	Analytical Methods		Monitoring Frequency
Alpha, gross		141.25(a)		141.26(a)
Beta, gross	141.16	141.25(a)	141.25(c)	141.26(b)
Cesium-134		141.25(a)	141.25(c)	
Iodine-131			141.25(c)	
Photon radioactivity	141.16			
Radium, total		141.25(a)		
Radium-226	141.15	141.25(a)		141.26(a)
Radium-228	141.15			141.26(a)
Strontium-89		141.25(a)	141.25(c)	141.26(b)
Strontium-90	141.16	141.25(a)	141.25(c)	
Tritium	141.16	141.25(a)	141.25(c)	
Uranium		141.25(a)		

4.1.6. TOTAL TRIHALOMETHANES SAMPLING, ANALYTICAL AND OTHER REQUIREMENTS (40CFR141.30)

4.1.7. ANALYSIS OF TRIHALOMETHANES (40CFR141-APPENDIX C)

- Part I - The analysis of trihalomethanes in drinking water by the purge and trap method
- Part II - The analysis of trihalomethanes in drinking water by liquid/liquid extraction
- Part III - Determination of maximum total trihalomethane potential

4.1.8. MICROBIOLOGICAL CONTAMINANTS

- Maximum contaminant level goals for microbiological contaminants (40CFR141.52)
- Maximum contaminant levels (MCLs) for microbiological contaminants (40CFR141.63)

4.2. MONITORING, SAMPLING AND ANALYTICAL METHODS IN EPA PUBLICATIONS

4.2.1. METHODS FOR THE DETERMINATION OF INORGANIC SUBSTANCES IN ENVIRONMENTAL SAMPLES

This manual (EPA-93/8) contains ten updated and revised automated, semi-automated or methods amenable to automation for the determination of a variety of inorganic substances in water and wastewater.

These methods include and address, in an expanded form, information concerning safety, quality control, pollution prevention, and waste management. Methods were selected which minimize the amount of hazardous reagents required and maximize sample throughput to allow expanded quality control.

Automated methods are included for nitrate-nitrite, phosphorus, and sulfate. Semi-automated methods cover cyanide, ammonia, total kjeldahl nitrogen (TKN), chemical oxygen demand (COD) and generic phenolics. Methods amenable to automation include turbidity and inorganic anions by ion chromatography.

Methods provided in the manual are as follows:

- Method 180.1   Determination of Turbidity by Nephelometry

- Method 300.0   Determination of Inorganic Anions by Ion Chromatography

- Method 335.4   Determination of Total CyanideMethod by Semi-Automated Colorimetry

- Method 350.1   Determination of Ammonia Nitrogen by Semi-Automated Colorimetry

- Method 351.2   Determination of Total Kjeldahl Nitrogen by Semi-Automated Colorimetry

- Method 353.2   Determination of Nitrate-Nitrite by Automated Colorimetry

- Method 365.1   Determination of Phosphorus Method by Automated Colorimetry

- Method 375.2   Determination of Sulfate by Automated Colorimetry

- Method 410.4   Determination of Chemical Oxygen Demand by Semi-Automated Colorimetry

- Method 420.4   Determination of Total Recoverable Phenolics by Semi-Automated Colorimetry

4.2.2. METHODS FOR THE DETERMINATION OF ORGANIC COMPOUNDS IN DRINKING WATER

Thirteen analytical methods (EPA-91/7) for the identification and measurement of organic compounds in drinking water are described in detail. Six of the methods are for volatile organic compounds (VOCs) and certain disinfection by-products, and these methods were cited in the Federal Register of July 8, 1987, under the National Primary Drinking Water Regulations. The other seven methods are designed for the determination of a variety of synthetic organic compounds and pesticides, and these methods were cited in proposed drinking water regulations in the Federal Register of May 22, 1989. Five of the methods utilize the inert gas purge-and-trap extraction procedure for VOCs, six methods employ a classical liquid-liquid extraction, one method uses a new liquid-solid extraction technique, and one method is for direct aqueous analysis. Of the 13 methods, 12 use either packed or capillary gas chromatography column separations followed by detection with mass spectrometry or a selective gas chromatography detector. One method is based on a high performance liquid chromatography separation.

Methods provided in the manual are as follows:

- Method 502.1   Volatile Halogenated Organic Compounds in Water by Purge and Trap Gas Chromatography

- Method 502.2   Volatile Organic Compounds in Water By Purge and Trap Capillary Column Gas Chromatography with Photoionization and Electrolytic Conductivity Detectors in Series

- Method 503.1   Volatile Aromatic and Unsaturated Organic Compounds in Water by Purge and Trap Gas Chromatography

- Method 504   1,2-Dibromoethane (EDB) and 1,2-Dibromo-3-Chloropropane (DBCP) in Water by Microextraction and Gas Chromatography

- Method 505   Analysis of Organohalide Pesticides and Commerical Polychlorinated Biphenyl Products in Water by Micro-Extraction and Gas Chromatography

- Method 507   Determination of Nitrogen - and Phosphorus-Containing Pesticides in Water by Gas Chromatography with a Nitrogen-Phosphorus Detector

- Method 508   Determination of Chlorinated Pesticides in Water by Gas Chromatography with An Electron Capture Detector

- Method 508A   Screening for Polychlorinated Biphenyls by Perchlorination and Gas Chromatography

- Method 515.1   Determination of Chlorinated Acids in Water by Gas Chromatography with an Electron Capture Detector

- Method 524.1   Measurement of Purgeable Organic Compounds in Water by Packed Column Gas Chromatography/Mass Spectrometry

- Method 524.2   Measurement of Purgeable Organic Compounds in Water by Capillary Column Gas Chromatography/Mass Spectrometry

- Method 525.1   Determination of Organic Compounds in Drinking Water by Liquid-Solid Extraction and Capillary Column Gas Chromatography/Mass Spectrometry

- Method 531.1   Measurement of N-Methylcarbamoyloximes and N-Methylcarbamates in Water by Direct Aqueous Injection HPLC with Post Column Derivatization

- Table 4.4.   Method cross reference for various analytes

SAMPLING, ANALYSIS, AND MONITORING METHODS: A GUIDE TO EPA REQUIREMENTS

TABLE 4.6. ANALYTE - METHOD CROSS-REFERENCE (EPA-91/7)

Method order		Analyte--alphabetical order	
502.1[1]	Bromobenzene	525.1	Acenaphthylene
502.1	Bromochloromethane	515.1	Acifluorfen
502.1	Bromodichloromethane	505	Alachlor
502.1	Bromoform	507	Alachlor
502.1	Bromomethane	525.1	Alachlor
502.1	Carbon tetrachloride	531.1	Aldicarb
502.1	Chlorobenzene	531.1	Aldicarb sulfone
502.1	Chloroethane	531.1	Aldicarb sulfoxide
502.1	Chloroform	505	Aldrin
502.1	Chloromethane	508	Aldrin
502.1	Chlorotoluene(2-)	525.1	Aldrin
502.1	Chlorotoluene(4-)	507	Ametryn
502.1	Dibromochloromethane	525.1	Anthracene
502.1	Dibromoethane(1,2-)	505	Aroclor 1016; 1221; 1232; 1242; 1248; 1254; 1260
502.1	Dibromomethane	508	Aroclor 1016; 1221; 1232; 1242; 1248; 1254; 1260
502.1	Dichlorobenzene(1,2-)	525.1	Aroclor 1016; 1221; 1232; 1242; 1248; 1254; 1260
502.1	Dichlorobenzene(1,3-)	508A	Aroclor (General screen)
502.1	Dichlorobenzene(1,4-)	525.1	Aroclor (General screen)
502.1	Dichlorodifluoromethane	507	Atraton
502.1	Dichloroethane(1,1-)	505	Atrazine
502.1	Dichloroethane(1,2-)	507	Atrazine
502.1	Dichloroethene(1,1-)	525.1	Atrazine
502.1	Dichloroethene(cis-1,2-)	531.1	Baygon
502.1	Dichloroethene(trans-1,2-)	515.1	Bentazon
502.1	Dichloropropane(1,2-)	525.1	Benz(a)anthracene
502.1	Dichloropropane(1,3-)	502.2	Benzene
502.1	Dichloropropane(2,2-)	503.1	Benzene
502.1	Dichloropropene(1,1-)	524.1	Benzene
502.1	Dichloropropene(cis-1,3-)	524.2	Benzene
502.1	Dichloropropene(trans-1,3-)	525.1	Benzo(b)fluoranthene
502.1	Methylene chloride	525.1	Benzo(k)fluoranthene
502.1	Tetrachloroethane(1,1,2,2-)	525.1	Benzo(ghi)perylene
502.1	Tetrachloroethane(1,1,1,2-)	525.1	Benzo(a)pyrene

4 - 12

TABLE 4.6. ANALYTE - METHOD CROSS-REFERENCE (EPA-91/7)

Method order		Analyte--alphabetical order	
502.1	Tetrachloroethene	507	Bromacil
502.1	Trichloroethane(1,1,1-)	502.1	Bromobenzene
502.1	Trichloroethane(1,1,2-)	502.2	Bromobenzene
502.1	Trichloroethene	503.1	Bromobenzene
502.1	Trichlorofluoromethane	524.1	Bromobenzene
502.1	Trichloropropane(1,2,3-)	524.2	Bromobenzene
502.1	Vinyl chloride	502.1	Bromochloromethane
502.2	Benzene	502.2	Bromochloromethane
502.2	Bromobenzene	524.1	Bromochloromethane
502.2	Bromochloromethane	524.2	Bromochloromethane
502.2	Bromodichloromethane	502.1	Bromodichloromethane
502.2	Bromoform	502.2	Bromodichloromethane
502.2	Bromomethane	524.1	Bromodichloromethane
502.2	Butylbenzene(n-)	524.2	Bromodichloromethane
502.2	Butylbenzene(sec-)	502.1	Bromoform
502.2	Butylbenzene(tert-)	502.2	Bromoform
502.2	Carbon tetrachloride	524.1	Bromoform
502.2	Chlorobenzene	524.2	Bromoform
502.2	Chloroethane	502.1	Bromomethane
502.2	Chloroform	502.2	Bromomethane
502.2	Chloromethane	524.1	Bromomethane
502.2	Chlorotoluene(2-)	524.2	Bromomethane
502.2	Chlorotoluene(4-)	507	Butachlor
502.2	Dibromo-3-chloropropane(1,2-)	507	Butylate
502.2	Dibromochloromethane	502.2	Butylbenzene(n-)
502.2	Dibromoethane(1,2-)	503.1	Butylbenzene(n-)
502.2	Dibromomethane	524.2	Butylbenzene(n-)
502.2	Dichlorobenzene(1,2-)	502.2	Butylbenzene(sec-)
502.2	Dichlorobenzene(1,3-)	503.1	Butylbenzene(sec-)
502.2	Dichlorobenzene(1,4-)	524.2	Butylbenzene(sec-)
502.2	Dichlorodifluoromethane	502.2	Butylbenzene(tert-)
502.2	Dichloroethane(1,1-)	503.1	Butylbenzene(tert-)
502.2	Dichloroethane(1,2-)	524.2	Butylbenzene(tert-)

TABLE 4.6. ANALYTE - METHOD CROSS-REFERENCE (EPA-91/7)			
Method order		Analyte--alphabetical order	
502.2	Dichloroethene(1,1-)	525.1	Butylbenzylphthalate
502.2	Dichloroethene(cis-1,2-)	531.1	Carbaryl
502.2	Dichloroethene(trans-1,2-)	531.1	Carbofuran
502.2	Dichloropropane(1,2-)	502.1	Carbon tetrachloride
502.2	Dichloropropane(1,3-)	502.2	Carbon tetrachloride
502.2	Dichloropropane(2,2-)	524.1	Carbon tetrachloride
502.2	Dichloropropene(1,1-)	524.2	Carbon tetrachloride
502.2	Dichloropropene(cis-1,3-)	507	Carboxin
502.2	Dichloropropene(trans-1,3-)	515.1	Chloramben
502.2	Ethylbenzene	505	Chlordane (Technical)
502.2	Hexachlorobutadiene	508	Chlordane (Technical)
502.2	Isopropylbenzene	505	Chlordane-alpha
502.2	Isopropyltoluene(4-)	508	Chlordane-alpha
502.2	Methylene chloride	525.1	Chlordane-alpha
502.2	Naphthalene	505	Chlordane-gamma
502.2	Propylbenzene	508	Chlordane-gamma
502.2	Styrene	525.1	Chlordane-gamma
502.2	Tetrachloroethane(1,1,2,2-)	508	Chlorneb
502.2	Tetrachloroethane(1,1,1,2-)	502.1	Chlorobenzene
502.2	Tetrachloroethene	502.2	Chlorobenzene
502.2	Toluene	503.1	Chlorobenzene
502.2	Trichlorobenzene(1,2,3-)	524.1	Chlorobenzene
502.2	Trichlorobenzene(1,2,4-)	524.2	Chlorobenzene
502.2	Trichloroethane(1,1,1-)	508	Chlorobenzilate
502.2	Trichloroethane(1,1,2-)	525.1	Chlorobiphenyl(2-)
502.2	Trichloroethene	502.1	Chloroethane
502.2	Trichlorofluoromethane	502.2	Chloroethane
502.2	Trichloropropane(1,2,3-)	524.1	Chloroethane
502.2	Trimethylbenzene(1,2,4-)	524.2	Chloroethane
502.2	Trimethylbenzene(1,3,5-)	502.1	Chloroform
502.2	Vinyl chloride	502.2	Chloroform
502.2	Xylene(m-)	524.1	Chloroform
502.2	Xylene(o-)	524.2	Chloroform

SECTION 4. DRINKING WATER PROGRAMS

TABLE 4.6. ANALYTE - METHOD CROSS-REFERENCE (EPA-91/7)			
Method order		Analyte--alphabetical order	
502.2	Xylene(p-)	502.1	Chloromethane
503.1	Benzene	502.2	Chloromethane
503.1	Bromobenzene	524.1	Chloromethane
503.1	Butylbenzene(n-)	524.2	Chloromethane
503.1	Butylbenzene(sec-)	508	Chlorothalonil
503.1	Butylbenzene(tert-)	502.1	Chlorotoluene(2-)
503.1	Chlorobenzene	502.2	Chlorotoluene(2-)
503.1	Chlorotoluene(2-)	503.1	Chlorotoluene(2-)
503.1	Chlorotoluene(4-)	524.1	Chlorotoluene(2-)
503.1	Dichlorobenzene(1,2-)	524.2	Chlorotoluene(2-)
503.1	Dichlorobenzene(1,3-)	502.1	Chlorotoluene(4-)
503.1	Dichlorobenzene(1,4-)	502.2	Chlorotoluene(4-)
503.1	Ethylbenzene	503.1	Chlorotoluene(4-)
503.1	Hexachlorobutadiene	524.1	Chlorotoluene(4-)
503.1	Isopropylbenzene	524.2	Chlorotoluene(4-)
503.1	Isopropyltoluene(4-)	507	Chlorpropham
503.1	Naphthalene	525.1	Chrysene
503.1	Propylbenzene(n-)	507	Cycloate
503.1	Styrene	515.1	Dalapon
503.1	Tetrachloroethene	508	Dimethyl-2,3,5,6-tetrachloroterephthalate; DCPA; Dacthal
503.1	Toluene	515.1	DCPA mono and diacid metabolites
503.1	Trichlorobenzene(1,2,3-)	515.1	Dichlorophenoxyacetic acid(2,4-); 2,4-D
503.1	Trichlorobenzene(1,2,4-)	515.1	Dichlorophenoxy(4-(2,4-))butyric acid; 2,4-DB
503.1	Trichloroethene	508	DDD(1,1-dichloro-2,2-bis(p-chlorophenyl)ethane)(4,4'-)
503.1	Trimethylbenzene(1,2,4-)	508	DDE(1,1-dichloro-2,2-bis(p-chlorophenyl)ethylene)(4,4'-)
503.1	Trimethylbenzene(1,3,5-)	508	DDT(1,1,1-trichloro-2,2-bis(p-chlorophenyl)ethane)(4,4'-)
503.1	Xylene(m-)	507	Diazinon
503.1	Xylene(o-)	525.1	Dibenz(ah)anthracene
503.1	Xylene(p-)	502.1	Dibromochloromethane
504	Dibromo-3-chloropropane(1,2-)	502.2	Dibromochloromethane

Method order		Analyte--alphabetical order	
504	Dibromoethane(1,2-)	524.1	Dibromochloromethane
505	Alachlor	524.2	Dibromochloromethane
505	Aldrin	502.2	Dibromo-3-chloropropane(1,2-)
505	Aroclor 1016; 1221; 1232; 1242; 1248; 1254; 1260	504	Dibromo-3-chloropropane(1,2-)
505	Atrazine	524.1	Dibromo-3-chloropropane(1,2-)
505	Chlordane (Technical)	524.2	Dibromo-3-chloropropane(1,2-)
505	Chlordane-alpha	502.1	Dibromomethane
505	Chlordane-gamma	502.2	Dibromomethane
505	Dieldrin	524.1	Dibromomethane
505	Endrin	524.2	Dibromomethane
505	Heptachlor	502.1	Dibromoethane(1,2-)
505	Heptachlor Epoxide	502.2	Dibromoethane(1,2-)
505	Hexachlorobenzene	504	Dibromoethane(1,2-)
505	Hexachlorocyclohexane(gamma-); Lindane; gamma-BHC; HCH-gamma	524.1	Dibromoethane(1,2-)
505	Hexachlorocyclopentadiene	524.2	Dibromoethane(1,2-)
505	Methoxychlor	525.1	Di-n-butylphthalate
505	Nonachlor(cis-)	515.1	Dicamba
505	nonachlor(trans-)	502.1	Dichlorobenzene(1,2-)
505	Simazine	502.2	Dichlorobenzene(1,2-)
505	Toxaphene (Technical)	503.1	Dichlorobenzene(1,2-)
507	Alachlor	524.1	Dichlorobenzene(1,2-)
507	Ametryn	524.2	Dichlorobenzene(1,2-)
507	Atraton	502.1	Dichlorobenzene(1,3-)
507	Atrazine	502.2	Dichlorobenzene(1,3-)
507	Bromacil	503.1	Dichlorobenzene(1,3-)
507	Butachlor	524.1	Dichlorobenzene(1,3-)
507	Butylate	524.2	Dichlorobenzene(1,3-)
507	Carboxin	502.1	Dichlorobenzene(1,4-)
507	Chlorpropham	502.2	Dichlorobenzene(1,4-)
507	Cycloate	503.1	Dichlorobenzene(1,4-)
507	Diazinon	524.1	Dichlorobenzene(1,4-)
507	Dichlorvos	524.2	Dichlorobenzene(1,4-)

TABLE 4.6. ANALYTE - METHOD CROSS-REFERENCE (EPA-91/7)

Method order		Analyte--alphabetical order	
507	Diphenamid	515.1	Dichlorobenzoic acid(3,5-)
507	Disulfoton	525.1	Dichlorobiphenyl(2,3-)
507	Disulfoton sulfone	502.1	Dichloroethane(1,1-)
507	Disulfoton sulfoxide	502.2	Dichloroethane(1,1-)
507	Ethoprop	524.1	Dichloroethane(1,1-)
507	Ethyl dipropylthiocarbamate(s-); EPTC	524.2	Dichloroethane(1,1-)
507	Ethylhexyl(n-(2-))bicyclo(2.2.1)-5-heptene-2,3-dicarboximide (MGK-264)	502.1	Dichloroethane(1,2-)
507	Fenamiphos	502.2	Dichloroethane(1,2-)
507	Fenarimol	524.1	Dichloroethane(1,2-)
507	Fluridone	524.2	Dichloroethane(1,2-)
507	Hexazinone	502.1	Dichloroethene(1,1-)
507	Merphos	502.2	Dichloroethene(1,1-)
507	Methyl paraoxon	524.1	Dichloroethene(1,1-)
507	Metolachlor	524.2	Dichloroethene(1,1-)
507	Metribuzin	502.1	Dichloroethene(cis-1,2-)
507	Mevinphos	502.2	Dichloroethene(cis-1,2-)
507	Molinate	524.1	Dichloroethene(cis-1,2-)
507	Napropamide	524.2	Dichloroethene(cis-1,2-)
507	Norflurazon	502.1	Dichloroethene(trans-1,2-)
507	Pebulate	502.2	Dichloroethene(trans-1,2-)
507	Prometon	524.1	Dichloroethene(trans-1,2-)
507	Prometryn	524.2	Dichloroethene(trans-1,2-)
507	Pronamide	502.1	Dichloropropane(1,2-)
507	Propazine	502.2	Dichloropropane(1,2-)
507	Simazine	524.1	Dichloropropane(1,2-)
507	Simetryn	524.2	Dichloropropane(1,2-)
507	Stirofos	502.1	Dichloropropane(1,3-)
507	Tebuthiuron	502.2	Dichloropropane(1,3-)
507	Terbacil	524.1	Dichloropropane(1,3-)
507	Terbufos	524.2	Dichloropropane(1,3-)
507	Terbutryn	502.1	Dichloropropane(2,2-)
507	Triademefon	502.2	Dichloropropane(2,2-)

TABLE 4.6. ANALYTE - METHOD CROSS-REFERENCE (EPA-91/7)			
Method order		Analyte--alphabetical order	
507	Tricyclazole	524.1	Dichloropropane(2,2-)
507	Vernolate	524.2	Dichloropropane(2,2-)
508	Aldrin	502.1	Dichloropropene(1,1-)
508A	Aroclor (General screen)	502.2	Dichloropropene(1,1-)
508	Aroclor 1016; 1221; 1232; 1242; 1248; 1254; 1260	524.1	Dichloropropene(1,1-)
508	Chlordane (Technical)	524.2	Dichloropropene(1,1-)
508	Chlordane-alpha	502.1	Dichloropropene(cis-1,3-)
508	Chlordane-gamma	502.2	Dichloropropene(cis-1,3-)
508	Chlorneb	524.1	Dichloropropene(cis-1,3-)
508	Chlorobenzilate	524.2	Dichloropropene(cis-1,3-)
508	Chlorothalonil	502.1	Dichloropropene(trans-1,3-)
508	DDD(1,1-dichloro-2,2-bis(p-chlorophenyl)ethane)(4,4'-)	502.2	Dichloropropene(trans-1,3-)
508	DDE(1,1-dichloro-2,2-bis(p-chlorophenyl)ethylene)(4,4'-)	524.1	Dichloropropene(trans-1,3-)
508	DDT(1,1,1-trichloro-2,2-bis(p-chlorophenyl)ethane)(4,4'-)	524.2	Dichloropropene(trans-1,3-)
508	Dieldrin	502.1	Dichlorodifluoromethane
508	Dimethyl-2,3,5,6-tetrachloroterephthalate; DCPA; Dacthal	502.2	Dichlorodifluoromethane
508	Endosulfan I	524.1	Dichlorodifluoromethane
508	Endosulfan II	524.2	Dichlorodifluoromethane
508	Endosulfan sulfate	515.1	Dichlorprop
508	Endrin	507	Dichlorvos
508	Endrin aldehyde	505	Dieldrin
508	Etridiazole	508	Dieldrin
508	Heptachlor	525.1	Diethylphthalate
508	Heptachlor Epoxide	525.1	Dimethylphthalate
508	Hexachlorobenzene	515.1	Dinoseb
508	Hexachlorocyclohexane(alpha-); alpha-BHC; HCH-alpha	507	Diphenamid
508	Hexachlorocyclohexane(beta-); beta-BHC; HCH-beta	507	Disulfoton sulfoxide
508	Hexachlorocyclohexane(delat-); delta-BHC; HCH-delta	507	Disulfoton sulfone

TABLE 4.6. ANALYTE - METHOD CROSS-REFERENCE (EPA-91/7)			
Method order		**Analyte--alphabetical order**	
508	Hexachlorocyclohexane(gamma-); Lindane; gamma-BHC; HCH-gamma	507	Disulfoton
508	Methoxychlor	508	Endosulfan sulfate
508	Permethrin(cis-)	508	Endosulfan I
508	Permethrin(trans-)	508	Endosulfan II
508A	Polychlorobiphenyls (General screen)	508	Endrin aldehyde
508	Propachlor	505	Endrin
508	Toxaphene (Technical)	508	Endrin
508	Trifluralin	525.1	Endrin
515.1	Acifluorfen	507	Ethyl dipropylthiocarbamate(s-); EPTC
515.1	Bentazon	507	Ethylhexyl(n-(2-))bicyclo(2.2.1)-5-heptene-2,3-dicarboximide (MGK-264)
515.1	Chloramben	507	Ethoprop
515.1	Dalapon	502.2	Ethylbenzene
515.1	DCPA mono and diacid metabolites	503.1	Ethylbenzene
515.1	Dicamba	524.1	Ethylbenzene
515.1	Dichlorobenzoic acid(3,5-)	524.2	Ethylbenzene
515.1	Dichlorophenoxy(4-(2,4-))butyric acid; 2,4-DB	525.1	bis(2-Ethylhexyl)adipate
515.1	Dichlorophenoxyacetic acid(2,4-); 2,4-D	525.1	bis(2-Ethylhexyl)phthalate
515.1	Dichlorprop	508	Etridiazole
515.1	Dinoseb	507	Fenamiphos
515.1	Hydroxydicamba(5-)	507	Fenarimol
515.1	Nitrophenol(4-)	525.1	Fluorene
515.1	Pentachlorophenol (PCP)	507	Fluridone
515.1	Picloram	508	Hexachlorocyclohexane(alpha-); alpha-BHC; HCH-alpha
515.1	Trichlorophenoxy(2-(2,4,5-))propionic acid; 2,4,5-TP; Silvex	508	Hexachlorocyclohexane(beta-); beta-BHC; HCH-beta
515.1	Trichlorophenoxyacetic acid(2,4,5-); 2,4,5-T	508	Hexachlorocyclohexane(delat-); delta-BHC; HCH-delta
524.1	Benzene	505	Hexachlorocyclohexane(gamma-); Lindane; gamma-BHC; HCH-gamma
524.1	Bromobenzene	508	Hexachlorocyclohexane(gamma-); Lindane; gamma-BHC; HCH-gamma
524.1	Bromochloromethane	525.1	Hexachlorocyclohexane(gamma-); Lindane; gamma-BHC; HCH-gamma

TABLE 4.6. ANALYTE - METHOD CROSS-REFERENCE (EPA-91/7)			
Method order		Analyte--alphabetical order	
524.1	Bromodichloromethane	505	Heptachlor
524.1	Bromoform	508	Heptachlor
524.1	Bromomethane	525.1	Heptachlor
524.1	Carbon tetrachloride	505	Heptachlor Epoxide
524.1	Chlorobenzene	508	Heptachlor Epoxide
524.1	Chloroethane	525.1	Heptachlor Epoxide
524.1	Chloroform	525.1	Heptachlorobiphenyl(2,2',3,3',4,4',6-)
524.1	Chloromethane	505	Hexachlorobenzene
524.1	Chlorotoluene(2-)	508	Hexachlorobenzene
524.1	Chlorotoluene(4-)	525.1	Hexachlorobenzene
524.1	Dibromo-3-chloropropane(1,2-)	525.1	Hexachlorobiphenyl(2,2',4,4',5,6'-)
524.1	Dibromochloromethane	502.2	Hexachlorobutadiene
524.1	Dibromoethane(1,2-)	503.1	Hexachlorobutadiene
524.1	Dibromomethane	524.2	Hexachlorobutadiene
524.1	Dichlorobenzene(1,2-)	505	Hexachlorocyclopentadiene
524.1	Dichlorobenzene(1,3-)	525.1	Hexachlorocyclopentadiene
524.1	Dichlorobenzene(1,4-)	507	Hexazinone
524.1	Dichlorodifluoromethane	531.1	Hydroxycarbofuran(3-)
524.1	Dichloroethane(1,1-)	515.1	Hydroxydicamba(5-)
524.1	Dichloroethane(1,2-)	525.1	Indeno(1,2,3,c,d)pyrene
524.1	Dichloroethene(1,1-)	502.2	Isopropylbenzene
524.1	Dichloroethene(cis-1,2-)	503.1	Isopropylbenzene
524.1	Dichloroethene(trans-1,2-)	524.2	Isopropylbenzene
524.1	Dichloropropane(1,2-)	502.2	Isopropyltoluene(4-)
524.1	Dichloropropane(1,3-)	503.1	Isopropyltoluene(4-)
524.1	Dichloropropane(2,2-)	524.2	Isopropyltoluene(4-)
524.1	Dichloropropene(1,1-)	507	Merphos
524.1	Dichloropropene(cis-1,3-)	531.1	Methiocarb
524.1	Dichloropropene(trans-1,3-)	531.1	Methomyl
524.1	Ethylbenzene	505	Methoxychlor
524.1	Methylene chloride	508	Methoxychlor
524.1	Styrene	525.1	Methoxychlor
524.1	Tetrachloroethane(1,1,2,2-)	507	Methyl paraoxon

TABLE 4.6. ANALYTE - METHOD CROSS-REFERENCE (EPA-91/7)			
Method order		Analyte--alphabetical order	
524.1	Tetrachloroethane(1,1,1,2-)	502.1	Methylene chloride
524.1	Tetrachloroethene	502.2	Methylene chloride
524.1	Toluene	524.1	Methylene chloride
524.1	Trichloroethane(1,1,1-)	524.2	Methylene chloride
524.1	Trichloroethane(1,1,2-)	507	Metolachlor
524.1	Trichloroethene	507	Metribuzin
524.1	Trichlorofluoromethane	507	Mevinphos
524.1	Trichloropropane(1,2,3-)	507	Molinate
524.1	Vinyl chloride	502.2	Naphthalene
524.1	Xylene(m-)	503.1	Naphthalene
524.1	Xylene(o-)	524.2	Naphthalene
524.1	Xylene(p-)	507	Napropamide
524.2	Benzene	515.1	Nitrophenol(4-)
524.2	Bromobenzene	505	Nonachlor(cis-)
524.2	Bromochloromethane	505	nonachlor(trans-)
524.2	Bromodichloromethane	525.1	nonachlor(trans-)
524.2	Bromoform	507	Norflurazon
524.2	Bromomethane	525.1	Octachlorobiphenyl(2,2',3,3',4,5',6,6'-)
524.2	Butylbenzene(n-)	531.1	Oxamyl
524.2	Butylbenzene(sec-)	515.1	Pentachlorophenol (PCP)
524.2	Butylbenzene(tert-)	525.1	Pentachlorophenol (PCP)
524.2	Carbon tetrachloride	507	Pebulate
524.2	Chlorobenzene	525.1	Pentachlorobiphenyl(2,2',3',4,6-)
524.2	Chloroethane	508	Permethrin(cis-)
524.2	Chloroform	508	Permethrin(trans-)
524.2	Chloromethane	525.1	Phenanthrene
524.2	Chlorotoluene(2-)	515.1	Picloram
524.2	Chlorotoluene(4-)	508A	Polychlorobiphenyls (General screen)
524.2	Dibromo-3-chloropropane(1,2-)	507	Prometon
524.2	Dibromochloromethane	507	Prometryn
524.2	Dibromoethane(1,2-)	507	Pronamide
524.2	Dibromomethane	508	Propachlor
524.2	Dichlorobenzene(1,2-)	507	Propazine

TABLE 4.6. ANALYTE - METHOD CROSS-REFERENCE (EPA-91/7)			
Method order		Analyte--alphabetical order	
524.2	Dichlorobenzene(1,3-)	502.2	Propylbenzene
524.2	Dichlorobenzene(1,4-)	503.1	Propylbenzene(n-)
524.2	Dichlorodifluoromethane	524.2	Propylbenzene(n-)
524.2	Dichloroethane(1,1-)	525.1	Pyrene
524.2	Dichloroethane(1,2-)	505	Simazine
524.2	Dichloroethene(1,1-)	507	Simazine
524.2	Dichloroethene(cis-1,2-)	525.1	Simazine
524.2	Dichloroethene(trans-1,2-)	507	Simetryn
524.2	Dichloropropane(1,2-)	507	Stirofos
524.2	Dichloropropane(1,3-)	502.2	Styrene
524.2	Dichloropropane(2,2-)	503.1	Styrene
524.2	Dichloropropene(1,1-)	524.1	Styrene
524.2	Dichloropropene(cis-1,3-)	524.2	Styrene
524.2	Dichloropropene(trans-1,3-)	525.1	Tetrachlorobiphenyl(2,2',4,4'-)
524.2	Ethylbenzene	515.1	Trichlorophenoxyacetic acid(2,4,5-); 2,4,5-T
524.2	Hexachlorobutadiene	515.1	Trichlorophenoxy(2-(2,4,5-))propionic acid; 2,4,5-TP; Silvex
524.2	Isopropylbenzene	507	Tebuthiuron
524.2	Isopropyltoluene(4-)	507	Terbacil
524.2	Methylene chloride	507	Terbufos
524.2	Naphthalene	507	Terbutryn
524.2	Propylbenzene(n-)	502.1	Tetrachloroethene
524.2	Styrene	502.2	Tetrachloroethene
524.2	Tetrachloroethane(1,1,1,2-)	503.1	Tetrachloroethene
524.2	Tetrachloroethane(1,1,2,2-)	524.1	Tetrachloroethene
524.2	Tetrachloroethene	524.2	Tetrachloroethene
524.2	Toluene	502.1	Tetrachloroethane(1,1,1,2-)
524.2	Trichlorobenzene(1,2,3-)	502.2	Tetrachloroethane(1,1,1,2-)
524.2	Trichlorobenzene(1,2,4-)	524.1	Tetrachloroethane(1,1,1,2-)
524.2	Trichloroethane(1,1,1-)	524.2	Tetrachloroethane(1,1,1,2-)
524.2	Trichloroethane(1,1,2-)	502.1	Tetrachloroethane(1,1,2,2-)
524.2	Trichloroethene	502.2	Tetrachloroethane(1,1,2,2-)
524.2	Trichlorofluoromethane	524.1	Tetrachloroethane(1,1,2,2-)

TABLE 4.6. ANALYTE - METHOD CROSS-REFERENCE (EPA-91/7)			
Method order		Analyte--alphabetical order	
524.2	Trichloropropane(1,2,3-)	524.2	Tetrachloroethane(1,1,2,2-)
524.2	Trimethylbenzene(1,2,4-)	502.2	Toluene
524.2	Trimethylbenzene(1,3,5-)	503.1	Toluene
524.2	Vinyl chloride	524.1	Toluene
524.2	Xylene(m-)	524.2	Toluene
524.2	Xylene(o-)	505	Toxaphene (Technical)
524.2	Xylene(p-)	508	Toxaphene (Technical)
525.1	Acenaphthylene	525.1	Toxaphene (Technical)
525.1	Alachlor	507	Triademefon
525.1	Aldrin	502.2	Trichlorobenzene(1,2,3-)
525.1	Anthracene	503.1	Trichlorobenzene(1,2,3-)
525.1	Aroclor (General screen)	524.2	Trichlorobenzene(1,2,3-)
525.1	Aroclor 1016; 1221; 1232; 1242; 1248; 1254; 1260	502.2	Trichlorobenzene(1,2,4-)
525.1	Atrazine	503.1	Trichlorobenzene(1,2,4-)
525.1	Benz(a)anthracene	524.2	Trichlorobenzene(1,2,4-)
525.1	Benzo(a)pyrene	525.1	Trichlorobiphenyl(2,4,5-)
525.1	Benzo(b)fluoranthene	502.1	Trichloroethene
525.1	Benzo(ghi)perylene	502.2	Trichloroethene
525.1	Benzo(k)fluoranthene	503.1	Trichloroethene
525.1	bis(2-Ethylhexyl)adipate	524.1	Trichloroethene
525.1	bis(2-Ethylhexyl)phthalate	524.2	Trichloroethene
525.1	Butylbenzylphthalate	502.1	Trichloroethane(1,1,1-)
525.1	Chlordane-alpha	502.2	Trichloroethane(1,1,1-)
525.1	Chlordane-gamma	524.1	Trichloroethane(1,1,1-)
525.1	Chlorobiphenyl(2-)	524.2	Trichloroethane(1,1,1-)
525.1	Chrysene	502.1	Trichloroethane(1,1,2-)
525.1	Di-n-butylphthalate	502.2	Trichloroethane(1,1,2-)
525.1	Dibenz(ah)anthracene	524.1	Trichloroethane(1,1,2-)
525.1	Dichlorobiphenyl(2,3-)	524.2	Trichloroethane(1,1,2-)
525.1	Diethylphthalate	502.1	Trichlorofluoromethane
525.1	Dimethylphthalate	502.2	Trichlorofluoromethane
525.1	Endrin	524.1	Trichlorofluoromethane

Method order		Analyte--alphabetical order	
525.1	Fluorene	524.2	Trichlorofluoromethane
525.1	Heptachlor	502.1	Trichloropropane(1,2,3-)
525.1	Heptachlor Epoxide	502.2	Trichloropropane(1,2,3-)
525.1	Heptachlorobiphenyl(2,2',3,3',4,4',6-)	524.1	Trichloropropane(1,2,3-)
525.1	Hexachlorobenzene	524.2	Trichloropropane(1,2,3-)
525.1	Hexachlorobiphenyl(2,2',4,4',5,6'-)	507	Tricyclazole
525.1	Hexachlorocyclohexane(gamma-); Lindane; gamma-BHC; HCH-gamma	508	Trifluralin
525.1	Hexachlorocyclopentadiene	502.2	Trimethylbenzene(1,2,4-)
525.1	Indeno(1,2,3,c,d)pyrene	503.1	Trimethylbenzene(1,2,4-)
525.1	Methoxychlor	524.2	Trimethylbenzene(1,2,4-)
525.1	nonachlor(trans-)	502.2	Trimethylbenzene(1,3,5-)
525.1	Octachlorobiphenyl(2,2',3,3',4,5',6,6'-)	503.1	Trimethylbenzene(1,3,5-)
525.1	Pentachlorobiphenyl(2,2',3',4,6-)	524.2	Trimethylbenzene(1,3,5-)
525.1	Pentachlorophenol (PCP)	507	Vernolate
525.1	Phenanthrene	502.1	Vinyl chloride
525.1	Pyrene	502.2	Vinyl chloride
525.1	Simazine	524.1	Vinyl chloride
525.1	Tetrachlorobiphenyl(2,2',4,4'-)	524.2	Vinyl chloride
525.1	Toxaphene (Technical)	502.2	Xylene(m-)
525.1	Trichlorobiphenyl(2,4,5-)	503.1	Xylene(m-)
531.1	Aldicarb	524.1	Xylene(m-)
531.1	Aldicarb sulfone	524.2	Xylene(m-)
531.1	Aldicarb sulfoxide	502.2	Xylene(o-)
531.1	Baygon	503.1	Xylene(o-)
531.1	Carbaryl	524.1	Xylene(o-)
531.1	Carbofuran	524.2	Xylene(o-)
531.1	Hydroxycarbofuran(3-)	503.1	Xylene(p-)
531.1	Methiocarb	524.1	Xylene(p-)
531.1	Methomyl	524.2	Xylene(p-)
531.1	Oxamyl	502.2	Xylene(p-)

TABLE 4.6. ANALYTE - METHOD CROSS-REFERENCE (EPA-91/7)

[1] Analysis methods provided in (EPA-91/7), "Methods for the Determination of Organic Compounds in Drinking Water," EPA600-4-88-039, July 1991.

4.2.3. METHODS FOR CHEMICAL ANALYSIS OF WATER AND WASTES

This manual (EPA-83/3) provides test procedures approved for the monitoring of water supplies waste discharges, and ambient waters under the Safe Drinking Water Act, the Nation Pollutant Discharge Elimination System and Ambient Monitoring Requirements of Section 106 and 208 of Public Law 92-500. The test methods have been selected to meet the needs of federal legislation and to provide guidance to laboratories engaged in the protection of human health and the aquatic environment. Tables 4.5A and 4.5B (EPA-83/3) provide methods for chemical analysis of water and wastes in an alphabetical order (parameter column) and a method number order (method column), respectively.

TABLE 4.7A. TEST METHODS--ALPHABETICAL ORDER BY PARAMETER			
Substance category	Method	Parameter	Description
Inorganic, non-metalls (300 series)	Method 305.1	Acidity	Titrimetric
Inorganic, non-metalls (300 series)	Method 305.2	Acidity	Titrimetric (acid rain)
Inorganic, non-metalls (300 series)	Method 310.2	Alkalinity	Colorimetric, Automated Methyl Orange
Inorganic, non-metalls (300 series)	Method 310.1	Alkalinity	Titrimetric (pH 4.5)
Metals (200 series)	Method 202.1	Aluminum	AA, Direct Aspiration
Metals (200 series)	Method 202.2	Aluminum	AA, Furnace
Metals (200 series)	Method 204.2	Antimony	AA, Furnace
Metals (200 series)	Method 204.1	Antimony	AA, Direct Aspiration
Metals (200 series)	Method 206.4	Arsenic	Spectrophotometric, SDDC
Metals (200 series)	Method 206.2	Arsenic	AA, Furnace
Metals (200 series)	Method 206.5	Arsenic	Digestion method for Hydride and SDDC
Metals (200 series)	Method 206.3	Arsenic	AA, Hydride
Metals (200 series)	Method 208.1	Barium	AA, Direct Aspiration
Metals (200 series)	Method 208.2	Barium	AA, Furnace
Metals (200 series)	Method 210.2	Beryllium	AA, Furnace
Metals (200 series)	Method 210.1	Beryllium	AA, Direct Aspiration
Inorganic, non-metalls (300 series)	Method 377.1	Biochemical Oxygen Demand	BOD (5 day, 20 C)
Metals (200 series)	Method 212.3	Boron	Colorimetric, Curcumin
Inorganic, non-metalls (300 series)	Method 320.1	Bromiide	Titrimetric
Metals (200 series)	Method 213.2	Cadmium	AA, Furnace
Metals (200 series)	Method 213.1	Cadmium	AA, Direct Aspiration
Metals (200 series)	Method 215.1	Calcium	AA, Direct Aspiration
Metals (200 series)	Method 215.2	Calcium	Titrimetric, EDTA
Inorganic, non-metalls (300 series)	Method 410.3	Chemical Oxygen Demand	Titrimetric, High-Level for Saline Waters
Inorganic, non-metalls (300 series)	Method 410.1	Chemical Oxygen Demand	Colorimetric, Mid-level
Inorganic, non-metalls (300 series)	Method 410.4	Chemical Oxygen Demand	Colorimetric, Automated; Manual

TABLE 4.7A. TEST METHODS--ALPHABETICAL ORDER BY PARAMETER			
Substance category	Method	Parameter	Description
Inorganic, non-metalls (300 series)	Method 410.2	Chemical Oxygen Demand	Titrimetric, Low-Level
Inorganic, non-metalls (300 series)	Method 325.3	Chloride	Titrimetric, Mercuric Nitrate
Inorganic, non-metalls (300 series)	Method 325.2	Chloride	Colorimetric, Automated Ferricyanide, AA II
Inorganic, non-metalls (300 series)	Method 325.1	Chloride	Colorimetric, Automated Ferricyanide, AA I
Inorganic, non-metalls (300 series)	Method 330.1	Chlorine, Total Residual	Titrimetric, Amperometric
Inorganic, non-metalls (300 series)	Method 330.4	Chlorine, Total Residual	Titrimetric, DPD-FAS
Inorganic, non-metalls (300 series)	Method 330.5	Chlorine, Total Residual	Spectrophotometric, DPD
Inorganic, non-metalls (300 series)	Method 330.3	Chlorine, Total Residual	Titrimetric, Iodometric
Inorganic, non-metalls (300 series)	Method 330.2	Chlorine, Total Residual	Titrimetric, Back-Iodometric
Metals (200 series)	Method 218.4	Chromium	Hexavalent, Chelation-Etraction
Metals (200 series)	Method 218.5	Chromium	Hexavalent, Dissolved
Metals (200 series)	Method 218.3	Chromium	Chelation-Etraction
Metals (200 series)	Method 218.2	Chromium	AA, Furnace
Metals (200 series)	Method 218.1	Chromium	AA, Direct Aspiration
Metals (200 series)	Method 219.2	Cobalt	AA, Furnace
Metals (200 series)	Method 219.1	Cobalt	AA, Direct Aspiration
Physical properties (100 series)	Method 110.1	Color	Colorimetric, ADMI
Physical properties (100 series)	Method 110.2	Color	Colorimetric, Platinum-cobalt
Physical properties (100 series)	Method 110.3	Color	Spectrophotometric
Physical properties (100 series)	Method 120.1	Conductance	Specific Conductance
Metals (200 series)	Method 220.1	Copper	AA, Direct Aspiration
Metals (200 series)	Method 220.2	Copper	AA, Furnace
Inorganic, non-metalls (300 series)	Method 335.1	Cyanide, Amenable to Chlorination	Titrimetric, Spectrophotometric
Inorganic, non-metalls (300 series)	Method 335.2	Cyanide, Total	Titrimetric Spectrophotometric
Inorganic, non-metalls (300 series)	Method 335.3	Cyanide, Total	Colorimetric, Automated UV
Inorganic, non-metalls (300 series)	Method 340.3	Fluoride	Colorimetric, Automated Complexone
Inorganic, non-metalls (300 series)	Method 340.1	Fluoride	Colorimetric SPADNS with Bellack distillation
Inorganic, non-metalls (300 series)	Method 340.2	Fluoride	Potentiometric, Ion Selective Electrode
Metals (200 series)	Method 231.2	Gold	AA, Furnace
Metals (200 series)	Method 231.1	Gold	AA, Direct Aspiration
Physical properties (100 series)	Method 130.1	Hartness, Total (mg/l as CaCO3)	Colorimetric, Automted EDTA

TABLE 4.7A. TEST METHODS--ALPHABETICAL ORDER BY PARAMETER			
Substance category	Method	Parameter	Description
Physical properties (100 series)	Method 130.2	Hartness, Total (mg/l as CaCO3)	Titrimetric, EDTA
Inorganic, non-metalls (300 series)	Method 345.1	Iodide	Titrimetric
Metals (200 series)	Method 235.1	Iridium	AA, Direct Aspiration
Metals (200 series)	Method 235.2	Iridium	AA, Furnace
Metals (200 series)	Method 236.1	Iron	AA, Direct Aspiration
Metals (200 series)	Method 236.2	Iron	AA, Furnace
Inorganic, non-metalls (300 series)	Method 351.2	Kjeldahl, Total	Colorimetric, Semi-Automated Block Digester AA II
Inorganic, non-metalls (300 series)	Method 351.1	Kjeldahl, Total	Colorimetric, Automated Phenate
Inorganic, non-metalls (300 series)	Method 351.3	Kjeldahl, Total	Colorimetric; Titrimetric; Potentiometric
Inorganic, non-metalls (300 series)	Method 351.4	Kjeldahl, Total	Potentiometric, Ion Selective Electrode
Metals (200 series)	Method 239.1	Lead	AA, Direct Aspiration
Metals (200 series)	Method 239.2	Lead	AA, Furnace
Metals (200 series)	Method 242.1	Magnesium	AA, Direct Aspiration
Metals (200 series)	Method 243.2	Mangnese	AA, Furnace
Metals (200 series)	Method 243.1	Mangnese	AA, Direct Aspiration
Metals (200 series)	Method 245.1	Mercury	Cold Vapor, Manual
Metals (200 series)	Method 245.5	Mercury	Cold Vapor, Sediments
Metals (200 series)	Method 24s.2	Mercury	Cold Vapor, Automtet
Metals (200 series)	Method 200.1	Metals	Inductively coupled Plasma
Metals (200 series)	Method 200.0	Metals	Atomic Absorption Methods
Inorganic, non-metalls (300 series)	Method 425.1	Methylene Blue Active Substances (MBAS)	Colorimetric
Metals (200 series)	Method 246.2	Molybdenum	AA, Furnace
Metals (200 series)	Method 246.1	Molybdenum	AA, Direct Aspiration
Metals (200 series)	Method 249.2	Nickel	AA, Furnace
Metals (200 series)	Method 249.1	Nickel	AA, Direct Aspiration
Inorganic, non-metalls (300 series)	Method 352.1	Nitrate	Colorimetric, Brucine
Inorganic, non-metalls (300 series)	Method 353.2	Nitrate-Nitrite	Colorimetric, Automated Cadmium Reduction
Inorganic, non-metalls (300 series)	Method 353.1	Nitrate-Nitrite	Colorimetric, Automated Hydrazine Reduction
Inorganic, non-metalls (300 series)	Method 353.3	Nitrate-Nitrite	Colorimetric, Manual Cadmium Reduction
Inorganic, non-metalls (300 series)	Method 354.1	Nitrite	Spectrophotometric

	TABLE 4.7A. TEST METHODS--ALPHABETICAL ORDER BY PARAMETER		
Substance category	Method	Parameter	Description
Inorganic, non-metalls (300 series)	Method 350.1	Nitrogen, Ammonia	Colorimetric, Automated Phenate
Inorganic, non-metalls (300 series)	Method 350.2	Nitrogen, Ammonia	Colorimetric; Titrimetric; Potentiometri Distillation Procedures
Inorganic, non-metalls (300 series)	Method 350.3	Nitrogen, Ammonia	Potentiometric, Ion Selective Electrode
Inorganic, non-metalls (300 series)	Method 430.1	NTA	Colorimetric, Manual, Zinc-Zincon
Inorganic, non-metalls (300 series)	Method 430.2	NTA	Colorimetric, Automated, Zinc-Zincon
Physical properties (100 series)	Method 140.1	Odor	Thresholt Odor (Consistent Series)
Inorganic, non-metalls (300 series)	Method 413.2	Oil and Grease, Total Recoverable	Spectrophotometric, Infrared
Inorganic, non-metalls (300 series)	Method 413.1	Oil and Grease, Total Recoverable	Gravimetric, Separatory Funnel Extraction
Inorganic, non-metalls (300 series)	Method 415.2	Organic Carbon, Total	UV Promoted, Persulfate Oxidation
Inorganic, non-metalls (300 series)	Method 415.1	Organic Carbon, Total	Combustion or Oxidation
Metals (200 series)	Method 252.1	Osmium	AA, Direct Aspiration
Metals (200 series)	Method 252.2	Osmium	AA, Furnace
Inorganic, non-metalls (300 series)	Method 360.2	Oxygen, Dissolved	Modified Winkler (Full Bottle Technique)
Inorganic, non-metalls (300 series)	Method 360.1	Oxygen, Dissolved	Membrane Electrode
Metals (200 series)	Method 253.2	Palladium	AA, Furnace
Metals (200 series)	Method 253.1	Palladium	AA, Direct Aspiration
Inorganic, non-metalls (300 series)	Method 418.1	Petroleum Hydrocarbon, Total, Recoverable	Spectrophotometric, Infrared
Physical properties (100 series)	Method 150.1	pH	Electrometric
Physical properties (100 series)	Method 150.2	pH	Electrometric (Continuous Monitoring)
Inorganic, non-metalls (300 series)	Method 420.2	Phenolics, Total, Recoverable	Colorimetric, Automated 4-AAP with distillation
Inorganic, non-metalls (300 series)	Method 420.3	Phenolics, Total, Recoverable	Spectrophotomctric, MBTH with distillation
Inorganic, non-metalls (300 series)	Method 420.1	Phenolics, Total, Recoverable	Spectrophotometric, Manual 4-AAP with distillation
Inorganic, non-metalls (300 series)	Method 365.1	Phosphorus, All Form	Colorimetric, Automated, Ascorbic Acid
Inorganic, non-metalls (300 series)	Method 365.2	Phosphorus, All Form	Colorimetric, Ascorbic Acid Single Reagent
Inorganic, non-metalls (300 series)	Method 365.3	Phosphorus, All Form	Colorimetric, Ascorbic Acid Two Reagent
Inorganic, non-metalls (300 series)	Method 365.4	Phosphorus, Total	Colorimetric, Automated, block digester, AA II
Metals (200 series)	Method 255.2	Platinum	AA, Furnace

TABLE 4.7A. TEST METHODS--ALPHABETICAL ORDER BY PARAMETER			
Substance category	Method	Parameter	Description
Metals (200 series)	Method 255.1	Platinum	AA, Direct Aspiration
Metals (200 series)	Method 258.1	Potassium	AA, Direct Aspiration
Physical properties (100 series)	Method 160.1	Residue Filterable	Gravimetric, Dried at 180 C
Physical properties (100 series)	Method 160.2	Residue Non-Filterable	Gravimetric, Dried at 103-105 C
Physical properties (100 series)	Method 160.5	Residue Settleable Matter	Volumetric, Imhoff cone
Physical properties (100 series)	Method 160.3	Residue Total	Gravimetric, Dried at 103-105 C
Physical properties (100 series)	Method 160.4	Residue Volatile	Gravimetric, ignition at 550 C
Metals (200 series)	Method 264.2	Rhenium	AA, Furnace
Metals (200 series)	Method 264.1	Rhenium	AA, Direct Aspiration
Metals (200 series)	Method 265.1	Rhodium	AA, Direct Aspiration
Metals (200 series)	Method 265.2	Rhodium	AA, Furnace
Metals (200 series)	Method 267.1	Ruthenium	AA, Direct Aspiration
Metals (200 series)	Method 267.2	Ruthenium	AA, Furnace
Metals (200 series)	Method 270.2	Selenium	AA, Direct Aspiration
Metals (200 series)	Method 270.3	Selenium	AA, Hydride
Inorganic, non-metalls (300 series)	Method 370.1	Silica, dissolved	Colorimetric
Metals (200 series)	Method 272.1	Silver	AA, Direct Aspiration
Metals (200 series)	Method 272.2	Silver	AA, Furnace
Metals (200 series)	Method 273.2	Sodium	AA, Furnace
Metals (200 series)	Method 273.1	Sodium	AA, Direct Aspiration
Inorganic, non-metalls (300 series)	Method 375.3	Sulfate	Gravimetric
Inorganic, non-metalls (300 series)	Method 375.2	Sulfate	Colorimetric, Automated Methyl thymol Blue, AA II
Inorganic, non-metalls (300 series)	Method 375.1	Sulfate	Colorimetric, Automated Chloranilate
Inorganic, non-metalls (300 series)	Method 375.4	Sulfate	Turbidimetric
Inorganic, non-metalls (300 series)	Method 376.1	Sulfide	Titrimetric, Iodine
Inorganic, non-metalls (300 series)	Method 376.2	Sulfide	Colorimetric, Methylene Blue
Inorganic, non-metalls (300 series)	Method 377.1	Sulfite	Titrimetric
Physical properties (100 series)	Method 170.1	Temperature	Thermometric
Metals (200 series)	Method 279.1	Thallium	AA, Direct Aspiration
Metals (200 series)	Method 279.2	Thallium	AA, Furnace
Metals (200 series)	Method 282.2	Tin	AA, Furnace
Metals (200 series)	Method 282.1	Tin	AA, Direct Aspiration

TABLE 4.7A. TEST METHODS--ALPHABETICAL ORDER BY PARAMETER			
Substance category	Method	Parameter	Description
Metals (200 series)	Method 283.2	Titanium	AA, Furnace
Metals (200 series)	Method 283.1	Titanium	AA, Direct Aspiration
Physical properties (100 series)	Method 180.1	Turbidity	Nephelometric
Metals (200 series)	Method 286.2	Vanadium	AA, Furnace
Metals (200 series)	Method 286.1	Vanadium	AA, Direct Aspiration
Metals (200 series)	Method 289.2	Zinc	AA, Furnace
Metals (200 series)	Method 289.1	Zinc	AA, Direct Aspiration

TABLE 4.7B. TEST METHODS--NUMBER ORDER BY METHOD			
Substance category	Method	Parameter	Description
Physical properties (100 series)	Method 110.1	Color	Colorimetric, ADMI
Physical properties (100 series)	Method 110.2	Color	Colorimetric, Platinum-cobalt
Physical properties (100 series)	Method 110.3	Color	Spectrophotometric
Physical properties (100 series)	Method 120.1	Conductance	Specific Conductance
Physical properties (100 series)	Method 130.1	Hartness, Total (mg/l as CaCO3)	Colorimetric, Automted EDTA
Physical properties (100 series)	Method 130.2	Hartness, Total (mg/l as CaCO3)	Titrimetric, EDTA
Physical properties (100 series)	Method 140.1	Odor	Thresholt Odor (Consistent Series)
Physical properties (100 series)	Method 150.1	pH	Electrometric
Physical properties (100 series)	Method 150.2	pH	Electrometric (Continuous Monitoring)
Physical properties (100 series)	Method 160.1	Residue Filterable	Gravimetric, Dried at 180 C
Physical properties (100 series)	Method 160.2	Residue Non-Filterable	Gravimetric, Dried at 103-105 C
Physical properties (100 series)	Method 160.3	Residue Total	Gravimetric, Dried at 103-105 C
Physical properties (100 series)	Method 160.4	Residue Volatile	Gravimetric, ignition at 550 C
Physical properties (100 series)	Method 160.5	Residue Settleablc Matter	Volumetric, Imhoff cone
Physical properties (100 series)	Method 170.1	Temperature	Thermometric
Physical properties (100 series)	Method 180.1	Turbidity	Nephelometric
Metals (200 series)	Method 200.0	Metals	Atomic Absorption Methods
Metals (200 series)	Method 200.1	Metals	Inductively coupled Plasma
Metals (200 series)	Method 202.1	Aluminum	AA, Direct Aspiration
Metals (200 series)	Method 202.2	Aluminum	AA, Furnace
Metals (200 series)	Method 204.1	Antimony	AA, Direct Aspiration
Metals (200 series)	Method 204.2	Antimony	AA, Furnace
Metals (200 series)	Method 206.2	Arsenic	AA, Furnace
Metals (200 series)	Method 206.3	Arsenic	AA, Hydride
Metals (200 series)	Method 206.4	Arsenic	Spectrophotometric, SDDC
Metals (200 series)	Method 206.5	Arsenic	Digestion method for Hydride and SDDC
Metals (200 series)	Method 208.1	Barium	AA, Direct Aspiration
Metals (200 series)	Method 208.2	Barium	AA, Furnace
Metals (200 series)	Method 210.1	Beryllium	AA, Direct Aspiration
Metals (200 series)	Method 210.2	Beryllium	AA, Furnace
Metals (200 series)	Method 212.3	Boron	Colorimetric, Curcumin
Metals (200 series)	Method 213.1	Cadmium	AA, Direct Aspiration

TABLE 4.7B. TEST METHODS--NUMBER ORDER BY METHOD			
Substance category	Method	Parameter	Description
Metals (200 series)	Method 213.2	Cadmium	AA, Furnace
Metals (200 series)	Method 215.1	Calcium	AA, Direct Aspiration
Metals (200 series)	Method 215.2	Calcium	Titrimetric, EDTA
Metals (200 series)	Method 218.1	Chromium	AA, Direct Aspiration
Metals (200 series)	Method 218.2	Chromium	AA, Furnace
Metals (200 series)	Method 218.3	Chromium	Chelation-Etraction
Metals (200 series)	Method 218.4	Chromium	Hexavalent, Chelation-Etraction
Metals (200 series)	Method 218.5	Chromium	Hexavalent, Dissolved
Metals (200 series)	Method 219.1	Cobalt	AA, Direct Aspiration
Metals (200 series)	Method 219.2	Cobalt	AA, Furnace
Metals (200 series)	Method 220.1	Copper	AA, Direct Aspiration
Metals (200 series)	Method 220.2	Copper	AA, Furnace
Metals (200 series)	Method 231.1	Gold	AA, Direct Aspiration
Metals (200 series)	Method 231.2	Gold	AA, Furnace
Metals (200 series)	Method 235.1	Iridium	AA, Direct Aspiration
Metals (200 series)	Method 235.2	Iridium	AA, Furnace
Metals (200 series)	Method 236.1	Iron	AA, Direct Aspiration
Metals (200 series)	Method 236.2	Iron	AA, Furnace
Metals (200 series)	Method 239.1	Lead	AA, Direct Aspiration
Metals (200 series)	Method 239.2	Lead	AA, Furnace
Metals (200 series)	Method 242.1	Magnesium	AA, Direct Aspiration
Metals (200 series)	Method 243.1	Mangnese	AA, Direct Aspiration
Metals (200 series)	Method 243.2	Mangnese	AA, Furnace
Metals (200 series)	Method 245.1	Mercury	Cold Vapor, Manual
Metals (200 series)	Method 245.5	Mercury	Cold Vapor, Sediments
Metals (200 series)	Method 246.1	Molybdenum	AA, Direct Aspiration
Metals (200 series)	Method 246.2	Molybdenum	AA, Furnace
Metals (200 series)	Method 249.1	Nickel	AA, Direct Aspiration
Metals (200 series)	Method 249.2	Nickel	AA, Furnace
Metals (200 series)	Method 24s.2	Mercury	Cold Vapor, Automtet
Metals (200 series)	Method 252.1	Osmium	AA, Direct Aspiration
Metals (200 series)	Method 252.2	Osmium	AA, Furnace
Metals (200 series)	Method 253.1	Palladium	AA, Direct Aspiration

TABLE 4.7B. TEST METHODS--NUMBER ORDER BY METHOD			
Substance category	Method	Parameter	Description
Metals (200 series)	Method 253.2	Palladium	AA, Furnace
Metals (200 series)	Method 255.1	Platinum	AA, Direct Aspiration
Metals (200 series)	Method 255.2	Platinum	AA, Furnace
Metals (200 series)	Method 258.1	Potassium	AA, Direct Aspiration
Metals (200 series)	Method 264.1	Rhenium	AA, Direct Aspiration
Metals (200 series)	Method 264.2	Rhenium	AA, Furnace
Metals (200 series)	Method 265.1	Rhodium	AA, Direct Aspiration
Metals (200 series)	Method 265.2	Rhodium	AA, Furnace
Metals (200 series)	Method 267.1	Ruthenium	AA, Direct Aspiration
Metals (200 series)	Method 267.2	Ruthenium	AA, Furnace
Metals (200 series)	Method 270.2	Selenium	AA, Direct Aspiration
Metals (200 series)	Method 270.3	Selenium	AA, Hydride
Metals (200 series)	Method 272.1	Silver	AA, Direct Aspiration
Metals (200 series)	Method 272.2	Silver	AA, Furnace
Metals (200 series)	Method 273.1	Sodium	AA, Direct Aspiration
Metals (200 series)	Method 273.2	Sodium	AA, Furnace
Metals (200 series)	Method 279.1	Thallium	AA, Direct Aspiration
Metals (200 series)	Method 279.2	Thallium	AA, Furnace
Metals (200 series)	Method 282.1	Tin	AA, Direct Aspiration
Metals (200 series)	Method 282.2	Tin	AA, Furnace
Metals (200 series)	Method 283.1	Titanium	AA, Direct Aspiration
Metals (200 series)	Method 283.2	Titanium	AA, Furnace
Metals (200 series)	Method 286.1	Vanadium	AA, Direct Aspiration
Metals (200 series)	Method 286.2	Vanadium	AA, Furnace
Metals (200 series)	Method 289.1	Zinc	AA, Direct Aspiration
Metals (200 series)	Method 289.2	Zinc	AA, Furnace
Inorganic, non-metalls (300 series)	Method 305.1	Acidity	Titrimetric
Inorganic, non-metalls (300 series)	Method 305.2	Acidity	Titrimetric (acid rain)
Inorganic, non-metalls (300 series)	Method 310.1	Alkalinity	Titrimetric (pH 4.5)
Inorganic, non-metalls (300 series)	Method 310.2	Alkalinity	Colorimetric, Automated Methyl Orange
Inorganic, non-metalls (300 series)	Method 320.1	Bromiide	Titrimetric
Inorganic, non-metalls (300 series)	Method 325.1	Chloride	Colorimetric, Automated Ferricyanide, AA I

TABLE 4.7B. TEST METHODS--NUMBER ORDER BY METHOD			
Substance category	Method	Parameter	Description
Inorganic, non-metalls (300 series)	Method 325.2	Chloride	Colorimetric, Automated Ferricyanide, AA II
Inorganic, non-metalls (300 series)	Method 325.3	Chloride	Titrimetric, Mercuric Nitrate
Inorganic, non-metalls (300 series)	Method 330.1	Chlorine, Total Residual	Titrimetric, Amperometric
Inorganic, non-metalls (300 series)	Method 330.2	Chlorine, Total Residual	Titrimetric, Back-Iodometric
Inorganic, non-metalls (300 series)	Method 330.3	Chlorine, Total Residual	Titrimetric, Iodometric
Inorganic, non-metalls (300 series)	Method 330.4	Chlorine, Total Residual	Titrimetric, DPD-FAS
Inorganic, non-metalls (300 series)	Method 330.5	Chlorine, Total Residual	Spectrophotometric, DPD
Inorganic, non-metalls (300 series)	Method 335.1	Cyanide, Amenable to Chlorination	Titrimetric, Spectrophotometric
Inorganic, non-metalls (300 series)	Method 335.2	Cyanide, Total	Titrimetric Spectrophotometric
Inorganic, non-metalls (300 series)	Method 335.3	Cyanide, Total	Colorimetric, Automated UV
Inorganic, non-metalls (300 series)	Method 340.1	Fluoride	Colorimetric SPADNS with Bellack distillation
Inorganic, non-metalls (300 series)	Method 340.2	Fluoride	Potentiometric, Ion Selective Electrode
Inorganic, non-metalls (300 series)	Method 340.3	Fluoride	Colorimetric, Automated Complexone
Inorganic, non-metalls (300 series)	Method 345.1	Iodide	Titrimetric
Inorganic, non-metalls (300 series)	Method 350.1	Nitrogen, Ammonia	Colorimetric, Automated Phenate
Inorganic, non-metalls (300 series)	Method 350.2	Nitrogen, Ammonia	Colorimetric; Titrimetric; Potentiometri Distillation Procedures
Inorganic, non-metalls (300 series)	Method 350.3	Nitrogen, Ammonia	Potentiometric, Ion Selective Electrode
Inorganic, non-metalls (300 series)	Method 351.1	Kjeldahl, Total	Colorimetric, Automated Phenate
Inorganic, non-metalls (300 series)	Method 351.2	Kjeldahl, Total	Colorimetric, Semi-Automated Block Digester AA II
Inorganic, non-metalls (300 series)	Method 351.3	Kjeldahl, Total	Colorimetric; Titrimetric; Potentiometric
Inorganic, non-metalls (300 series)	Method 351.4	Kjeldahl, Total	Potentiometric, Ion Selective Electrode
Inorganic, non-metalls (300 series)	Method 352.1	Nitrate	Colorimetric, Brucine
Inorganic, non-metalls (300 series)	Method 353.1	Nitrate-Nitrite	Colorimetric, Automated Hydrazine Reduction
Inorganic, non-metalls (300 series)	Method 353.2	Nitrate-Nitrite	Colorimetric, Automated Cadmium Reduction
Inorganic, non-metalls (300 series)	Method 353.3	Nitrate-Nitrite	Colorimetric, Manual Cadmium Reduction
Inorganic, non-metalls (300 series)	Method 354.1	Nitrite	Spectrophotometric
Inorganic, non-metalls (300 series)	Method 360.1	Oxygen, Dissolved	Membrane Electrode
Inorganic, non-metalls (300 series)	Method 360.2	Oxygen, Dissolved	Modified Winkler (Full Bottle Technique)

TABLE 4.7B. TEST METHODS--NUMBER ORDER BY METHOD			
Substance category	Method	Parameter	Description
Inorganic, non-metalls (300 series)	Method 365.1	Phosphorus, All Form	Colorimetric, Automated, Ascorbic Acid
Inorganic, non-metalls (300 series)	Method 365.2	Phosphorus, All Form	Colorimetric, Ascorbic Acid Single Reagent
Inorganic, non-metalls (300 series)	Method 365.3	Phosphorus, All Form	Colorimetric, Ascorbic Acid Two Reagent
Inorganic, non-metalls (300 series)	Method 365.4	Phosphorus, Total	Colorimetric, Automated, block digester, AA II
Inorganic, non-metalls (300 series)	Method 370.1	Silica, dissolved	Colorimetric
Inorganic, non-metalls (300 series)	Method 375.1	Sulfate	Colorimetric, Automated Chloranilate
Inorganic, non-metalls (300 series)	Method 375.2	Sulfate	Colorimetric, Automated Methyl thymol Blue, AA II
Inorganic, non-metalls (300 series)	Method 375.3	Sulfate	Gravimetric
Inorganic, non-metalls (300 series)	Method 375.4	Sulfate	Turbidimetric
Inorganic, non-metalls (300 series)	Method 376.1	Sulfide	Titrimetric, Iodine
Inorganic, non-metalls (300 series)	Method 376.2	Sulfide	Colorimetric, Methylene Blue
Inorganic, non-metalls (300 series)	Method 377.1	Sulfite	Titrimetric
Inorganic, non-metalls (300 series)	Method 377.1	Biochemical Oxygen Demand	BOD (5 day, 20 C)
Inorganic, non-metalls (300 series)	Method 410.1	Chemical Oxygen Demand	Colorimetric, Mid-level
Inorganic, non-metalls (300 series)	Method 410.2	Chemical Oxygen Demand	Titrimetric, Low-Level
Inorganic, non-metalls (300 series)	Method 410.3	Chemical Oxygen Demand	Titrimetric, High-Level for Saline Waters
Inorganic, non-metalls (300 series)	Method 410.4	Chemical Oxygen Demand	Colorimetric, Automated; Manual
Inorganic, non-metalls (300 series)	Method 413.1	Oil and Grease, Total Recoverable	Gravimetric, Separatory Funnel Extraction
Inorganic, non-metalls (300 series)	Method 413.2	Oil and Grease, Total Recoverable	Spectrophotometric, Infrared
Inorganic, non-metalls (300 series)	Method 415.1	Organic Carbon, Total	Combustion or Oxidation
Inorganic, non-metalls (300 series)	Method 415.2	Organic Carbon, Total	UV Promoted, Persulfate Oxidation
Inorganic, non-metalls (300 series)	Method 418.1	Petroleum Hydrocarbon, Total, Recoverable	Spectrophotometric, Infrared
Inorganic, non-metalls (300 series)	Method 420.1	Phenolics, Total, Recoverable	Spectrophotometric, Manual 4-AAP with distillation
Inorganic, non-metalls (300 series)	Method 420.2	Phenolics, Total, Recoverable	Colorimetric, Automated 4-AAP with distillation
Inorganic, non-metalls (300 series)	Method 420.3	Phenolics, Total, Recoverable	Spectrophotomctric, MBTH with distillation
Inorganic, non-metalls (300 series)	Method 425.1	Methylene Blue Active Substances (MBAS)	Colorimetric

TABLE 4.7B. TEST METHODS--NUMBER ORDER BY METHOD			
Substance category	Method	Parameter	Description
Inorganic, non-metalls (300 series)	Method 430.1	NTA	Colorimetric, Manual, Zinc-Zincon
Inorganic, non-metalls (300 series)	Method 430.2	NTA	Colorimetric, Automated, Zinc-Zincon

4.3. REFERENCES

(EPA-83/3), "Methods for Chemical Analysis of Water and Wastes," EPA600-4-79-020, March 1983.

(EPA-91/7), "Methods for the Determination of Organic Compounds in Drinking Water," EPA600-4-88-039, July 1991.

(EPA-93/8), "Methods for the Determination of Inorganic Substances in e Environmental Samples," EPA600-R-93-100, August 1993.

WATER PROGRAMS

5.1. NPDES PERMIT APPLICATION TESTING REQUIREMENTS (40CFR122, APPENDIX D) 2

TABLE 5.1. TESTING REQUIREMENTS FOR ORGANIC TOXIC POLLUTANTS 2

TABLE 5.2. ORGANIC TOXIC POLLUTANTS IN EACH OF FOUR FRACTIONS 4

TABLE 5.3. OTHER TOXIC POLLUTANTS (METALS AND CYANIDE) AND
TOTAL PHENOLS 6

TABLE 5.4. CONVENTIONAL AND NONCONVENTIONAL POLLUTANTS
REQUIRED TO BE TESTED 6

TABLE 5.5. TOXIC POLLUTANTS AND HAZARDOUS SUBSTANCES
REQUIRED TO BE IDENTIFIED 7

5.2. IDENTIFICATION OF TEST PROCEDURES (40CFR136.3) 10

TABLE 5.6. LIST OF APPROVED BIOLOGICAL TEST PROCEDURES
(40CFR136.3 TABLE IA) 10

TABLE 5.7. LIST OF APPROVED BIOLOGICAL TEST PROCEDURES
(40CFR136.3 TABLE IB) 11

TABLE 5.8. LIST OF APPROVED TEST PROCEDURES
FOR NON-PESTICIDE ORGANIC COMPOUNDS 25

TABLE 5.9. LIST OF APPROVED TEST PROCEDURES FOR PESTICIDES 29

TABLE 5.10. LIST OF APPROVED RADIOLOGIC TEST PROCEDURES
(40CFR136.3 TABLE IE) 33

TABLE 5.11. REQUIRED CONTAINERS, PRESERVATION TECHNIQUES,
AND HOLDING TIMES 34

5.3. METHODS FOR ORGANICS CHEMICAL ANALYSIS OF MUNICIPAL AND
INDUSTRIAL WASTEWATERS (40CFR136, APPENDIX A) 38

5.4. REFERENCES (40CFR136.3.b) 38

WATER PROGRAMS

5.1. NPDES PERMIT APPLICATION TESTING REQUIREMENTS (40CFR122 APPENDIX D)

TABLE 5.1. TESTING REQUIREMENTS FOR ORGANIC TOXIC POLLUTANTS (40CFR122 APPENDIX D TABLE I)				
	GC/MS Fraction[1]			
NPDES Primary Industry Category	Volatile	Acid	Base/neutral	Pesticide
Adhesives and Sealants	2	2	2	
Aluminum Forming	2	2	2	
Auto and Other Laundries	2	2	2	2
Battery Manufacturing	2		2	
Coal Mining	2	2	2	2
Coil Coating	2	2	2	
Copper Forming	2	2	2	
Electric and Electronic Components	2	2	2	2
Electroplating	2	2	2	
Explosives Manufacturing		2	2	
Foundries	2	2	2	
Gum and Wood Chemicals	2	2	2	2
Inorganic Chemicals Manufacturing	2	2	2	
Iron and Steel Manufacturing	2	2	2	
Leather Tanning and Finishing	2	2	2	2
Mechanical Products Manufacturing	2	2	2	
Nonferrous Metals Manufacturing	2	2	2	2
Ore Mining	2	2	2	2
Organic Chemicals Manufacturing	2	2	2	2
Paint and Ink Formulation	2	2	2	2
Pesticides	2	2	2	2
Petroleum Refining	2	2	2	2
Pharmaceutical Preparations	2	2	2	
Photographic Equipment and Supplies	2	2	2	2
Plastic and Synthetic Materials Manufacturing	2	2	2	2
Plastic Processing	2			
Porcelain Enameling	2		2	2
Printing and Publishing	2	2	2	2
Pulp and Paper Mills	2	2	2	2
Rubber Processing	2	2	2	

NPDES Primary Industry Category	GC/MS Fraction[1]			
	Volatile	Acid	Base/neutral	Pesticide
Soap and Detergent Manufacturing	2	2	2	
Steam Electric Power Plants	2	2	2	
Textile Mills	2	2	2	2
Timber Products Processing	2	2	2	2

TABLE 5.1. TESTING REQUIREMENTS FOR ORGANIC TOXIC POLLUTANTS (40CFR122 APPENDIX D TABLE I)

[1] The toxic pollutants in each fraction are listed in 40CFR122 Appendix D Table II and in the following section 3.1.2.
[2] Testing required

5.1.2. ORGANIC TOXIC POLLUTANTS IN ANALYSIS BY GAS CHROMATOGRAPHY/MASS SPECTROSCOPY (GS/MS)

TABLE 5.2. ORGANIC TOXIC POLLUTANTS IN EACH OF FOUR FRACTIONS
IN ANALYSIS BY GS/MS (40CFR122 APPENDIX D TABLE II)
===

VOLATILES

1V   acrolein
2V   acrylonitrile
3V   benzene
5V   bromoform
6V   carbon tetrachloride
7V   chlorobenzene
8V   chlorodibromomethane
9V   chloroethane
10V 2-chloroethylvinyl ether
11V chloroform
12V dichlorobromomethane
14V 1,1-dichloroethane
15V 1,2-dichloroethane
16V 1,1-dichloroethylene
17V 1,2-dichloropropane
18V 1,3-dichloropropylene
19V ethylbenzene
20V methyl bromide
21V methyl chloride
22V methylene chloride
23V 1,1,2,2-tetrachloroethane
24V tetrachloroethylene
25V toluene
26V 1,2-trans-dichloroethylene
27V 1,1,1-trichloroethane
28V 1,1,2-trichloroethane
29V trichloroethylene
31V vinyl chloride

ACID COMPOUNDS

1A   2-chlorophenol
2A   2,4-dichlorophenol
3A   2,4-dimethylphenol
4A   4,6-dinitro-o-cresol
5A   2,4-dinitrophenol
6A   2-nitrophenol
7A   4-nitrophenol
8A   p-chloro-m-cresol
9A   pentachlorophenol
10A phenol
11A 2,4,6-trichlorophenol

BASE/NEUTRAL

1B   acenaphthene
2B   acenaphthylene
3B   anthracene
4B   benzidine
5B   benzo(a)anthracene
6B   benzo(a)pyrene
7B   3,4-benzofluoranthene
8B   benzo(ghi)perylene

9B   benzo(k)fluoranthene
10B bis(2-chloroethoxy)methane
11B bis(2-chloroethyl)ether
12B bis(2-chloroisopropyl)ether
13B bis (2-ethylhexyl)phthalate
14B 4-bromophenyl phenyl ether
15B butylbenzyl phthalate
16B 2-chloronaphthalene
17B 4-chlorophenyl phenyl ether
18B chrysene
19B dibenzo(a,h)anthracene
20B 1,2-dichlorobenzene
21B 1,3-dichlorobenzene
22B 1,4-dichlorobenzene
23B 3,3'-dichlorobenzidine
24B diethyl phthalate
25B dimethyl phthalate
26B di-n-butyl phthalate
27B 2,4-dinitrotoluene
28B 2,6-dinitrotoluene
29B di-n-octyl phthalate
30B 1,2-diphenylhydrazine (as azobenzene)
31B fluroranthene
32B fluorene
33B hexachlorobenzene
34B hexachlorobutadiene
35B hexachlorocyclopentadiene
36B hexachloroethane
37B indeno(1,2,3-cd)pyrene
38B isophorone
39B napthalene
40B nitrobenzene
41B N-nitrosodimethylamine
42B N-nitrosodi-n-propylamine
43B N-nitrosodiphenylamine
44B phenanthrene
45B pyrene
46B 1,2,4-trichlorobenzene

PESTICIDES

1P   aldrin
2P   alpha-BHC
3P   beta-BHC
4P   gamma-BHC
5P   delta-BHC
6P   chlordane
7P   4,4'-DDT
8P   4,4'-DDE
9P   4,4'-DDD
10P dieldrin
11P alpha-endosulfan
12P beta-endosulfan
13P endosulfan sulfate
14P endrin
15P endrin aldehyde
16P heptachlor
17P heptachlor epoxide
18P PCB-1242
19P PCB-1254
20P PCB-1221

21P PCB-1232
22P PCB-1248
23P PCB-1260
24P PCB-1016
25P toxaphene

==

TABLE 5.3. OTHER TOXIC POLLUTANTS (METALS AND CYANIDE) AND TOTAL PHENOLs
(40CFR122 APPENDIX D TABLE III)

==

Antimony,	Total
Arsenic,	Total
Beryllium,	Total
Cadmium,	Total
Chromium,	Total
Copper,	Total
Lead,	Total
Mercury,	Total
Nickel,	Total
Selenium,	Total
Silver,	Total
Thallium,	Total
Zinc,	Total
Cyanide,	Total
Phenols,	Total

==

TABLE 5.4. CONVENTIONAL AND NONCONVENTIONAL POLLUTANTS REQUIRED TO BE TESTED
BY EXISTING DISCHARGERS IF EXPECTED TO BE PRESENT (40CFR122 APPENDIX D TABLE IV)

==

Bromide
Chlorine, Total Residual
Color
Fecal Coliform
Fluoride
Nitrate-Nitrite
Nitrogen, Total Organic
Oil and Grease
Phosphorus, Total
Radioactivity
Sulfate
Sulfide
Sulfite
Surfactants
Aluminum, Total
Barium, Total
Boron, Total
Cobalt, Total
Iron, Total
Magnesium, Total
Molybdenum, Total
Manganese, Total
Tin, Total
Titanium, Total

==

TABLE 5.5. TOXIC POLLUTANTS AND HAZARDOUS SUBSTANCES REQUIRED TO BE IDENTIFIED
BY EXISTING DISCHARGERS IF EXPECTED TO BE PRESENT (40CFR122 APPENDIX D TABLE V)
==

TOXIC POLLUTANTS

    Asbestos

HAZARDOUS SUBSTANCES

    Acetaldehyde
    Allyl alcohol
    Allyl chloride
    Amyl acetate
    Aniline
    Benzonitrile
    Benzyl chloride
    Butyl acetate
    Butylamine
    Captan
    Carbaryl
    Carbofuran
    Carbon disulfide
    Chlorpyrifos
    Coumaphos
    Cresol
    Crotonaldehyde
    Cyclohexane
    2,4-D (2,4-Dichlorophenoxy acetic acid)
    Diazinon
    Dicamba
    Dichlobenil
    Dichlone
    2,2-Dichloropropionic acid
    Dichlorvos
    Diethyl amine
    Dimethyl amine
    Dintrobenzene
    Diquat
    Disulfoton
    Diuron
    Epichlorohydrin
    Ethion
    Ethylene diamine
    Ethylene dibromide
    Formaldehyde
    Furfural
    Guthion
    Isoprene
    Isopropanolamine Dodecylbenzenesulfonate
    Kelthane
    Kepone
    Malathion
    Mercaptodimethur
    Methoxychlor
    Methyl mercaptan
    Methyl methacrylate
    Methyl parathion
    Mevinphos
    Mexacarbate
    Monoethyl amine

Monomethyl amine
Naled
Napthenic acid
Nitrotoluene
Parathion
Phenolsulfanate
Phosgene
Propargite
Propylene oxide
Pyrethrins
Quinoline
Resorcinol
Strontium
Strychnine
Styrene
2,4,5-T (2,4,5-Trichlorophenoxy acetic acid)
TDE (Tetrachlorodiphenylethane)
2,4,5-TP [2-(2,4,5-Trichlorophenoxy) propanoic acid]
Trichlorofan
Triethanolamine dodecylbenzenesulfonate
Triethylamine
Trimethylamine
Uranium
Vanadium
Vinyl acetate
Xylene
Xylenol
Zirconium

===

[Note 1: The Environmental Protection Agency has suspended the requirements of 40CFR122.21(g)(7)(ii)(A) and Table I of Appendix D as they apply to certain industrial categories. The suspensions are as follows:

a. At 46 FR 2046, Jan. 8, 1981, the Environmental Protection Agency suspended until further notice ^U 122.21(g)(7)(ii)(A) as it applies to coal mines.

b. At 46 FR 22585, Apr. 20, 1981, the Environmental Protection Agency suspended until further notice ^U 122.21(g)(7)(ii)(A) and the corresponding portions of Item V-C of the NPDES application Form 2c as they apply to:

1. Testing and reporting for all four organic fractions in the Greige Mills Subcategory of the Textile Mills industry (Subpart C-Low water use processing of 40 CFR part 410), and testing and reporting for the pesticide fraction in all other subcategories of this industrial category.

2. Testing and reporting for the volatile, base/neutral and pesticide fractions in the Base and Precious Metals Subcategory of the Ore Mining and Dressing industry (subpart B of 40 CFR part 440), and testing and reporting for all four fractions in all other subcategories of this industrial category.

3. Testing and reporting for all four GC/MS fractions in the Porcelain Enameling industry.

c. At 46 FR 35090, July 1, 1981, the Environmental Protection Agency suspended until further notice ^U 122.21(g)(7)(ii)(A) and the corresponding portions of Item V-C of the NPDES application Form 2c as they apply to:

1. Testing and reporting for the pesticide fraction in the Tall Oil Rosin Subcategory (subpart D) and Rosin-Based Derivatives Subcategory (subpart F) of the Gum and Wood Chemicals industry (40 CFR part 454), and testing and reporting for the pesticide and base/netural fractions in all other subcategories of this industrial category.

2. Testing and reporting for the pesticide fraction in the Leather Tanning and Finishing, Paint and Ink Formulation, and Photographic Supplies industrial categories.

3. Testing and reporting for the acid, base/neutral and pesticide fractions in the Petroleum Refining industrial category.

4. Testing and reporting for the pesticide fraction in the Papergrade Sulfite subcategories (subparts J and U) of the Pulp and Paper industry

(40 CFR part 430); testing and reporting for the base/neutral and pesticide fractions in the following subcategories: Deink (subpart Q), Dissolving Kraft (subpart F), and Paperboard from Waste Paper (subpart E); testing and reporting for the volatile, base/neutral and pesticide fractions in the following subcategories: BCT Bleached Kraft (subpart H), Semi-Chemical (subparts B and C), and Nonintegrated-Fine Papers (subpart R); and testing and reporting for the acid, base/neutral, and pesticide fractions in the following subcategories: Fine Bleached Kraft (subpart I), Dissolving Sulfite Pulp (subpart K), Groundwood-Fine Papers (subpart O), Market Bleached Kraft (subpart G), Tissue from Wastepaper (subpart T), and Nonintegrated-Tissue Papers (subpart S).

5. Testing and reporting for the base/neutral fraction in the Once-Through Cooling Water, Fly Ash and Bottom Ash Transport Water process wastestreams of the Steam Electric Power Plant industrial category.

5.2. IDENTIFICATION OF TEST PROCEDURES (40CFR136.3)

TABLE 5.6. LIST OF APPROVED BIOLOGICAL TEST PROCEDURES (40CFR136.3 TABLE IA)					
Parameter, units and methods	Method[1]	Reference (method No. or page)			
		EPA[2]	Std methods 18th ed	ASTM	USGS[3]
Bacteria:					
1. Coliform (fecal), number per 100 ml.	MPN, 5 tube, dilution or, membrane filter (MF)[4], single step	p. 132	9221C		B-0050-85
		p. 124	9222D		
2. Coliform (fecal) in presence of chlorine, number per 100 ml.	MPN, 5 tube, 3 dilution; or, MF[4], single step[5]	p. 132	9221C		
		p. 124	9222D		
3. Coliform (total, number per 100 ml.)	MPN, 5 tube, 3 dilution; or,	p. 114	9221B		
	MF[4], single step or two step	p. 108	9222B		
4. Coliform (total), in. presence of chlorine, number per 100 ml.	MPN, 5 tube, dilution; or	p. 114	9221B		
	MF[4] with enrichment	p. 111	9222B+B.5C		
5. Fecal streptococci, number per 100 ml.	MPN, 5 tube, 3 dilution; MF[4]; or, plate count	p. 139	9230B		
		p. 136	9230C		
		p. 143			

[1] The method used must be specified when results are reported.

[2] Bordner, R.H., and J.A. Winter, eds. 1978. "Microbiological Methods for Monitoring the Environment, Water and Waste." Environmental Monitoring Systems Laboratory, U.S. Environmental Protection Agency. EPA-600/8-78-017.

[3] Britton, L.J., and P.E. Greeson, eds., 1989. "Methods for Collection and Analysis of Aquatic Biological and Microbological Samples," Techniques of Water Resources Investigations of the U.S. Geological Survey, Techniques of Water Resources Investigations, Book 5, Chapter A4, Laboratory Analysis, U.S. Geological Survey, U.S. Department of Interior, Reston, Virginia.

[4] A 0.45 μm membrane filter (MF) or other pore size certified by the manufacturer to fully retain organisms to be cultivated, and to be free of extractables which could interfere with their growth.

[5] Because the MF technique usually yields low and variable recovery from chlorinated wastewaters, the Most Probable Number method will be required to resolve any controversies.

TABLE 5.7. LIST OF APPROVED BIOLOGICAL TEST PROCEDURES (40CFR136.3 TABLE IB)

Parameter, units and methods	Method Reference (method No. or page)				
	EPA[1,35]	Std method 18th ed	ASTM	USGS[2]	Other
1. Acidity, as CaCO3, mg/L:					
Electrometric endpoint or phenolphthalein endpoint.	305.1	2310 B(4a)	D1067-92		
2. Alkalinity, as CaCO3, mg/L:	310.1	2320 B	D1067-92	I-1030-85	973.43[3]
Electrometric or Colorimetric titration to pH 4.5, manual or automated.	310.2			I-2030-85	
3. Aluminum-Total[4], mg/L; Digestion[4] followed by:					
AA direct aspiration[36]	202.1	3111 D		3051-85	
AA furnace	202.2	3113 B			
Inductively Coupled Plasma/Atomic Emission Spectrometry (ICP/AES)[36]	[5]200.7	3120 B			
Direct Current Plasma (DCP)[36]			D4190-82(88)		Note 34.
Colorimetric (Eriochrome cyanine R).		3500-A1 D			
4. Ammonia (as N), mg/L:					
Manual distillation (at pH 9.5)[6], followed by.	350.2	4500-NH3 B			973.49[3]
Nesslerization	350.2	4500-NH3 C	D1426-89(A)	I-3520-85	973.49[3]
Titration	350.2	4500-NH3 E			
Electrode	350.3	4500-NH3 F or G	D1426-89(B)		
Automated phenate, or	350.1	4500-NH3 H		I-4523-85	
Automated electrode					Note 7.
5. Antimony-Total[4], mg/L; Digestion[4] followed by:					
AA direct aspiration[36]	204.1	3111 B			
AA furnace	204.2	3113 B			
ICP/AES[36]	[5]200.7	3120 B			
6. Arsenic-Total[4], mg/L:					
Digestion[4] followed by	206.5				
AA gaseous hydride	206.3	314 B 4.d	D2972-88(B)	I-3062-85	
AA furnace	206.2	3113 B	D2972-88(C)		
ICP/AES[36], or	[5]200.7	3120 B			
Colorimetric (SDDC)	206.4	3500-As C	D2972-88(A)	I-3060-85	

TABLE 5.7. LIST OF APPROVED BIOLOGICAL TEST PROCEDURES (40CFR136.3 TABLE IB)					
Parameter, units and methods	Method Reference (method No. or page)				
	EPA[1,35]	Std method 18th ed	ASTM	USGS[2]	Other
7. Barium-Total[4], mg/L; Digestion[4]followed by:					
AA direct aspiration[36]	208.1	3111 D		I-3084-85	
AA furnace	208.2	3113 B	D4382-91		
ICP/AES[36]	[5]200.7	3120 B			
DCP[36]					Note 34.
8. Beryllium-Total[4], mg/L; Digestion[4]followed by:					
AA direct aspiration	210.1	3111 D	D3645-84(88)(A)	I-3095-85	
AA furnace	210.2	3113 B	D3645(88)(B)		
ICP/AES	[5]200.7	3120 B			
DCP, or			D4190-82(88)		Note 34.
Colorimetric (aluminon)		3500-Be D			
9. Biochemical oxygen demand (BOD5), mg/L:					
Dissolved Oxygen Depletion	405.1	5210 B		I-1578-78[8]	973.44[3], p.17[9]
10. Boron[37]-Total, mg/L:					
Colorimetric (curcumin)	212.3	4500-B B		I-3112-85	
ICP/AES, or	[5]200.7	3120 B			
DCP			D4190-82(88)		Note 34.
11. Bromide, mg/L:					
Titrimetric, or	320.1		D1246-82(88)(C)	I-1125-85	p.S44[10]
12. Cadmium-Total[4], mg/L; Digestion[4]followed by:					
AA direct aspiration[36]	213.1	3111 B or C	D3557-90(A or B)	I-3135-85 or I-3136-85	974.27[3], p.37[9]
AA furnace	213.2	3113 B	D3557-90(D)		
ICP/AES[36]	[5]200.7	3120 B		I-1472-85	
DCP[36]			D4190-82(88)		Note 34.
Voltametry[11], or			D3557-90(C)		
Colorimetric (Dithizone)		3500-Cd D			
13. Calcium-Total[4], mg/L; Digestion[4]followed by:					
AA direct aspiration	215.1	3111 B	D511-92(B)	I-3152-85	

TABLE 5.7. LIST OF APPROVED BIOLOGICAL TEST PROCEDURES (40CFR136.3 TABLE IB)					
Parameter, units and methods	Method Reference (method No. or page)				
	EPA[1,35]	Std method 18th ed	ASTM	USGS[2]	Other
ICP/AES	[5]200.7	3120 B			
DCP, or					Note 34.
Titrimetric (EDTA)	215.2	3500-Ca D	D511-92(A)		
14. Carbonaceous biochemical oxygen demand (CBOD5), mg/L[12]:					
Dissolved Oxygen Depletion with nitrification inhibitor		5210 B			
15. Chemical oxygen demand (COD), mg/L; Titrimetric, or	410.1	5220 C	D1252-88(A)	I-3560-85	973.46[3], p. 17[9]
	410.2			I-3562-85	
	410.3				
Spectrophotometric, manual or automated.	410.4	5220 D	D1252-88(B)	I-3561-85	Notes 13 or 14
16. Chloride, mg/L:					
Titrimetric (silver nitrate)		4500-C1 B	D512-89(B)	I-1183-85	
Or (Mercuric nitrate)	325.3	4500-C1 C	D512-89(A)	I-1184-85	973.51[3]
Colorimetric, manual or				I-1187-85	
Automated (Ferricyanide)	325.1 or	4500-C1 E		I-2187-85	
	325.2				
17. Chlorine-Total residual, mg/L; Titrimetric:					
Amperometric direct	330.1	4500-C1 D	D1253-86(92)		
Iodometric direct	330.3	4500-C1 B			
Black titration ether end-point[15]or	330.2	4500-C1 C			
DPD-FAS	330.4	4500-C1 F			
Spectrophotometric, DPD	330.5	4500-C1 G			
Or Electrode					Note 16
18. Chromium VI dissolved, mg/L; 0.45 micron filtration followed by:					
AA chelation-extraction or	218.4	3111 C		I-1232-85	
Colorimetric (Diphenylcarbazide)		3500-Cr D	D1687-92(A)	I-1230-85	
19. Chromium-Total[4], mg/L; Digestion[4]followed by:					
AA direct aspiration[36]	218.1	3111 B	D1687-92(B)	I-3236-85	974.27[3]

TABLE 5.7. LIST OF APPROVED BIOLOGICAL TEST PROCEDURES (40CFR136.3 TABLE IB)					
Parameter, units and methods	Method Reference (method No. or page)				
	EPA[1,35]	Std method 18th ed	ASTM	USGS[2]	Other
AA chelation-extraction	218.3	3111 C			
AA furnace	218.2	3113 B	D1687-92(C)		
ICP/AES[36]	[5]200.7	3120 B			
DCP[36], or			D4190-82(88)		Note 34.
Colorimetric (Diphenylcarbazide)		3500-Cr D			
20. Cobalt-Total[4], mg/L; Digestion[4] followed by:					
AA direct aspiration	219.1	3111 B or C	D3558-90(A or B)	I-3239-85	p.37[9]
AA furnace	219.2	3113 B	D3558-90(C)		
ICP/AES	[5]200.7	3120 B			
DCP			D4190-82(88)		Note 34
21. Color platinum cobalt units or dominant wavelength, hue, luminance purity:					
Colorimetric (ADMI), or	110.1	2120 E			Note 18
(Platinum cobalt), or	110.2	2120 B		I-1250-85	
Spectrophotometric	110.3	2120C			
22. Copper-Total[4], mg/L; Digestion[4] followed by:					
AA direct aspiration[36]	220.1	3111 B or C	D1688-90(A or B)	I-3270-85 or I-3271-85	974.27[3], p.37[9]
AA furnace	220.2	3113 B	D1688-90(C)		
ICP/AES[36]	[5]200.7	3120 B			
DCP[36], or			D4190-82(88)		Note 34
Colorimetric (Neocuproine) or		3500-Cu D			
(Bicinchoninate)		or E			Note 19
23. Cyanide-Total, mg/L:					
Manual distillation with MgCl2 followed by		4500-CN C			
Titrimetric, or		4500-CN D			p. 22[9]
Spectrophotometric, manual or	[31]335.3	4500-CN E	D2036-91(A)	I-3300-85	
Automoted[20]	[31]335.3				
24. Cyanide amenable to chlorination, mg/L:					

TABLE 5.7. LIST OF APPROVED BIOLOGICAL TEST PROCEDURES (40CFR136.3 TABLE IB)					
Parameter, units and methods	Method Reference (method No. or page)				
	EPA[1,35]	Std method 18th ed	ASTM	USGS[2]	Other
Manual distillation with MgCl2 followed by titrimetric or	335.1	4500-CN G	2036-91(B)		
Spectrophotometric					
25. Fluoride-Total, mg/L:					
Manual distillation[6] followed by		4500-F - B			
Electrode, manual or	340.2	4500-F - C	D1179-88(B)		
Automated				I-4327-85	
Colorimetric (SPADNS)	340.1	4500-F - D	D1179-88(A)		
Or Automated complexone or	340.3	4500-F - E			
26. Gold-total[4], mg/L; Digestion[4] followed by:					
AA direct aspiration	231.1	3111 B			
AA furnace, or	231.2				
DCP					Note 34
27. Hardness-Total, as CaCO3, mg/L:					
Automated colorimetric	130.1				
Titrimetric (EDTA), or Ca plus Mg as their carbonates, by inductively coupled plasma or AA direct aspiration. (Parameters 13 and 33)	130.2	2340 C	D1126-86(92)	I-1338-85	973.52B[3]
28. Hydrogen ion (pH), pH units:					
Electrometric measurement, or	150.1	4500-H$^+$ B	D1293-84(90) (A or B)	I-1586-85	973.41[3]
Automated electrode					Note 21
29. Iridium-Total[4], mg/L; Digestion[4] followed by:					
AA direct aspiration or	235.1	3111 B			
AA furnace	235.2				
30. Iron-Total[4], mg/L; Digestion[4] followed by:					
AA direct aspiration[36]	236.1	3111 B or C	D1068-90(A or B)	I-3381-85	974.27[3]
AA furnace	236.2	3113 B	D1068-90(C)		
ICP/AES[36]	[5]200.7	3120 B			
DCP[36], or			D4190-82(88)		Note 34
Colorimetric (Phenanthroline)		3500-Fe D	D1068-90(D)		Note 22

TABLE 5.7. LIST OF APPROVED BIOLOGICAL TEST PROCEDURES (40CFR136.3 TABLE IB)					
Parameter, units and methods	Method Reference (method No. or page)				
	EPA[1,35]	Std method 18th ed	ASTM	USGS[2]	Other
31. Kjeldahl Nitrogen-Total, (as N), mg/L:					
Digestion and distillation followed by	351.3	4500-NH 3 B or C	D3590-89(A)		
Titration	351.3	4500-NH 3 E	D3590-89(A)		973.48[3]
Nesslerization	51.3	4500-NH 3 C	D3590-89(A)		
Electrode	351.3	4500-NH 3 F or G			
Automated phenate colorimetric	351.1			I-4551-78[8]	
Semi-automated block digestor colorimetric, or	351.2		3590-89(B)		
Manual or block digestor Potentiometric	351.4		D359-89(A)		
32. Lead-Total[4], mg/L; Digestion[4]followed by:					
AA direct aspiration[36]	239.1	3111 B or C	D3559-90(A or B)	I-3399-85	974.27[3]
AA furnace	239.2	3113 B	D3559-90(D)		
ICP/AES[36]	[5]200.7	3120 B			
DCP[36]			D4190-82(88)		Note 34
Voltametry[11], or			D3559-90(C)		
Colorimtric (Dithizone)		3500-Pb D			
33. Magnesium-Total[4], mg/L; Digestion[4]followed by:					
AA direct aspiration	242.1	3111 B	D511-92(B)	I-3447-85	974.27[3]
ICP/AES	[5]200.7	3120 B			
DCP, or					Note 34
Gravimetric		3500-Mg D			
34. Manganese-Total[4], mg/L; Digestion[4]followed by:					
AA direct aspiration[36]	243.1	3111 B	D858-90(A or B)	I-3454-85	974.27[3]
AA furnace	243.2	3113 B	D858-90(C)		
ICP/AES[36]	[5]200.7	3120 B			
DCP[36]or			D4190-82(88)		Note 34
Colorimetric (Persulfate), or		3500-Mn D			920.203[3]
(Periodate)					Note 23
35. Mercury-Total[4], mg/L:					

TABLE 5.7. LIST OF APPROVED BIOLOGICAL TEST PROCEDURES (40CFR136.3 TABLE IB)					
Parameter, units and methods	Method Reference (method No. or page)				
	EPA[1,35]	Std method 18th ed	ASTM	USGS[2]	Other
Cold vapor, manual or	245.1	3112 B	D3223-91	I-3462-85	977.22[3]
Automated	245.2				
36. Molybdenum-Total[4], mg/L; Digestion[4]followed by:					
AA direct aspiration	246.1	3111 D		I-3490-85	
AA furnace	246.2	3113 B			
ICP/AES	[5]200.7	3120 B			
DCP					Note 34
37. Nickel-Total[4], mg/L; Digestion[4]followed by:					
AA direct aspiration	249.1	3111 B or C	D1886-90(A or B)	I-3499-85	
AA furnace	249.2	3113 B	D1886-90(C)		
ICP/AES[36]	[5]200.7	3120 B			
DCP[36], or			D4190-82(88)		Note 34
Colorimetric (heptoxime)		3500-Ni D			
38. Nitrate (as N), mg/L:					
Colorimetric (Brucine sulfate), or Nitrate-nitrite N minus Nitrite N (See parameters 39 and 40)	352.1				973.50[3], 419 D[17], p.28[9]
39. Nitrate-nitrite (as N), mg/L:					
Cadmium reduction, Manual or	353.3	4500-NO 3 E	D3867-90(B)		
Automated, or	353.2	4500-NO 3 F	D3867-90(A)	I-4545-85	
Automated hydrazine	353.1	4500-NO 3 H			
40. Nitrite (as N), mg/L; Spectrophotometric:					
Manual or	354.1				Note 25
Automated (Diazotization)				4500-No2	
41. Oil and grease-Total recoverable, mg/L:					
Gravimetric (extraction)	413.1	5520 B			
42. Organic carbon-Total (TOC), mg/L:					
Combustion or oxidation	[17]415.1	5310 B, C, or D	D2579-85(A or B)		973.47[3]p.14[24]
43. Organic nitrogen (as N), mg/L:					

TABLE 5.7. LIST OF APPROVED BIOLOGICAL TEST PROCEDURES (40CFR136.3 TABLE IB)					
Parameter, units and methods	Method Reference (method No. or page)				
	EPA[1,35]	Std method 18th ed	ASTM	USGS[2]	Other
Total Kjeldahl N (Parameter 31) minus ammonia N (Parameter 4)					
44. Orthophosphate (as P), mg/L Ascorbic acid method:					
Automated, or	365.1	4500-P F		I-4601-85	973.56[3]
Manual single reagent	365.2	4500-P E	D515-88(A)		973.55[3]
Manual two reagent	365.3				
45. Osmium-Total,[4]mg/L; Digestion[4]followed by:					
AA direct aspiration, or	252.1	3111 D			
AA furnace	252.2				
46. Oxygen, dissolved, mg/L:					
Winkler (Azide modification), or	360.2	4500-0 C	D888-92(A)	[8]I-1575-78	973.45B[3]
Electrode	360.1	4500-0-G	D888-92(B)	[8]I-1576-78	
47. Palladium-Total,[4]mg/L; Digestion[4]followed by:					
AA direct aspiration, or	253.1	3111 B			p.S27[10]
AA furnace	253.2				p.S28[10]
DCP					Note 34
48. Phenols, mg/L:					
Manual distillation[26]	420.1				Note 27
Followed by:					
Colorimetric (4AAP) manual, or	420.1				Note 27
Automated[19]	420.2				
49. Phosphorus (elemental), mg/L:					
Gas-liquid chromatography					Note 28
50. Phosphorus-Total, mg/L:					
Persulfate digestion followed by	365.2	4500-P B,5			973.55[3]
Manual or	365.2 or	4500-P E	D515-88(A)		
	365.3				
Automated ascorbic acid reduction	365.1	4500-P F		I4600-85	973.56[3]
Semi-automated block digestor	365.4		D515-88(B)		

TABLE 5.7. LIST OF APPROVED BIOLOGICAL TEST PROCEDURES (40CFR136.3 TABLE IB)					
Parameter, units and methods	Method Reference (method No. or page)				
	EPA[1,35]	Std method 18th ed	ASTM	USGS[2]	Other
51. Platinum-Total,[4]mg/L; Digestion[4]followed by:					
AA direct aspiration	255.1	3111 B			
AA furnace	255.2				
DCP					Note 34
52. Potassium-Total,[4]mg/L; Digestion[4]followed by:					
AA direct aspiration	258.1	3111 B		I-3630-85	973.53[3]
ICP/AES	[5]200.7	3120 B			
Flame photometric, or		3500-K D			
Colormetric					317 B[17]
53. Residue-Total, mg/L:					
Gravimetric, 103-105°	160.3	2540 B		I-3750-85	
54. Residue-filterable, mg/L:					
Gravimetric, 180°	160.1	2540 C		I-1750-85	
55. Residue-nonfilterable (TSS), mg/L:					
Gravimetric, 103-105° post washing of residue	160.2	2540 D		I-3765-85	
56. Residue-settleable, mg/L:					
Volumetric, (Imhoff cone), or gravimetric	160.5	2540 F			
57. Residue-Volatile, mg/L:					
Gravimetric, 550°	160.4			I-3753-85	
58. Rhodium-Total,[4]mg/L; Digestion[4]followed by:					
AA direct aspiration, or	265.1	3111 B			
AA furnace	265.2				
59. Ruthenium-Total,[4]mg/L; Digestion[4]followed by:					
AA direct aspiration, or	267.1	3111 B			
AA furnace	267.2				
60. Selenium-Total,[4]mg/L; Digestion[4]followed by:					
AA furnace	270.2	3113 B			

TABLE 5.7. LIST OF APPROVED BIOLOGICAL TEST PROCEDURES (40CFR136.3 TABLE IB)					
Parameter, units and methods	Method Reference (method No. or page)				
	EPA[1,35]	Std method 18th ed	ASTM	USGS[2]	Other
ICP/AES,[36]or	[5]200.7	3120 B			
AA gaseous hydride		3114 B	D3859-88(A)	I-3667-85	
61. Silica[37]-Dissolved, mg/L; 0.45 micron filtration followed by:					
Colorimetric, Manual or	370.1	4500-Si D	D859-88	I-1700-85	
Automated (Molybdosilicate), or				I-2700-85	
ICP	[5]200.7	3120 B			
62. Silver-Total,[4]mg/L; Digestion[4,29]followed by:					
AA direct aspiration	272.1	3111 B or C		I-3720-85	974.27[3], p. 37[9]
AA furnace	272.2	3113 B			
ICP/AES	[5]200.7	3120 B			
DCP					Note 34
63. Sodium-Total,[4]mg/L; Digestion[4]followed by:					
AA direct aspiration	273.I	3111 B		I-3735-85	973.54[3]
ICP/AES	[5]200.7	3120 B			
DCP, or					Note 34
Flame photometric		3500 Na D			
64. Specific conductance, micromhos/cm at 25°C:					
Wheatstone bridge	120.1	2510 B	D1125-91(A)	I-1780-85	973.40[3]
65. Sulfate (as SO4), mg/L:					
Automated colorimetric (barium chloranilate)	375.1				
Gravimetric	375.3	4500-SO4-2 C or D			925.54[3]
Turbidimetric, or	375.4		516-90		426C[30]
66. Sulfide (as S), mg/L:					
Titrimetric (iodine), or	376.1	4500-S-2 E		I-3840-85	
Colorimetric (methylene blue)	376.2	4500-S-2 D			
67. Sulfite (as SO3), mg/L:					
Titrimetric (iodine-iodate)	377.1	4500-SO3-2 B			
68. Surfactants, mg/L:					

TABLE 5.7. LIST OF APPROVED BIOLOGICAL TEST PROCEDURES (40CFR136.3 TABLE IB)					
Parameter, units and methods	Method Reference (method No. or page)				
	EPA[1,35]	Std method 18th ed	ASTM	USGS[2]	Other
Colorimetric (methylene blue)	425.1	5540 C	D2330-88		
69. Temperature, °C:					
Thermometric	170.1	2550 B			Note 32
70. Thallium-Total,[4] mg/L; Digestion[4] followed by:					
AA direct aspiration	279.1	3111 B			
AA furnace	279.2				
ICP/AES, or	[5]200.7	3120 B			
71. Tin-Total,[4] mg/L; Digestion[4] followed by:					
AA direct aspiration	282.1	3111 B		I-3850-78[8]	
AA furnace, or	282.2	3113 B			
ICP/AES	[5]200.7				
72. Titanium-Total,[4] mg/L; Digestion[4] followed by:					
AA direct aspiration	283.1	3111 D			
AA furnace	283.2				
DCP					Note 34
73. Turbidity, NTU:					
Nephelometric	180.1	2130 B	D1889-88(A)	I-3860-85	
74. Vanadium-Total,[4] mg/L; Digestion[4] followed by:					
AA direct aspiration	286.1	3111 D			
AA furnace	286.2				
ICP/AES	[5]200.7	3120 B			
DCP, or			D4190-82(88)		Note 34
Colorimetric (Gallic acid)		3500-V D			
75. Zinc-Total[4], mg/L; Digestion[4] followed by:					
AA direct aspiration[36]	289.1	3111 B or C	D1691-90 (A or B)	1-3900-85	974.27[3]p.37[9]
AA furnace	289.2				
ICP/AES[36]	[5]200.7	3120 B			
DCP,[36] or			D4190-82(88)		Note 34

TABLE 5.7. LIST OF APPROVED BIOLOGICAL TEST PROCEDURES (40CFR136.3 TABLE IB)					
Parameter, units and methods	Method Reference (method No. or page)				
	EPA[1,35]	Std method 18th ed	ASTM	USGS[2]	Other
Colorimetric (Dithizone) or		3500-Zn E			
(Zincon)		3500-Zn F			Note 33

Notes:

[1] "Methods for Chemical Analysis of Water and Wastes," Environmental Protection Agency, Environmental Monitoring Systems Laboratory-Cincinnati (EMSL-CI), EPA-600/4-79-020, Revised March 1983 and 1979 where applicable.

[2] "Fishman, M.J., et al, "Methods for Analysis of Inorganic Substances in Water and Fluvial Sediments," U.S. Department of the Interior, Techniques of Water-Resource Investigations of the U.S. Geological Survey, Denver, CO, Revised 1989, unless otherwise stated.

[3] "Official Methods of Analysis of the Association of Official Analytical Chemists," methods manual, 15th ed. (1990).

[4] For the determination of total metals the sample is not filtered before processing. A digestion procedure is required to solubilize suspended material and to destroy possible organic-metal complexes. Two digestion procedures are given in "Methods for Chemical Analysis of Water and Wastes, 1979 and 1983." One (section 4.1.3), is a vigorous digestion using nitric acid. A less vigorous digestion using nitric and hydrochloric acids (section 4.1.4) is preferred; however, the analyst should be cautioned that this mild digestion may not suffice for all samples types. Particularly, if a colorimetric procedure is to be employed, it is necessary to ensure that all organo-metallic bonds be broken so that the metal is in a reactive state. In those situations, the vigorous digestion is to be preferred making certain that at no time does the sample go to dryness. Samples containing large amounts of organic materials may also benefit by this vigorous digestion, however, vigorous digestion with concentrated nitric acid will convert antimony and tin to insoluble oxides and render them unavailable for analysis. Use of ICP/AES as well as determinations for certain elements such as antimony, arsenic, the noble metals, mercury, selenium, silver, tin, and titanium require a modified sample digestion procedure and in all cases the method write-up should be consulted for specific instructions and/or cautions. Note: If the digestion procedure for direct aspiration AA included in one of the other approved references is different than the above, the EPA procedure must be used. Dissolved metals are defined as those constituents which will pass through a 0.45 micron membrane filter. Following filtration of the sample, the referenced procedure for total metals must be followed. Sample digestion of the filtrate for dissolved metals (or digestion of the original sample solution for total metals) may be omitted for AA (direct aspiration or graphite furnace) and ICP analyses, provided the sample solution to be analyzed meets the following criteria:
 a. Has a low COD (<20),
 b. Is visibly transparent with a turbidity measurement of 1 NTU or less,
 c. Is colorless with no perceptible odor, and
 d. Is of one liquid phase and free of particulate or suspended matter following acidification.

[5] The full text of Method 200.7, "Inductively Coupled Plasma Atomic Emission Spectrometric Method for Trace Element Analysis of Water and Wastes," is given at Appendix C of this Part 136.

[6] Manual distillation is not required if comparability data on representative effluent samples are on company file to show that this preliminary distillation step is not necessary: however, manual distillation will be required to resolve any controversies.

[7] Ammonia, Automated Electrode Method, Industrial Method Number 379-75 WE, dated February 19, 1976, Bran & Luebbe (Technicon) Auto Analyzer II, Bran & Luebbe Analyzing Technologies, Inc., Elmsford, N.Y. 10523.

[8] The approved method is that cited in "Methods for Determination of Inorganic Substances in Water and Fluvial Sediments", USGS TWRI, Book 5, Chapter A1 (1979).

[9] American National Standard on Photographic Processing Effluents, Apr. 2, 1975. Available from ANSI, 1430 Broadway, New York, NY 10018.

[10] "Selected Analytical Methods Approved and Cited by the United States Environmental Protection Agency," Supplement to the Fifteenth Edition of Standard Methods for the Examination of Water and Wastewater (1981).

[11] The use of normal and differential pulse voltage ramps to increase sensitivity and resolution is acceptable.

[12] Carbonaceous biochemical oxygen demand (CBOD5) must not be confused with the traditional BOD5 test which measures "total BOD".

The addition of the nitrification inhibitor is not a procedural option, but must be included to report the CBOD5 parameter. A discharger whose permit requires reporting the traditional BOD5 may not use a nitrification inhibitor in the procedure for reporting the results. Only when a discharger's permit specifically states CBOD5 is required can the permittee report data using the nitrification inhibitor.

[13] OIC Chemical Oxygen Demand Method, Oceanography International Corporation, 1978, 512 West Loop, P.O. Box 2980, College Station, TX 77840.

[14] Chemical Oxygen Demand, Method 8000, Hach Handbook of Water Analysis, 1979, Hach Chemical Company, P.O. Box 389, Loveland, CO 80537.

[15] The back titration method will be used to resolve controversy.

[16] Orion Research Instruction Manual, Residual Chlorine Electrode Model 97-70, 1977, Orion Research Incorporated, 840 Memorial Drive, Cambridge, MA 02138. The calibration graph for the Orion residual chlorine method must be derived using a reagent blank and three standard solutions, containing 0.2, 1.0, and 5.0 ml 0.00281 N potassium iodate/100 ml solution, respectively.

[17] The approved method is that cited in Standard Methods for the Examination of Water and Wastewater, 14th Edition, 1976.

[18] National Council of the Paper Industry for Air and Stream Improvement, (Inc.) Technical Bulletin 253, December 1971.

[19] Copper, Biocinchoinate Method, Method 8506, Hach Handbook of Water Analysis, 1979, Hach Chemical Company, P.O. Box 389, Loveland, CO 80537.

[20] After the manual distillation is completed, the autoanalyzer manifolds in EPA Methods 335.3 (cyanide) or 420.2 (phenols) are simplified by connecting the re-sample line directly to the sampler. When using the manifold setup shown in Method 335.3, the buffer 6.2 should be replaced with the buffer 7.6 found in Method 335.2.

[21] Hydrogen ion (pH) Automated Electrode Method, Industrial Method Number 378-75WA, October 1976, Bran & Luebbe (Technicon) Autoanalyzer II. Bran & Luebbe Analyzing Technologies, Inc., Elmsford, N.Y. 10523.

[22] Iron, 1,10-Phenanthroline Method, Method 8008, 1980, Hach Chemical Company, P.O. Box 389, Loveland, CO 80537.

[23] Manganese, Periodate Oxidation Method, Method 8034, Hach Handbook of Wastewater Analysis, 1979, pages 2-113 and 2-117, Hach Chemical Company, Loveland, CO 80537.

[24] Wershaw, R.L., et al, "Methods for Analysis of Organic Substances in Water," Techniques of Water-Resources Investigation of the U.S. Geological Survey, Book 5, Chapter A3, (1972 Revised 1987) p. 14.

[25] Nitrogen, Nitrite, Method 8507, Hach Chemical Company, P.O. Box 389, Loveland, CO 80537.

[26] Just prior to distillation, adjust the sulfuric-acid-preserved sample to pH 4 with 19 NaOH.

[27] The approved method is cited in Standard Methods for the Examination of Water and Wastewater, 14th Edition. The colorimetric reaction is conducted at a pH of 10.0±0.2. The approved methods are given on pp. 576-81 of the 14th Edition: Method 510A for distillation, Method 510B for the manual colorimetric procedure, or Method 510C for the manual spectrophotometric procedure.

[28] R. F. Addison and R. G. Ackman, "Direct Determination of Elemental Phosphorus by Gas-Liquid Chromatography," Journal of Chromatography, vol. 47, No. 3, pp. 421-426, 197.

[29] Approved methods for the analysis of silver in industrial wastewaters at concentrations of 1 mg/L and above are inadequate where silver exists as an inorganic halide. Silver halides such as the bromide and chloride are relatively insoluble in reagents such as nitric acid but are readily soluble in an aqueous buffer of sodium thiosulfate and sodium hydroxide to pH of 12. Therefore, for levels of silver above 1 mg/L, 20 mL of sample should be diluted to 100 mL by adding 40 mL each of 2 M Na2S2O3 and NaOH. Standards should be prepared in the same manner. For levels of silver below 1 mg/L the approved method is satisfactory.

[30] The approved method is that cited in Standard Methods for the Examination of Water and Wastewater, 15th Edition.

[31] EPA Methods 335.2 and 335.3 require the NaOH absorber solution final concentration to be adjusted to 0.25 N before colorimetric determination of total cyanide.

[32] Stevens, H. H., Ficke, J. F., and Smoot, G. F., "Water Temperature-Influential Factors, Field Measurement and Data Presentation",

Techniques of Water-Resources Investigations of the U.S. Geological Survey, Book 1, Chapter D1, 1975.

[33] Zinc, Zincon Method, Method 8009, Hach Handbook of Water Analysis, 1979, pages 2-231 and 2-333, Hach Chemical Company, Loveland, CO 80537.

[34] "Direct Current Plasma (DCP) Optical Emission Spectrometric Method for Trade Elemental Analysis of Water and Wastes, Method AES0029," 1986-Revised 1991, Applied Research Laboratories, Inc., 24911 Avenue Stanford, Valencia, CA 91355.

[35] Precision and recovery statements for the atomic absorption direct aspiration and graphite furnace methods, and for the spectrophotometric SDDC method for arsenic are provided in Appendix D of this part titled, "Precision and Recovery Statements for Methods for Measuring Metals".

[36] "Closed Vessel Microwave Digestion of Wastewater Samples for Determination of Metals," CEM Corporation, P.O. Box 200, Matthews, NC 28106-0200, April 16, 1992. Available from the CEM Corporation.

[37] When determining boron and silica, only plastic, PTFE, or quartz sampling and laboratory ware must be used from time of collection until completion of analysis.

TABLE 5.8. LIST OF APPROVED TEST PROCEDURES FOR NON-PESTICIDE ORGANIC COMPOUNDS
(40CFR136.3 TABLE IC)

Parameter[1]	EPA Method number[2,7]				ASTM	Other
	GC	GC/MS	HPLC	Std method 18th ed.		
1. Acenaphthene	610	625, 1625	610	6410 B, 6440 B	D4657-92	
2. Acenaphthylene	610	625, 1625	610	6410 B, 6440 B	D4657-92	
3. Acrolein	603	[4]604, 1624				
4. Acrylonitrile	603	[4]624, 1624	610			
5. Anthracene	610	625, 1625	610	6410 B, 6440 B	D4657-92	
6. Benzene	602	624, 1624		6210 B, 6220 B		
7. Benzidine		[5]625, 1625	605			Note 3, p.1.
8. Benzo(a)anthracene	610	625, 1625	610	6410 B, 6440 B	D4657-92	
9. Benzo(a)pyrene	610	625, 1625	610	6410 B, 6440 B	D4657-92	
10. Benzo(b)fluoranthene	610	625, 1625	610	6410 B, 6440 B	D4657-92	
11. Benzo(g,h,i)perylene	610	625, 1625	610	6410 B, 6440 B	D4657-92	
12. Benzo(k)fluoranthene	610	625, 1625	610	6410 B, 6440 B	D4657-92	
13. Benzyl chloride						Note 3, p.130: Note 6, p. S102
14. Benzyl butyl phthalate	606	625, 1625		6410 B		
15. Bis(2-chloroethoxy) methane	611	625, 1625		6410 B		
16. Bis(2-chloroethyl) ether	611	625, 1625		6410 B		
17. Bis (2-ethylhexyl) phthalate	606	625, 1625		6410 B, 6230 B		
18. Bromodichloromethane	601	624, 1624		6210 B, 6230 B		
19. Bromoform	601	624, 1624		6210 B, 6230 B		
20. Bromomethane	601	624, 1624		6210 B, 6230 B		
21. 4-Bromophenylphenyl ether	611	625, 1625		6410 B		
22. Carbon tetrachloride	601	624, 1624		6230 B, 6410 B		Note 3, p.130.
23. 4-Chloro-3-methylphenol	604	625, 1625		6410 B, 6420 B		
24. Chlorobenzene	601, 602	624, 1624		6210 B, 6220 B, 6230 B		Note 3, p.130.
25. Chloroethane	601	624, 1624		6210 B, 6230 B		
26. 2-Chloroethylvinyl ether	601	624, 1624		6210 B, 6230 B		
27. Chloraform	601	624, 1624		6210 B, 6230 B		Note, p.130.
28. Chloromethane	601	624, 1624		6210 B. 6230 B		
29. 2-Chloronaphthalene	612	625, 1625		6410 B		

TABLE 5.8. LIST OF APPROVED TEST PROCEDURES FOR NON-PESTICIDE ORGANIC COMPOUNDS
(40CFR136.3 TABLE IC)

Parameter[1]	EPA Method number[2,7]				ASTM	Other
	GC	GC/MS	HPLC	Std method 18th ed.		
30. 2-Chlorophenol	604	625, 1625		6410 B, 6420 B		
31. 4-Chlorophenylphenyl ether	611	625, 1625		6410 B		
32. Chrysene	610	625, 1625	610	6410 B, 6440 B	D4657-92	
33. Dibenzo(a,h)anthracene	610	625, 1625	610	6410 B, 6440 B	D4657-92	
34. Dibromochloromethane	601	624, 1624		6210 B, 6230 B		
35. 1, 2-Dichlorobenzene	601, 602, 612	624, 625, 1625		6410 B, 6230 B, 6220 B		
36. 1, 3-Dichlorobenzene	601, 602, 612	624, 625, 1625		6410 B, 6230 B, 6220 B		
37. 1, 4-Dichlorobenzene	601, 602, 612	624, 625, 1625		6410 B, 6220 B, 6230 B		
38. 3, 3-Dichlorobenzidine		625, 1625	605	6410 B		
39. Dichlorodifluoromethane	601			6230 B		
40. 1, 1-Dichloroethane	601	624, 1624		6230 B, 6210 B		
41. 1, 2-Dichloroethane	601	624, 1624		6230 B, 6210 B		
42. 1, 1-Dichloroethene	601	624, 1624		6230 B, 6210 B		
43. trans-1, 2-Dichloroethene	601	624, 1624		6230 B, 6210 B		
44. 2, 4-Dichlorophenol	604	625, 1625		6420 B, 6410 B		
45. 1, 2-Dichloropropane	601	624, 1624		6230 B, 6210 B		
46. cis-1, 3-Dichloropropene	601	624, 1624		6230 B, 6210 B		
47. trans-1, 3-Dichloropropene	601	624, 1624		6230 B, 6210 B		
48. Diethyl phthalate	606	625, 1625		6410 B		
49. 2, 4-Dimethylphenol	604	625, 1625		6420 B, 6410 B		
50. Dimethyl phthalate	606	625, 1625		6410 B		
51. Di-n-butyl phthalate	606	625, 1625		6410 B		
52. Di-n-octyl phthalate	606	625, 1625		6410 B		
53. 2,3-Dinitrophenol	604	625, 1625		6420 B, 6410 B		
54. 2,4-Dinitrotoluene	609	625, 1625		6410 B		
55. 2,6-Dinitrotoluene	609	625, 1625		6410 B		
56. Epichlorohydrin						Note 3, p.130 Note 6, p.S102
57. Ethylbenzene	602	624, 1624		6220 B, 6210 B		
58. Fluoranthene	610	625, 1625	610	6410 B, 6440 B	D4657-92	

TABLE 5.8. LIST OF APPROVED TEST PROCEDURES FOR NON-PESTICIDE ORGANIC COMPOUNDS
(40CFR136.3 TABLE IC)

Parameter[1]	EPA Method number[2,7]				ASTM	Other
	GC	GC/MS	HPLC	Std method 18th ed.		
59. Fluorene	610	625, 1625	610	6410 B, 6440 B	D4657-92	
60. Hexachlorobenzene	612	625, 1625		6410 B		
61. Hexachlorobutadiene	612	625, 1625		6410 B		
62. Hexachlorocyclopentadiene	612	[5]625, 1625		6410 B		
63. Hexachloroethane	616	625, 1625		6410 B		
64. Ideno(1,2,3-cd) pyrene	610	625, 1625	610	6410 B, 6440 B	D4657-92	
65. Isophorone	609	625, 1625		6410 B		
66. Methylene chloride	601	624, 1624		6230 B		Note 3, p.130.
67. 2-Methyl-4,6-dinitrophenol	604	625, 1625		6420 B, 6410 B		
68. Naphthalene	610	625, 1625	610	6410 B, 6440 B		
69. Nitrobenezene	609	625, 1625		6410 B		
70. 2-Nitrophenol	604	625, 1625		6410 B, 6420 B		
71. 4-Nitrophenol	604	625, 1625		6410 B, 6420 B		
72. N-Nitrosodimethylamine	607	625, 1625		6410 B		
73. N-Nitrosodi-n-propylamine	607	[5]625, 1625		6410 B		
74. N-Nitrosodiphenylamine	607	[5]625, 1625		6410 B		
75. 2,2-Oxybis(1-chloropropane)	611	625, 1625		6410 B		
76. PCB-1016	608	625		6410 B		Note 3, p.43
77. PCB-1221	608	625		6410 B		Note 3, p.43
78. PCB-1232	608	625		6410 B		Note 3, p.43
79. PCB-1242	608	625		6410 B		Note 3, p.43
80. PCB-1248	608	625				
81. PCB-1254	608	625		6410 B		Note 3, p.43
82. PCB-1260	608	625		6410 B, 6630 B		Note 3, p.43.
83. Pentachlorophenol	604	625, 1625		6410 B, 6630 B		Note 3, p.140.
84. Phenanthrene	610	625, 1625	610	6410 B, 6440 B	D4657-92	
85. Phenol	604	625, 1625		6420 B, 6410 B		
86. Pyrene	610	625, 1625	610	6410 B, 6440 B	D4675-92	
87. 2,3,7,8-Tetrachlorodibenzo-p-dioxin		5a 613				
88. 1,1,2,2-Tetrachloroethane	601	624, 1624		6230 B, 6210 B		Note 3, p.130.
89. Tetrachloroethene	601	624, 1624		6230 B, 6210 B		Note 3, p.130.

TABLE 5.8. LIST OF APPROVED TEST PROCEDURES FOR NON-PESTICIDE ORGANIC COMPOUNDS
(40CFR136.3 TABLE IC)

Parameter[1]	EPA Method number[2,7]				ASTM	Other
	GC	GC/MS	HPLC	Std method 18th ed.		
90. Toluene	602	624, 1624		6210 B, 6220 B		
91. 1,2,4-Trichlorobenzene	612	625, 1625		6410 B		Note 3, p.130.
92. 1,1,1-Trichloroethane	601	624, 1624		6210 B, 6230 B		
93. 1,1,2-Trichloroethane	601	624, 1624		6210 B, 6230 B		Note 3, p.130.
94. Trichloroethene	601	624, 1624		6210 B, 6230 B		
95. Trichlorofluoromethane	601	624		6210 B, 6230 B		
96. 2,4,6-Trichlorophenol	604	625, 1625		6410 B, 6240 B		
97. Vinyl chloride	601	624, 1624		6210 B, 6230 B		

Notes:

[1] All parameters are expressed in micrograms per liter (μg/L).

[2] The full text of Methods 601-613, 624, 625, 1624, and 1625, are given at appendix A, "Test Procedures for Analysis of Organic Pollutants," of this part 136. The standardized test procedure to be used to determine the method detection limit (MDL) for these test procedures is given at appendix B, "Definition and Procedure for the Determination of the Method Detection Limit" of this part 136.

[3] "Methods for Benzidine: Chlorinated Organic Compounds, Pentachlorophenol and Pesticides in Water and Wastewater," U.S. Environmental Protection Agency, September, 1978.

[4] Method 624 may be extended to screen samples for Acrolein and Acrylonitrile. However, when they are known to be present, the preferred method for these two compounds is Method 603 or Method 1624.

[5] Method 625 may be extended to include benzidine, hexachlorocyclopentadiene, N-nitrosodimethylamine, and N-nitrosodiphenylamine. However, when they are known to be present, Methods 605, 607, and 612, or Method 1625, are preferred methods for these compounds.

[5] [6a] 625, Screening only.

[6] "Selected Analytical Methods Approved and Cited by the United States Environmental Protection Agency," Supplement to the Fifteenth Edition of Standard Methods for the Examination of Water and Wastewater (1981).

[7] Each Analyst must make an initial, one-time demonstration of their ability to generate acceptable precision and accuracy with Methods 601-603, 624, 625, 1624, and 1625 (See Appendix A of this Part 136) in accordance with procedures each in section 8.2 of each of these Methods. Additionally, each laboratory, on an on-going basis must spike and analyze 10% (5% for Methods 624 and 625 and 100% for methods 1624 and 1625) of all samples to monitor and evaluate laboratory data quality in accordance with sections 8.3 and 8.4 of these Methods. When the recovery of any parameter falls outside the warning limits, the analytical results for that parameter in the unspiked sample are suspect and cannot be reported to demonstrate regulatory compliance.

Note: These warning limits are promulgated as an "interim final action with a request for comments."

TABLE 5.9. LIST OF APPROVED TEST PROCEDURES FOR PESTICIDES[1]
(40CFR136.3 TABLE ID)

Parameter	EPA method number			ASTM	Other
	Method	EPA[2,7]	Std methods 18th Ed.		
1. Aldrin	GC	608	6630 B & C	D3086-90	Note 3, p. 7; Note 4, p. 30
	GC/MS	625	6410 B		
2. Ametryn	GC				Note 3, p. 83; Note 6, p. S68
3. Aminocarb	TLC				Note 3, p. 94; Note 6, p. S16
4. Atraton	GC				Note 3, p. 83; Note 6, p. S68
5. Atrazine	GC				Note 3, p. 83; Note 6, p. S68
6. Azinphos methyl	GC				Note 3, p. 25; Note 6, p. S51
7. Barban	TLC				Note 3, p. 104; Note 6, p. S64
8. alpha=BHC	GC	608	6630 B & C	D3086-90	Note 3, p. 7
	GC/MS	[5]625	6410 B		
9. beta=BHC	GC	608	6630	D3086-90	
	GC/MS	[5]625	6410 B		
10. delta=BHC	GC	608	6630 B & C	D3086-90	
	GC/MS	[5]625	6410 B		
11. gamma-BHC (Lindane)	GC	608	6630 B & C	D3086-90	Note 3, p. 7; Note 4, p. 30
	GC/MS	625	6410 B		
12. Captan	GC		6630 B	D3086-90	Note 3, p. 7
13. Carbaryl	TLC				Note 3, p. 94: Note 6, p. S60
14. Carbophenothion	GC				Note 4, p. 30; Note 6, p. S73
15. Chlordane	GC	608	6630 B & C	D3086-90	Note 3, p. 7
	GC/MS	625	6410 B		
16. Chloropropham	TLC				Note 3, p. 104; Note 6, p. S64
17. 2,4-D	GC		6640 B		Note 3, p. 115; Note 4, p. 35
18. 4,4'-D-DDD	GC	608	6630 B & C	D3086-90	Note 3, p. 7; Note 4, p. 30
	GC-MS	625	6410B		
19. 4,4'-DDE	GC	608	6630 B & C	D3086-90	Note 3, p. 7; Note 4, p. 30
	GC-MS	625	6410 B		
20. 4,4'-DDT	GC	608	6630 B & C	D3086-90	Note 3, p. 7; Note 4, p. 30
	GC-MS	625	6410 B		
21. Demeton-O	GC				Note 3, p. 25; Note 6, p. S51
22. Dementon-S	GC				Note 3, p. 25; Note 6, p. S51

TABLE 5.9. LIST OF APPROVED TEST PROCEDURES FOR PESTICIDES[1] (40CFR136.3 TABLE ID)					
Parameter	EPA method number			ASTM	Other
	Method	EPA[2,7]	Std methods 18th Ed.		
23. Diazinon	GC				Note 3, p. 25; Note 4, p. 30; Note 6, p. S51
24. Dicamba	GC				Note 3, p. 115
25. Dichlofenthion.	GC				Note 4, p. 30; Note 6, p. S73
26. Dichloran	GC		6630 B & C		Note 3, p. 7
27. Dicofol	GC			D3086-90	
28. Dieldrin	GC	608	6630 B & C		Note 3, p. 7; Note 4, p. 30
	GC-MS	625	6410 B		
29. Dioxathion	GC				Note 4, p. 30; Note 6, p. S73
30. Disulfoton	GC				Note 3, p. 25; Note 6, p. S51
31.Diuron	TLC				Note 3, p. 104; Note 6, p. S64
32. Endosulfan I	GC	608	6630 B & C	D3086-90	Note 3, p. 7
	GC-MS	[5]625	6410 B		
33. Endosulfan II	GC	608	6630 B & C	D3086-90	Note 3, p. 7
	GC-MS	[5]625	6410 B		
34. Endosulfan Sulfate	GC	608	6630 C		
	GC-MS	625	6410 B		
35. Endrin	GC	608	6630 B & C	D3086-90	Note 3, p. 7; Note 4, p. 30
	GC-MS	[5]625	6410 B		
36. Endrin aldehyde	GC	608			
37. Ethion	GC				Note 4, p. 30; Note 6, p. S73
38. Fenuron.	TLC				Note 3, p. 104; Note 6, p. S64
39. Fenuron-TCA	TLC				Note 3, p. 104; Note 6, p. S64
40. Heptachlor	GC	608	6630 B & C	D3086-90	Note 3, p. 7; Note 4, p. 30
	GC/MS	625	6410 B		
41. Heptachlor epoxide	GC	608	6630 B	D3086-90	Note 3, p. 7; Note 4, p. 30
	GC/MS	625	6410 B		Note 6, p. S73.
42. Isodrin	GC				Note 4, p. 30; Note 6, p. S73
43. Linuron	GC				Note 3, p. 104; Note 6, p. S64
44. Malathion	GC		6630 C		Note 3, p. 25; Note 4, p.
					30; Note 6, p. S51

Parameter	EPA method number			ASTM	Other
	Method	EPA[2,7]	Std methods 18th Ed.		
45. Methiocarb	TLC				Note 3, p. 94; Note 6, p. S60
46. Methoxychlo	GC		6630 B & C	D3086-90	Note 3, p. 7; Note 4, p. 30
47. Mexacarbate	TLC				Note 3, p. 94; Note 6, p. S60
48. Mirex	GC		6630 B & C		Note 3, p. 7
49. Monuron	TLC				Note 3, p. 104; Note 6, p. S64
50. Monuron	TLC				Note 3, p. 104; Note 6, p. S64
51. Nuburon	TLC				Note 3, p. 104; Note 6, p. S64
52. Parathion methyl	GC		6630 C		Note 3, p. 25; Note 4, p. 30
53. Parathion ethyl	GC		6630 C		Note 3, p. 25
54. PCNB	GC		6630 B & C		Note 3, p. 7
55. Perthane	GC			D3086-90	
56. Prometron	GC				Note 3, p. 83; Note 6, p. S68
57. Prometryn	GC				Note 3, p. 83; Note 6, p. S68
58. Propazine	GC				Note 3, p. 83; Note 6, p. S68
59. Propham	TLC				Note 3, p. 104; Note 6, p. S64
60. Propoxur	TLC				Note 3, p. 94; Note 6, p. S60
61. Secbumeton	TLC				Note 3, p. 83; Note 6, p. S68
62. Siduron	TLC				Note 3, p. 104; Note 6, p. S64
63. Simazine	GC				Note 3, p. 83; Note 6, p. S68
64. Strobane	GC		6630 B & C		Note 3, p. 7
65. Swep	TLC				Note 3, p. 104; Note 6, p. S64
66. 2,4,5-T.	GC		6640 B		Note 3, p. 115; Note 4, p. 35
67. 2,4,5-TP (Silvex)	GC		6640 B		Note 3, p. 115
68. Terbuthylazine	GC				Note 3, p. 83; Note 6, p. S68
69. Toxaphene	GC	608	6630 B & C	D3086-90	Note 3, p. 7; Note 4, p. 30
	GC/MS	625	6410 B		
70. Trifluralin	GC		6630 B		Note 3, p. 7

Notes:

[1] Pesticides are listed in this table by common name for the convenience of the reader. Additional pesticides may be found under Table 1C, where entries are listed by chemical name.

[2] The full text of Methods 608 and 625 are given at appendix A. "Test Procedures for Analysis of Organic Pollutants," of this part 136. The

standardized test procedure to be used to determine the method detection limit (MDL) for these test procedures is given at appendix B. "Definition and Procedure for the Determination of the Method Detection Limit", of this part 136.

[3] "Methods for Benzidine, Chlorinated Organic Compounds, Pentachlorophenol and Pesticides in Water and Wastewater," U.S. Environmental Protection Agency, September, 1978. This EPA publication includes thin-layer chromatography (TLC) methods.

[4] "Methods for Analysis of Organic Substances in Water and Fluvial Sediments," Techniques of Water-Resources Investigations of the U.S. Geological Survey, Book 5, Chapter A3 (1987).

[5] The method may be extended to include α-BHC, δ-BHC, endosulfan I, endosulfan II, and endrin, However, when they are known to exist, Method 608 is the preferred method.

[6] "Selected Analytical Methods Approved and Cited by the United States Environmental Protection Agency," Supplement to the Fifteenth Edition of Standard Methods for the Examination of Water and Wastewater (1981).

[7] Each analyst must make an initial, one-time, demonstration of their ability to generate acceptable precision and accuracy with Methods 608 and 625 (See appendix A of this part 136) in accordance with procedures given in section 8.2 of each of these methods. Additionally, each laboratory, on an on-going basis, must spike and analyze 10% of all samples analyzed with Method 608 or 5% of all samples analyzed with Method 625 to monitor and evaluate laboratory data quality in accordance with Sections 8.3 and 8.4 of these methods. When the recovery of any parameter falls outside the warning limits, the analytical results for that parameter in the unspiked sample are suspect and cannot be reported to demonstrat regulatory compliance. These quality control requirements also apply to the Standard Methods, ASTM Methods, and other Methods cited.

Note: These warning limits are promulgated as an "Interim final action with a request for comments."

TABLE 5.10. LIST OF APPROVED RADIOLOGIC TEST PROCEDURES (40CFR136.3 TABLE IE)						
Parameter and units	Method	Reference (method No. or page)				
		EPA[1]	Std methods 18th ed.	ASTM	USGS[2]	
1. Alpha-Total, pCi per liter	Proportional or scintillation counter	900	7110 B	D1943-90	pp. 75 and 78[3]	
2. Alpha-Counting error, pCi per liter	Proportional or scintillation counter	Appendix B	7110 B	D1943-90	P. 79	
3. Beta-Total, pCi per liter	Proportional counter	900.0	7110 B	D1890-90	pp. 75 and 78[3]	
4. Beta-Counting error, pCi	Proportional counter	Appendix B	7110 B	D1890-90	p. 79	
5. (a) Radium Total pCi per liter.	Proportional counter	903.0	7500Ra B	D2460-90		
(b) Ra, pCi per liter	Scintillation counter	903.1	7500Ra C	D3454-91	p. 81	

Notes:

[1] "Prescribed Procedures for Measurement of Radioactivity in Drinking Water," EPA-600/4-80-032 (1980), U.S. Environmental Protection Agency, August 1980.

[2] Fishman, M.J. and Brown, Eugene, " Selected Methods of the U.S. Geological Survey of Analysis of Wastewaters," U.S. Geological Survey, Open-File Report 76-177 (1976).

[3] The method found on p. 75 measures only the dissolved portion while the method on p. 78 measures only the suspended portion. Therefore, the two results must be added to obtain the "total".

Parameter No/name	Container	Preservation{2,3}	Maximum holding time[4]
TABLE 5.11. REQUIRED CONTAINERS, PRESERVATION TECHNIQUES, AND HOLDING TIMES (40CFR136.3 TABLE II)			
Table IA-Bacterial Tests:			
1-4.coliform, fecal and total	P,G	Cool, 4°C, 0.008% Na2S2O3[5]	6 hours
5. Fecal streptococci	P,G	Do	Do
Table IB-Inorganic Tests:			
1. Acidity	P,G	Cool, 4°C	14 days
2. Alkalinity	P,G	Do	Do
4. Ammonia	P,G	Cool, 4°C, H2SO4 to pH<2	28 days
9. Biochemical oxygen demand	P,G	Cool, 4°C	48 hours
10. Boron	P (PFTE), or Quartz	HNO3 TO pH<2	6 months
11. Bromide	P,G	None required.	28 days
14. Biochemical oxygen demand, carbonaceous	P,G	Cool, 4°C	48 hours
15.chemical oxygen demand	P,G	Cool, 4°C, H2SO4 to pH<2	28 days
16.chloride	P,G	None required.	Do
17.chlorine, total residual	P,G	Do	Analyze immediately.
21.color	P,G	Cool, 4°C	48 hours
23-24.cyanide, total and amenable to chlorination	P,G	Cool, 4°C, NaOH to pH>12, 0. 6g ascorbic acid[5]	14 days6
25. Fluoride	P	None required	28 days
27. Hardness	P,G	HNO3 to pH<2, H2SO4 to pH<2	6 months
28. Hydrogen ion (pH)	P,G	None required	Analyze immediately.
31, 43. Kjeldahl and organic nitrogen Metals[7]:	P,G	Cool, 4°C, H2SO4 to pH<2	28 days
18. Chromium VI	P,G	cool, 4°C	24 hours
35. Mercury	P,G	HNO3 to pH<2	28 days
3, 5-8, 12, 13, 19, 20, 22, 26, 29, 30, 32-34, 36, 37, 45, 47, 51, 52, 58-60, 62, 63, 70-72, 74, 75. Metals, except boron, chromium VI and mercury.			
38. Nitrate	P,G	Cool, 4°C	48 hours
39. Nitrate-nitrite	P,G	Cool, 4°C, H2SO4 to pH<2	28 days
40. Nitrite	P,G	Cool, 4°C	48 hours
41. Oil and grease	G	Cool to 4°C, HCl or H2SO4 to pH<2	28 days

TABLE 5.11. REQUIRED CONTAINERS, PRESERVATION TECHNIQUES, AND HOLDING TIMES
(40CFR136.3 TABLE II)

Parameter No/name	Container	Preservation {2,3}	Maximum holding time[4]
42. Organiccarbon	G	Cool to 4 °C HC1 or H2SO4 or H3PO4, to pH<2	28 days
44. Orthophosphate	P,G	Filter immediately,cool, 4°C	48 hours
46. Oxygen, DissolvedProbe	G Bottle and top	None required	Analyze immediately.
47. Winkler	Do	Fix on site and store in dark	8 hours
48. Phenols	G only	Cool, 4°C, H2SO4 to pH<2	28 days
49. Phosphorus (elemental)	G	Cool, 4°C	48 hours
50. Phosphorus, total	P,G	Cool, 4°C, H2SO4 to pH<2	28 days
53. Residue, total	P,G	Cool, 4°C	7 days
54. Residue, Filterable	P,G	Do	7 days
55. Residue, Nonfilterable (TSS)	P,G	Do	7 days
56. Residue, Settleable	P,G	Do	48 hours
57. Residue, volatile	P,G	Do	7 days
61. Silica	P (PFTE), or Quartz	Cool, 4 °C	28 days
64. Specific conductance	P,G	Do	Do
65. Sulfate	P,G	Do	Do
66. Sulfide	P,G	Cool, 4°C add zinc acetate plus sodium hydroxide to pH> 9	7 days
67. Sulfite	P,G	None required.	Analyze immediately.
68. Surfactants	P,G	Cool, 4°C	48 hours
69. Temperature	P,G	None required	Analyze
73. Turbidity	P,G	Cool, 4°C	48 hours
Table IC-Organic Tests[8]			
13, 18-20, 22, 24-28, 34-37, 39-43, 45-47, 56, 66, 88, 89, 92-95, 97.Purgeable Halocarbons	G, Telflon-lined septum	Cool, 4°C, 0.008% Na2S2O3[5]	14 days
6, 57, 90.Purgeable aromatic hydrocarbons	Do	Cool, 4°C, 0.008% Na2S2O3[5], HC1 to pH2[9]	Do
3, 4, Acrolein and acrylonitrile.	Do	Cool, 4°C, 0.008% Na2S2O3[5]; Adjust pH to 4-5[10]	Do
23, 30, 44, 49, 53,67, 70, 71, 83, 85, 96.Phenols[11]	G, Teflon-lined cap	Cool, 4°C, 0.008% Na2S2O3[5]	7 days until extraction, 40 days after extraction

TABLE 5.11. REQUIRED CONTAINERS, PRESERVATION TECHNIQUES, AND HOLDING TIMES
(40CFR136.3 TABLE II)

Parameter No/name	Container	Preservation {2,3}	Maximum holding time[4]
7, 38. Benzidines[11]	Do	Do	7 days until extraction[13]
14, 17, 48, 50-52.Phthalate esters[11]	Do	Cool, 4°C	7 days until extraction, 40 days after extraction
72-74. Nitrosamines{11, 14}	Do	Cool, 4°C, store in dark, 0. 008% Na2S2O35	Do
76-82.PCBs[11]acrylonitrile	Do	Cool, 4°C	Do
54, 55,65,69. Nitroaromatics and isophorone[11]	Do	Cool, 4°C, 0.008% Na2S2O35 store in dark	Do
1, 2, 5, 8-12, 32,33, 58, 59,64, 68, 84, 86.Polynuclear aromatic hydrocarbons[11]	Do	Do	Do
15, 16, 21, 31, 75. Haloethers[11]	Do	Cool, 4°C, 0.008% Na2S2O35	Do
29, 35-37,60-63, 91.chlorinated hydrocarbons[11]	Do	Cool, 4°C	Do
87. TCDD[11]	Do	Cool, 4°C, 0.008% Na2S2O35	Do
Table ID-Pesticides Tests:			
1-70.Pesticides[11]	Do	Cool, 4°C, pH 5-915	Do
Table IE-Radiological Tests:			
1-5. Alpha, beta and radium	P,G	HNO3 to pH<2	6 months

Notes:

[1] Polyethylene (P) or Glass (G).

[2] Sample preservation should be performed immediately upon sample collection. For composite chemical samples each aliquot should be preserved at the time of collection. When use of an automated sampler makes it impossible to preserve each aliquot, then chemical samples may be preserved by maintaining at 4°C until compositing and sample splitting is completed.

[3] When any sample is to be shipped by common carrier or sent through the United States Mails, it must comply with the Department of Transportation Hazardous Materials regulations (49 CFR part 172). The person offering such material for transportation is responsible for ensuring such compliance. For the preservation requirements of Table II, the Office of Hazardous Materials, Materials Transportation Bureau, Department of Transportation has determined that the Hazardous Materials Regulations do not apply to the following materials: Hydrochloric acid (HCl) in water solutions at concentrations of 0.04% by weight or less (pH about 1.96 or greater); Nitric acid (HNO3) in water solutions at concentrations of 0.15% by weight or less (pH about 1.62 or greater); Sulfuric acid (H2SO4) in water solutions at concentrations of 0.35% by weight or less (pH about 1.15 or greater); and Sodium hydroxide (NaOH) in water solutions at concentrations of 0.080% by weight or less (pH about 12.30 or less).

[4] Samples should be analyzed as soon as possible after collection. The times listed are the maximum times that samples may be held before analysis and still be considered valid. Samples may be held for longer periods only if the permittee, or monitoring laboratory, has data on file to show that for the specific types of samples under study, the analytes are stable for the longer time, and has received a variance from the Regional Administrator under 40CFR136.3(e). Some samples may not be stable for the maximum time period given in the table. A permittee, or monitoring laboratory, is obligated to hold the sample for a shorter time if knowledge exists to show that this is necessary to maintain sample stability. See 40CFR136.3(e) for details. The term "analyze immediately" usually means within 15 minutes or less of sample collection.

[5] Should only be used in the presence of residual chlorine.

[6] Maximum holding time is 24 hours when sulfide is present. Optionally all samples may be tested with lead acetate paper before pH adjustments in order to determine if sulfide is present. If sulfide is present, it can be removed by the addition of cadmium nitrate powder until a negative spot test is obtained. The sample is filtered and then NaOH is added to pH 12.

[7] Samples should be filtered immediately on-site before adding preservative for dissolved metals.

[8] Guidance applies to samples to be analyzed by GC, LC, or GC/MS for specific compounds.

[9] Sample receiving no pH adjustment must be analyzed within seven days of sampling.

[10] The pH adjustment is not required if acrolein will not be measured. Samples for acrolein receiving no pH adjustment must be analyzed within 3 days of sampling.

[11] When the extractable analytes of concern fall within a single chemical category, the specified preservative and maximum holding times should be observed for optimum safeguard of sample integrity. When the analytes of concern fall within two or more chemical categories, the sample may be preserved by cooling to 4°C, reducing residual chlorine with 0.008% sodium thiosulfate, storing in the dark, and adjusting the pH to 6-9; samples preserved in this manner may be held for seven days before extraction and for forty days after extraction. Exceptions to this optional preservation and holding time procedure are noted in footnote 5 (re the requirement for thiosulfate reduction of residual chlorine), and footnotes 12, 13 (re the analysis of benzidine).

[12] If 1,2-diphenylhydrazine is likely to be present, adjust the pH of the sample to 4.0 ± 0.2 to prevent rearrangement to benzidine.

[13] Extracts may be stored up to 7 days before analysis if storage is conducted under an inert (oxidant-free) atmosphere.

[14] For the analysis of diphenylnitrosamine, add 0.008% $Na_2S_2O_3$ and adjust pH to 7-10 with NaOH within 24 hours of sampling.

[15] The pH adjustment may be performed upon receipt at the laboratory and may be omitted if the samples are extracted within 72 hours of collection. For the analysis of aldrin, add 0.008% $Na_2S_2O_3$.

5.3. METHODS FOR ORGANICS CHEMICAL ANALYSIS OF MUNICIPAL AND INDUSTRIAL WASTEWATERS (40CFR136 APPENDIX A)

- Method 601 - Purgeable halocarbons
- Method 602 - Purgeable aromatics
- Method 603 - Acrolein and acrylonitrile
- Method 604 - Phenols
- Method 605 - Benzidincs
- Method 606 - Phthalate
- Method 607 - Nitrosamines
- Method 608 - Organochlorine pesticides and PCBs
- Method 609 - Nitroaromatics and isophorone
- Method 610 - Polynuclear aromatic hydrocarbons
- Method 611 - Haloethers
- Method 612 - Chlorinated hydrocarbons
- Method 613 - 2,3,7,8-Tetrachlorodibenzo-p-dioxins
- Method 624 - Purgeables
- Method 625 - Base/neutrals and acids
- Method 1624 revision B - Volatile organic compounds by isotope dilution GC/MS
- Method 1625 revision B - Semivolatile organic compounds by isotope dilution GC/MS

5.4. REFERENCES (40CFR136.3.b)

(1) The full text of Methods 601-613, 624, 625, 1624, and 1625 are printed in appendix A of this part 136. The full text for determining the method detection limit when using the test procedures is given in appendix B of this part 136. The full text of Method 200.7 is printed in appendix C of this part 136. Cited in: Table IB, Note 5; Table IC, Note 2; and Table ID, Note 2.

(2) "Microbiological Methods for Monitoring the Environment, Water and Wastes," U.S. Environmental Protection Agency, EPA-600/8-78-017, 1978. Available from: ORD Publications, CERI, U.S. Environmental Protection Agency, Cincinnati, Ohio 45268.
Table IA, Note 2.

(3) "Methods for Chemical Analysis of Water and Wastes," U.S. Environmental Protection Agency, EPA-600/4-79-020, March 1979, or "Methods for Chemical Analysis of Water and Wastes," U.S. Environmental Protection Agency, EPA-600/4-79-020, Revised March 1983. Available from: ORD Publications, CERI, U.S. Environmental Protection Agency, Cincinnati, Ohio 45268, Table IB, Note 1.

(4) "Methods for Benzidine, Chlorinated Organic Compounds, Pentachlorophenol and Pesticides in Water and Wastewater," U.S. Environmental Protection Agency, 1978. Available from: ORD Publications, CERI, U.S. Environmental Protection Agency, Cincinnati, Ohio 45268, Table IC, Note 3; Table D, Note 3.

(5) "Prescribed Procedures for Measurement of Radioactivity in Drinking Water," U.S. Environmental Protection Agency, EPA-600/4-80-032, 1980. Available from: ORD Publications, CERI, U.S. Environmental Protection Agency, Cincinnati, Ohio 45268, Table IE, Note 1.

(6) "Standard Methods for the Examination of Water and Wastewater," Joint Editorial Board, American Public Health Association, American Water Works Association, and Water Environment Federation, 18th Edition, 1992. Available from: American Public Health Association, 1015 Fifteenth Street, NW., Washington, DC 20005. Cost $160.00. Tables IA, IB, IC, ID and IE.

(7) Ibid, 15th Edition, 1980. Table IB, Note 30; Table ID.

(8) Ibid, 14th Edition, 1975. Table IB, Notes 17 and 27.

(9) "Selected Analytical Methods Approved and Cited by the United States Environmental Protection Agency," Supplement to the 15th Edition of Standard Methods for the Examination of Water and Wastewater, 1981. Available from: American Public Health Association, 1015 Fifteenth Street NW., Washington, DC 20036. Cost available from publisher. Table IB, Note 10; Table IC, Note 6; Table ID, Note 6.

(10) Annual Book of ASTM Standards, Water and Environmental Technology, Section 11, Volumes 11.01 and 11.02, 1994 in 40 CFR 136.3, Tables IB, IC, ID and IE.

(11) "Methods for Collection and Analysis of Aquatic Biological and Microbiological Samples," edited by Britton, L.J. and P.E. Greason, Techniques of Water Resources Investigations, of the U.S. Geological Survey, Book 5, Chapter A4 (1989). Available from: U.S. Geological Survey, Denver Federal Center, Box 25425, Denver, CO 80225. Cost: $9.25 (subject to change). Table IA.

(12) "Methods for Determination of Inorganic Substances in Water and Fluvial Sediments," by M.J. Fishman and Linda C. Friedman, Techniques of Water-Resources Investigations of the U.S. Geological Survey, Book 5 Chapter A1 (1989). Available from: U.S. Geological Survey, Denver Federal Center, Box 25425, Denver, CO 80225. Cost: $108.75 (subject to change). Table IB, Note 2.

(13) "Methods for Determination of Inorganic Substances in Water and Fluvial Sediments," N.W. Skougstad and others, editors. Techniques of Water-Resources Investigations of the U.S. Geological Survey, Book 5, Chapter A1 (1979). Available from: U.S. Geological Survey, Denver Federal Center, Box 25425, Denver, CO 80225. Cost: $10.00 (subject to change), Table IB, Note 8.

(14) "Methods for the Determination of Organic Substances in Water and Fluvial Sediments," Wershaw, R.L., et al, Techniques of Water-Resources Investigations of the U.S. Geological Survey, Book 5, Chapter A3 (1987). Available from: U.S. Geological Survey, Denver Federal Center, Box 25425, Denver, CO 80225. Cost: $0.90 (subject to change). Table IB, Note 24; Table ID, Note 4. (15) "Water Temperature-Influential Factors, Field Measurement and Data Presentation," by H.H. Stevens, Jr., J. Ficke, and G.F. Smoot, Techniques of Water-Resources Investigations of the U.S. Geological Survey, Book 1, Chapter D1, 1975. Available from: U.S. Geological Survey, Denver Federal Center, Box 25425, Denver, CO 80225. Cost: $1.60 (subject to change). Table IB, Note 32.

(16) "Selected Methods of the U.S. Geological Survey of Analysis of Wastewaters," by M.J. Fishman and Eugene Brown; U.S. Geological Survey Open File Report 76-77 (1976). Available from: U.S. Geological Survey, Branch of Distribution, 1200 South Eads Street, Arlington, VA 22202. Cost: $13.50 (subject to change). Table IE, Note 2.

(17) "Official Methods of Analysis of the Association of Official Analytical Chemicals," Methods manual, 15th Edition (1990). Price: $240.00. Available from: The Association of Official Analytical Chemists, 2200 Wilson Boulevard, Suite 400, Arlington, VA 22201. Table IB, Note 3.

(18) "American National Standard on Photographic Processing Effluents," April 2, 1975. Available from: American National Standards Institute, 1430 Broadway, New York, New York 10018. Table IB, Note 9.

(19) "An Investigation of Improved Procedures for Measurement of Mill Effluent and Receiving Water Color," NCASI Technical Bulletin No. 253, December 1971. Available from: National Council of the Paper Industry for Air and Stream Improvements, Inc., 260 Madison Avenue, New York, NY 10016. Cost available from publisher. Table IB, Note 18.

(20) Ammonia, Automated Electrode Method, Industrial Method Number 379-75WE, dated February 19, 1976. Technicon Auto Analyzer II. Method and price available from Technicon Industrial Systems, Tarrytown, New York 10591. Table IB, Note 7.

(21) Chemical Oxygen Demand, Method 8000, Hach Handbook of Water Analysis, 1979. Method price available from Hach Chemical Company, P.O. Box 389, Loveland, Colorado 80537. Table IB, Note 14.

(22) OIC Chemical Oxygen Demand Method, 1978. Method and price available from Oceanography International Corporation, 512 West Loop, P.O. Box 2980, College Station, Texas 77840. Table IB, Note 13.

(23) ORION Research Instruction Manual, Residual Chlorine Electrode Model 97-70, 1977. Method and price available from ORION Research Incorporation, 840 Memorial Drive, Cambridge, Massachusetts 02138. Table IB, Note 16.

(24) Bicinchoninate Method for Copper. Method 8506, Hach Handbook of Water Analysis, 1979, Method and price available from Hach Chemical Company, P.O. Box 300, Loveland, Colorado 80537. Table IB, Note 19.

(25) Hydrogen Ion (pH) Automated Electrode Method, Industrial Method Number 378-75WA. October 1976. Bran & Luebbe (Technicon) Auto Analyzer II. Method and price available from Bran & Luebbe Analyzing Technologies, Inc. Elmsford, N.Y. 10523. Table IB, Note 21.

(26) 1,10-Phenanthroline Method using FerroVer Iron Reagent for Water, Hach Method 8008, 1980. Method and price available from Hach Chemical Company, P.O. Box 389 Loveland, Colorado 80537. Table IB, Note 22.

(27) Periodate Oxidation Method for Manganese, Method 8034, Hach Handbook for Water Analysis, 1979. Method and price available from Hach Chemical Company, P.O. Box 389, Loveland, Colorado 80537. Table IB, Note 23.

(28) Nitrogen, Nitrite-Low Range, Diazotization Method for Water and Wastewater, Hach Method 8507, 1979. Method and price available from Hach Chemical Company, P.O. Box 389, Loveland, Colorado 80537. Table IB, Note 25.

(29) Zincon Method for Zinc, Method 8009. Hach Handbook for Water Analysis, 1979. Method and price available from Hach Chemical Company, P.O. Box 389, Loveland, Colorado 80537. Table IB, Note 33.

(30) "Direct Determination of Elemental Phosphorus by Gas-Liquid Chromatography," by R.F. Addison and R.G. Ackman, Journal of Chromatography, Volume 47, No. 3, pp. 421-426, 1970. Available in most public libraries. Back volumes of the Journal of Chromatography are

available from Elsevier/North-Holland, Inc., Journal Information Centre, 52 Vanderbilt Avenue, New York, NY 10164. Cost available from publisher. Table IB, Note 28.

(31) "Direct Current Plasma (DCP) Optical Emission Spectrometric Method for Trace Elemental Analysis of Water and Wastes", Method AES 0029, 1986-Revised 1991, Fison Instruments, Inc., 32 Commerce Center, Cherry Hill Drive, Danvers, MA 01923. Table B, Note 34.

(32) "Closed Vessel Microwave Digestion of Wastewater Samples for Determination of Metals, CEM Corporation, P.O. Box 200, Matthews, North Carolina 28106-0200, April 16, 1992. Available from the CEM Corporation. Table IB, Note 36.

Section 6

AIR PROGRAMS

6.l. NATIONAL PRIMARY AND SECONDARY AMBIENT AIR QUALITY STANDARDS
    (40CFR50) 2
    6.1.1. NATIONAL STANDARDS 2
    6.1.2. REFERENCE METHODS FOR CRITERIA POLLUTANTS 2
    TABLE 6.lA. REFERENCE METHODS--ALPHABETICAL ORDER BY
        CRITERIA POLLUTANTS 2
    TABLE 6.lB. REFERENCE METHODS--NUMBER ORDER BY METHOD 3

6.2. CONTINUOUS EMISSION MONITORS FOR S0₂ (40CFR52, APPENDIX D) 4

6.3. NATIONAL STANDARDS FOR NEW STATIONARY SOURCES 4
    TABLE 6.2A. NATIONAL STANDARDS--ALPHABETICAL ORDER BY
        INDUSTRY CATEGORY 4
    TABLE 6.2B. NATIONAL STANDARDS--40CFR60 ORDER BY SUBPART 8

6.4. REFERENCE METHODS FOR STATIONARY SOURCE STANDARDS 12
    TABLE 6.3A. REFERENCE METHODS--ALPHABETICAL ORDER BY
        METHOD PARAMETER 12
    TABLE 6.3B. REFERENCE METHODS--NUMBER ORDER BY METHOD 15

6.5 NATIONAL EMISSION STANDARDS FOR HAZARDOUS AIR POLLUTANTS (HAPS)
    (40CFR60) 18
    TABLE 6.4. HAZARDOUS AIR POLLUTANTS AND THEIR STANDARDS 18
    TABLE 6.5A TEST METHODS FOR HAPS--ALPHABETICAL ORDER BY POLLUTANT ... 25
    TABLE 6.5B. TEST METHODS FOR HAPS--NUMBER ORDER BY METHOD 26

6.6. NATIONAL EMISSION STANDARDS FOR HAPs FOR SOURCE CATEGORIES (40CFR63) 27
    TABLE 6.6. TEST METHODS FOR HAPs 27

6.7. TESTS METHODS FOR FUELS AND FUEL ADDITIVES (40CFR80, APPENDIXES A-G) 27

AIR PROGRAMS

6.1. NATIONAL PRIMARY AND SECONDARY AMBIENT AIR QUALITY STANDARDS (40CFR50)

6.1.1. NATIONAL STANDARDS

- National primary ambient air quality standards for sulfur oxides (sulfur dioxide) (40CFR50.4)
- National secondary ambient air quality standards for sulfur oxides (sulfur dioxide) (40CFR50.5)
- National primary and secondary ambient air quality standards for particulate matter (40CFR 50.6)
- National primary ambient air quality standards for carbon monoxide (40CFR50.8)
- National primary and secondary ambient air quality standards for ozone (40CFR50.9)
- National primary and secondary ambient air quality standards for nitrogen dioxide (40CFR50.11)
- National primary and secondary ambient air quality standards for lead (40CFR50.12)

6.1.2. REFERENCE METHODS FOR CRITERIA POLLUTANTS

<table>
<tr><td colspan="4" align="center">TABLE 6.1A. REFERENCE METHODS--ALPHABETICAL ORDER BY CRITERIA POLLUTANTS
(REFERENCE METHODS FOR NATIONAL PRIMARY AND SECONDARY AMBIENT AIR QUALITY STANDARDS)
(40CFR50 APPENDIXES A TO K)</td></tr>
<tr><th>Pollutant</th><th>Method</th><th>Source</th><th>Description</th></tr>
<tr><td>Carbon monoxide</td><td>Appendix-C</td><td>40CFR50</td><td>Measurement principle and calibration procedure for the measurement of carbon monoxide in the atmosphere (non-dispersive infrared photometry).</td></tr>
<tr><td>Hydrocarbons</td><td>Appendix-E</td><td>40CFR50</td><td>Reference method for the determination of hydrocarbons corrected methane.</td></tr>
<tr><td>Lead</td><td>Appendix-G</td><td>40CFR50</td><td>Reference method for the determination of lead in suspended particulate matter collected from ambient air.</td></tr>
<tr><td>Nitrogen dioxide</td><td>Appendix-F</td><td>40CFR50</td><td>Measurement principle and calibration procedure for the measurement of nitrogen dioxide in the atmosphere (gas phase chemiluminescence).</td></tr>
<tr><td>Ozone</td><td>Appendix-H</td><td>40CFR50</td><td>Interpretation of the National Ambient Air Quality Standards for ozone.</td></tr>
<tr><td>Ozone</td><td>Appendix-D</td><td>40CFR50</td><td>Measurement principle and calibration procedure for the measurement of ozone in the atmosphere.</td></tr>
<tr><td>Particulate matter</td><td>Method 202</td><td>40CFR51-AM[1,2]</td><td>Determination of condensible particulate emissions from stationary sources.</td></tr>
<tr><td>Particulate matter</td><td>Appendix-K</td><td>40CFR50</td><td>40CFR50 Appendix-K: interpretation of the National Ambient Air Quality Standards for particulate matter.</td></tr>
<tr><td>Particulate matter</td><td>Appendix-B</td><td>40CFR50</td><td>Reference method for the determination of suspended particulate matter in the atmosphere (high volume method).</td></tr>
<tr><td>PM10</td><td>Appendix-J</td><td>40CFR50</td><td>Reference method for the determination of particulate matter as PM10 in the atmosphere.</td></tr>
<tr><td>PM10</td><td>Method 201A</td><td>40CFR51-AM</td><td>Determination of PM10 Emissions (Constant Sampling Rate Procedure).</td></tr>
<tr><td>PM10</td><td>Method 201</td><td>40CFR51-AM</td><td>Determination of PM10 Emissions (Exhaust Gas Recycle Procedure).</td></tr>
<tr><td>Sulfur dioxide</td><td>Appendix-A</td><td>40CFR50</td><td>Reference method for the determination of sulfur dioxide in the atmosphere (pararosaniline method).</td></tr>
</table>

TABLE 6.1B. REFERENCE METHODS--NUMBER ORDER BY METHOD
(REFERENCE METHODS FOR NATIONAL PRIMARY AND SECONDARY AMBIENT AIR QUALITY STANDARDS)
(40CFR50 APPENDIXES A TO K)

Pollutant	Method	Source	Description
Sulfur dioxide	Appendix-A	40CFR50	Reference method for the determination of sulfur dioxide in the atmosphere (pararosaniline method).
Particulate matter	Appendix-B	40CFR50	Reference method for the determination of suspended particulate matter in the atmosphere (high volume method).
Carbon monoxide	Appendix-C	40CFR50	Measurement principle and calibration procedure for the measurement of carbon monoxide in the atmosphere (non-dispersive infrared photometry).
Ozone	Appendix-D	40CFR50	Measurement principle and calibration procedure for the measurement of ozone in the atmosphere.
Hydrocarbons	Appendix-E	40CFR50	Reference method for the determination of hydrocarbons corrected methane.
Nitrogen dioxide	Appendix-F	40CFR50	Measurement principle and calibration procedure for the measurement of nitrogen dioxide in the atmosphere (gas phase chemiluminescence).
Lead	Appendix-G	40CFR50	Reference method for the determination of lead in suspended particulate matter collected from ambient air.
Ozone	Appendix-H	40CFR50	Interpretation of the National Ambient Air Quality Standards for ozone.
PM10	Appendix-J	40CFR50	Reference method for the determination of particulate matter as PM10 in the atmosphere.
Particulate matter	Appendix-K	40CFR50	40CFR50 Appendix-K: interpretation of the National Ambient Air Quality Standards for particulate matter.
PM10	Method 201	40CFR51-AM[1,2]	Determination of PM10 Emissions (Exhaust Gas Recycle Procedure).
PM10	Method 201A	40CFR51-AM	Determination of PM10 Emissions (Constant Sampling Rate Procedure).
Particulate matter	Method 202	40CFR51-AM	Determination of condensible particulate emissions from stationary sources.

Notes for tables 6.1A and 6.1B:

[1] 40CFR51-AM: 40CFR51-Appendix M

[2] Presented herein are recommended test methods for measuring air pollutants emanating from an emission source. They are provided for States to use in their plans to meet the requirements of subpart K-Source Surveillance.

The State may also choose to adopt other methods to meet the requirements of subpart K of this part, subject to the normal plan review process.

The State may also meet the requirements of subpart K of this part by adopting, again subject to the normal plan review process, any of the relevant methods in appendix A to 40 CFR part 60.

6.2. CONTINUOUS EMISSION MONITORS FOR SO$_2$ (40CFR52 APPENDIX D)

6.3. NATIONAL STANDARDS FOR NEW STATIONARY SOURCES

TABLE 6.2A. NATIONAL STANDARDS--ALPHABETICAL ORDER BY INDUSTRY CATEGORY (40CFR60 Subparts Ca-VVV)		
Industry category	Source	Description
Aluminum Reduction Plants, Primary	Subpart S	Standards of Performance for Primary Aluminum Reduction Plants
Ammonium Sulfate	Subpart PP	Standards of Performance for Ammonium Sulfate Manufacture
Asphalt concrete plants	Subpart I	Standards of Performance for Asphalt concrete plants
Asphalt Processing	Subpart UU	Standards of Performance for Asphalt Processing and Asphalt Roofing Manufacture
Basic oxygen Process, Secondary	Subpart Na	Standards of Performance for Secondary Emissions From Basic oxygen Process Steelmaking Facilities for Which Construction Is Commenced After January 20, 1983
Basic Oxygen Furnaces, Primary	Subpart N	Standards of Performance for Primary Emissions from Basic Oxygen Process Furnaces for Which Construction is Commenced After June 11, 1973
Brass and Bronze Plants, Secondary	Subpart M	Standards of Performance for Secondary Brass and Bronze Production Plants
Calciners and Dryers	Subpart UUU	Standards of Performance for Calciners and Dryers in Mineral Industries
Coal Preparation Plants	Subpart Y	Standards of Performance for Coal Preparation Plants
Coating and Printing, Vinyl and Urethane	Subpart FFF	Standards of Performance for Flexible Vinyl and Urethane Coating and Printing
Coating, automobile Surface	Subpart MM	Standards of Performance for Automobile and Light Duty Truck Surface Coating Operations
Coating, Beverage Can Surface	Subpart WW	Standards of Performance for the Beverage Can Surface Coating Industry
Coating, Industrial Surface	Subpart SS	Standards of Performance for Industrial Surface Coating: Large Appliances
Coating, Industrial Surface	Subpart TTT	Standards of Performance for Industrial Surface Coating: Surface Coating of Plastic Parts for Business Machines
Coating, Magnetic Tape	Subpart SSS	Standards of Performance for Magnetic Tape Coating Facilities
Coating, Metal Coil Surface	Subpart TT	Standards of Performance for Metal Coil Surface Coating
Coating, Metal Surface	Subpart EE	Standards of Performance for Surface Coating of Metal Furniture
Coating, Polymeric	Subpart VVV	Standards of Performance for Polymeric Coating of Supporting Substrates Facilities
Coating, Tape and Label Surface	Subpart RR	Standards of Performance for Pressure Sensitive Tape and Label Surface Coating Operations
Copper Smelters, Primary	Subpart P	Standards of Performance for Primary Copper Smelters
Dry Cleaners	Subpart JJJ	Standards of Performance for Petroleum Dry Cleaners
Ferroalloy Production	Subpart Z	Standards of Performance for Ferroalloy Production Facilities

TABLE 6.2A. NATIONAL STANDARDS--ALPHABETICAL ORDER BY INDUSTRY CATEGORY
(40CFR60 Subparts Ca-VVV)

Industry category	Source	Description
Gas Turbines, Stationary	Subpart GG	Standards of Performance for Stationary Gas Turbines
Gasoline Terminals, Bulk	Subpart XX	Standards of Performance for Bulk Gasoline Terminals
Glass Manufacturing Plants	Subpart CC	Standards of Performance for Glass Manufacturing Plants
Grain Elevators	Subpart DD	Standards of Performance for Grain Elevators
Graphic Arts Industry	Subpart QQ	Standards of Performance for the Graphic Arts Industry: Publication Rotogravure Printing
Incinerators	Subpart E	Standards of Performance for Incinerators, see also municipal waste combustors
Kraft Pulp Mills	Subpart BB	Standards of Performance for Kraft Pulp Mills
Lead Smelters, Secondary	Subpart L	Standards of Performance for Secondary Lead Smelters
Lead Smelters, Primary	Subpart R	Standards of Performance for Primary Lead Smelters
Lead-Acid Battery Plants	Subpart KK	Standards of Performance for Lead-Acid Battery Manufacturing Plants
Lime Manufacturing Plants	Subpart HH	Standards of Performance for Lime Manufacturing Plants
Mineral plants, Metallic	Subpart LL	Standards of Performance for Metallic Mineral Processing plants
Mineral Processing Plants, Nonmetallic	Subpart OOO	Standards of Performance for Nonmetallic Mineral Processing Plants
Municipal Waste Combustors	Subpart Ca	Emissions Guidelines and Compliance Times for Municipal Waste Combustors, See also incinerators
Municipal Waste Combustors	Subpart Ea	Standards of Performance for Municipal Waste Combustors
Nitric Acid Plants	Subpart G	Standards of Performance for Nitric Acid Plants
Onshore Natural Gas	Subpart KKK	Standards of Performance for Equipment Leaks of VOC From Onshore Natural Gas Processing Plants.
Onshore Natural Gas	Subpart LLL	Standards of Performance for Onshore Natural Gas Processing: SO2 Emissions
Petroleum Refineries	Subpart GGG	Standards of Performance for Equipment Leaks of VOC in Petroleum Refineries
Petroleum Refineries	Subpart J	Standards of Performance for Petroleum Refineries
Petroleum Wastewater Systems	Subpart QQQ	Standards of Performance for VOC Emissions From Petroleum Refinery Wastewater Systems
Phosphate Fertilizer	Subpart T	Standards of Performance for the Phosphate Fertilizer Industry: Wet-Process Phosphoric Acid Plants
Phosphate Fertilizer	Subpart U	Standards of Performance for the Phosphate Fertilizer Industry: Superphosphoric Acid Plants
Phosphate Fertilizer	Subpart V	Standards of Performance for the Phosphate Fertilizer Industry: Diammonium Phosphate Plants

TABLE 6.2A. NATIONAL STANDARDS--ALPHABETICAL ORDER BY INDUSTRY CATEGORY (40CFR60 Subparts Ca-VVV)		
Industry category	Source	Description
Phosphate Fertilizer	Subpart W	Standards of Performance for the Phosphate Fertilizer Industry: Triple Superphosphate Plants
Phosphate Fertilizer	Subpart X	Standards of Performance for the Phosphate Fertilizer Industry: Granular Triple Superphosphate Storage Facilities
Phosphate Rock Plants	Subpart NN	Standards of Performance for Phosphate Rock Plants
Polymer Manufacturing	Subpart DDD	Standards of Performance for Volatile Organic Compound (VOC) Emissions from the Polymer Manufacturing Industry
Portland Cement Plants	Subpart F	Standards of Performance for Portland Cement Plants
Rubber Tire Manufacturing	Subpart BBB	Standards of Performance for the Rubber Tire Manufacturing Industry
Sewage Treatment Plants	Subpart O	Standards of Performance for Sewage Treatment Plants
Steam Generating Units	Subpart Da	Standards of Performance for Electric Utility Steam Generating Units
Steam Generating Units	Subpart Db	Standards of Performance for Industrial-Commercial-Institutional Steam Generating Units
Steam Generating Units	Subpart Dc	Standards of Performance for Small Industrial-Commercial-Institutional Steam Generating Units
Steel Plants: Electric Arc	Subpart AA	Standards of Performance for Steel Plants: Electric Arc Furnaces Constructed After October 21, 1974, and On or Before August 17, 1983
Steel Plants: Electric Arc	Subpart AAa	Standards of Performance for Steel Plants: Electric Arc Furnaces and Argon-Oxygen Decarburization Vessels Constructed After August 7, 1983
Storage Vessels for Petroleum Liquids	Subpart K	Standards of Performance for Storage Vessels for Petroleum Liquids for Which Construction, Reconstruction, or Modification Commenced After June 11, 1973, and Prior to May 19, 1978
Storage Vessels for Petroleum Liquids	Subpart Ka	Standards of Performance for Storage Vessels for Petroleum Liquids for Which Construction, Reconstruction, or Modification Commenced After May 18, 1978, and Prior to July 23, 1984
Storage Vessels, Volatile Organic Liquid	Subpart Kb	Standards of Performance for Volatile Organic Liquid Storage Vessels (Including Petroleum Liquid Storage Vessels) for Which Construction, Reconstruction, or Modification Commenced after July 23, 1984
Sulfuric Acid Plants	Subpart H	Standards of Performance for Sulfuric Acid Plants
Sulfuric Acid Units	Subpart Cb	Emission Guidelines and Compliance Times for Sulfuric Acid Production Units
Synthetic Fiber Production	Subpart HHH	Standards of Performance for Synthetic Fiber Production Facilities
Synthetic Organic Chemical	Subpart III	Standards of Performance for Volatile Organic Compound (VOC) Emissions From the Synthetic Organic Chemical Manufacturing Industry (SOCMI) Air Oxidation Unit Processes
Synthetic Organic Chemical	Subpart NNN	Standards of Performance for Volatile Organic Compound (VOC) Emissions From Synthetic Organic Chemical Manufacturing Industry (SOCMI) Distillation Operations

TABLE 6.2A. NATIONAL STANDARDS--ALPHABETICAL ORDER BY INDUSTRY CATEGORY
(40CFR60 Subparts Ca-VVV)

Industry category	Source	Description
Synthetic Organic Chemical	Subpart RRR	Standards of Performance for Volatile Organic Compound Emissions From Synthetic Organic Chemical Manufacturing Industry (SOCMI) Reactor Processes
Synthetic Organic Chemicals	Subpart VV	Standards of Performance for Equipment Leaks of VOC in the Synthetic Organic Chemicals Manufacturing Industry
Wood Heaters, Residential	Subpart AAA	Standards of Performance for New Residential Wood Heaters
Wool Fiberglass Insulation	Subpart PPP	Standard of Performance for Wool Fiberglass Insulation Manufacturing Plants
Zinc Smelters, Primary	Subpart Q	Standards of Performance for Primary Zinc Smelters
	Subpart CCC	[Reserved]
	Subpart EEE	[Reserved]
	Subpart FF	[Reserved]
	Subpart MMM	[Reserved]

Industry	Source	Description
TABLE 6.2B. NATIONAL STANDARDS--40CFR60 ORDER BY SUBPART (40CFR60 Subparts Ca-VVV)		
Municipal Waste Combustors	Subpart Ca	Emissions Guidelines and Compliance Times for Municipal Waste Combustors, See also incinerators
Sulfuric Acid Units	Subpart Cb	Emission Guidelines and Compliance Times for Sulfuric Acid Production Units
Steam Generating Units	Subpart Da	Standards of Performance for Electric Utility Steam Generating Units
Steam Generating Units	Subpart Db	Standards of Performance for Industrial-Commercial-Institutional Steam Generating Units
Steam Generating Units	Subpart Dc	Standards of Performance for Small Industrial-Commercial-Institutional Steam Generating Units
Incinerators	Subpart E	Standards of Performance for Incinerators, see also municipal waste combustors
Municipal Waste Combustors	Subpart Ea	Standards of Performance for Municipal Waste Combustors
Portland Cement Plants	Subpart F	Standards of Performance for Portland Cement Plants
Nitric Acid Plants	Subpart G	Standards of Performance for Nitric Acid Plants
Sulfuric Acid Plants	Subpart H	Standards of Performance for Sulfuric Acid Plants
Asphalt concrete plants	Subpart I	Standards of Performance for Asphalt concrete plants
Petroleum Refineries	Subpart J	Standards of Performance for Petroleum Refineries
Storage Vessels for Petroleum Liquids	Subpart K	Standards of Performance for Storage Vessels for Petroleum Liquids for Which Construction, Reconstruction, or Modification Commenced After June 11, 1973, and Prior to May 19, 1978
Storage Vessels for Petroleum Liquids	Subpart Ka	Standards of Performance for Storage Vessels for Petroleum Liquids for Which Construction, Reconstruction, or Modification Commenced After May 18, 1978, and Prior to July 23, 1984
Storage Vessels, Volatile Organic Liquid	Subpart Kb	Standards of Performance for Volatile Organic Liquid Storage Vessels (Including Petroleum Liquid Storage Vessels) for Which Construction, Reconstruction, or Modification Commenced after July 23, 1984
Lead Smelters, Secondary	Subpart L	Standards of Performance for Secondary Lead Smelters
Brass and Bronze Plants, Secondary	Subpart M	Standards of Performance for Secondary Brass and Bronze Production Plants
Basic Oxygen Furnaces, Primary	Subpart N	Standards of Performance for Primary Emissions from Basic Oxygen Process Furnaces for Which Construction is Commenced After June 11, 1973
Basic oxygen Process, Secondary	Subpart Na	Standards of Performance for Secondary Emissions From Basic oxygen Process Steelmaking Facilities for Which Construction Is Commenced After January 20, 1983
Sewage Treatment Plants	Subpart O	Standards of Performance for Sewage Treatment Plants
Copper Smelters, Primary	Subpart P	Standards of Performance for Primary Copper Smelters
Zinc Smelters, Primary	Subpart Q	Standards of Performance for Primary Zinc Smelters
Lead Smelters, Primary	Subpart R	Standards of Performance for Primary Lead Smelters

TABLE 6.2B. NATIONAL STANDARDS--40CFR60 ORDER BY SUBPART
(40CFR60 Subparts Ca-VVV)

Industry	Source	Description
Aluminum Reduction Plants, Primary	Subpart S	Standards of Performance for Primary Aluminum Reduction Plants
Phosphate Fertilizer	Subpart T	Standards of Performance for the Phosphate Fertilizer Industry: Wet-Process Phosphoric Acid Plants
Phosphate Fertilizer	Subpart U	Standards of Performance for the Phosphate Fertilizer Industry: Superphosphoric Acid Plants
Phosphate Fertilizer	Subpart V	Standards of Performance for the Phosphate Fertilizer Industry: Diammonium Phosphate Plants
Phosphate Fertilizer	Subpart W	Standards of Performance for the Phosphate Fertilizer Industry: Triple Superphosphate Plants
Phosphate Fertilizer	Subpart X	Standards of Performance for the Phosphate Fertilizer Industry: Granular Triple Superphosphate Storage Facilities
Coal Preparation Plants	Subpart Y	Standards of Performance for Coal Preparation Plants
Ferroalloy Production	Subpart Z	Standards of Performance for Ferroalloy Production Facilities
Steel Plants: Electric Arc	Subpart AA	Standards of Performance for Steel Plants: Electric Arc Furnaces Constructed After October 21, 1974, and On or Before August 17, 1983
Steel Plants: Electric Arc	Subpart AAa	Standards of Performance for Steel Plants: Electric Arc Furnaces and Argon-Oxygen Decarburization Vessels Constructed After August 7, 1983
Kraft Pulp Mills	Subpart BB	Standards of Performance for Kraft Pulp Mills
Glass Manufacturing Plants	Subpart CC	Standards of Performance for Glass Manufacturing Plants
Grain Elevators	Subpart DD	Standards of Performance for Grain Elevators
Coating, Metal Surface	Subpart EE	Standards of Performance for Surface Coating of Metal Furniture
	Subpart FF	[Reserved]
Gas Turbines, Stationary	Subpart GG	Standards of Performance for Stationary Gas Turbines
Lime Manufacturing Plants	Subpart HH	Standards of Performance for Lime Manufacturing Plants
Lead-Acid Battery Plants	Subpart KK	Standards of Performance for Lead-Acid Battery Manufacturing Plants
Mineral plants, Metallic	Subpart LL	Standards of Performance for Metallic Mineral Processing plants
Coating, automobile Surface	Subpart MM	Standards of Performance for Automobile and Light Duty Truck Surface Coating Operations
Phosphate Rock Plants	Subpart NN	Standards of Performance for Phosphate Rock Plants
Ammonium Sulfate	Subpart PP	Standards of Performance for Ammonium Sulfate Manufacture
Graphic Arts Industry	Subpart QQ	Standards of Performance for the Graphic Arts Industry: Publication Rotogravure Printing
Coating, Tape and Label Surface	Subpart RR	Standards of Performance for Pressure Sensitive Tape and Label Surface Coating Operations

		TABLE 6.2B. NATIONAL STANDARDS--40CFR60 ORDER BY SUBPART (40CFR60 Subparts Ca-VVV)	
Industry	Source	Description
Coating, Industrial Surface	Subpart SS	Standards of Performance for Industrial Surface Coating: Large Appliances
Coating, Metal Coil Surface	Subpart TT	Standards of Performance for Metal Coil Surface Coating
Asphalt Processing	Subpart UU	Standards of Performance for Asphalt Processing and Asphalt Roofing Manufacture
Synthetic Organic Chemicals	Subpart VV	Standards of Performance for Equipment Leaks of VOC in the Synthetic Organic Chemicals Manufacturing Industry
Coating, Beverage Can Surface	Subpart WW	Standards of Performance for the Beverage Can Surface Coating Industry
Gasoline Terminals, Bulk	Subpart XX	Standards of Performance for Bulk Gasoline Terminals
Wood Heaters, Residential	Subpart AAA	Standards of Performance for New Residential Wood Heaters
Rubber Tire Manufacturing	Subpart BBB	Standards of Performance for the Rubber Tire Manufacturing Industry
	Subpart CCC	[Reserved]
Polymer Manufacturing	Subpart DDD	Standards of Performance for Volatile Organic Compound (VOC) Emissions from the Polymer Manufacturing Industry
	Subpart EEE	[Reserved]
Coating and Printing, Vinyl and Urethane	Subpart FFF	Standards of Performance for Flexible Vinyl and Urethane Coating and Printing
Petroleum Refineries	Subpart GGG	Standards of Performance for Equipment Leaks of VOC in Petroleum Refineries
Synthetic Fiber Production	Subpart HHH	Standards of Performance for Synthetic Fiber Production Facilities
Synthetic Organic Chemical	Subpart III	Standards of Performance for Volatile Organic Compound (VOC) Emissions From the Synthetic Organic Chemical Manufacturing Industry (SOCMI) Air Oxidation Unit Processes
Dry Cleaners	Subpart JJJ	Standards of Performance for Petroleum Dry Cleaners
Onshore Natural Gas	Subpart KKK	Standards of Performance for Equipment Leaks of VOC From Onshore Natural Gas Processing Plants.
Onshore Natural Gas	Subpart LLL	Standards of Performance for Onshore Natural Gas Processing: SO2 Emissions
	Subpart MMM	[Reserved]
Synthetic Organic Chemical	Subpart NNN	Standards of Performance for Volatile Organic Compound (VOC) Emissions From Synthetic Organic Chemical Manufacturing Industry (SOCMI) Distillation Operations
Mineral Processing Plants, Nonmetallic	Subpart OOO	Standards of Performance for Nonmetallic Mineral Processing Plants
Wool Fiberglass Insulation	Subpart PPP	Standard of Performance for Wool Fiberglass Insulation Manufacturing Plants
Petroleum Wastewater Systems	Subpart QQQ	Standards of Performance for VOC Emissions From Petroleum Refinery Wastewater Systems

	TABLE 6.2B. NATIONAL STANDARDS--40CFR60 ORDER BY SUBPART (40CFR60 Subparts Ca-VVV)	
Industry	Source	Description
Synthetic Organic Chemical	Subpart RRR	Standards of Performance for Volatile Organic Compound Emissions From Synthetic Organic Chemical Manufacturing Industry (SOCMI) Reactor Processes
Coating, Magnetic Tape	Subpart SSS	Standards of Performance for Magnetic Tape Coating Facilities
Coating, Industrial Surface	Subpart TTT	Standards of Performance for Industrial Surface Coating: Surface Coating of Plastic Parts for Business Machines
Calciners and Dryers	Subpart UUU	Standards of Performance for Calciners and Dryers in Mineral Industries
Coating, Polymeric	Subpart VVV	Standards of Performance for Polymeric Coating of Supporting Substrates Facilities

6.4. REFERENCE METHODS FOR STATIONARY SOURCE STANDARDS

			TABLE 6.3A. REFERENCE METHODS--ALPHABETICAL ORDER BY METHOD PARAMETER
Parameter	Method	Source	Description
Air to fuel ratio	Method 28A	40CFR60-AA	Measurement of air to fuel ratio and minimum achievable burn rates for wood-fired appliances
Auditing	Method 28	40CFR60-AA	Certification and auditing of wood heaters
Carbon monoxide	Method 1OA	40CFR60-AA	Determination of carbon monoxide emissions in certifying continuous emission monitoring systems at petroleum refineries
Carbon monoxide	Method 10	40CFR60-AA	Determination of carbon monoxide emissions from stationary sources
Carbon monoxide	Method 1OB	40CFR60-AA	Determination of carbon monoxide emissions from stationary sources
CO_2, O_2	Method 3	40CFR60-AA	Gas analysis for carbon dioxide, oxygen, excess air, and dry molecular weight
CO_2, O_2	Method 3A	40CFR60-AA	Determination of Oxygen and Carbon Dioxide Concentrations in Emissions From Stationary Sources (Instrumental Analyzer Procedure)
Flouride	Method 13B	40CFR60-AA	Determination of total fluoride emissions from stationary sources -- Specific ion electrode method
Flouride	Method 14	40CFR60-AA	Determination of fluoride emissions from potroom roof monitors for primary aluminum plants
Flouride	Method 13A	40CFR60-AA	Determination of total fluoride emissions from stationary sources -- SPADNS zirconium lake method
Fugitive emission	Method 22	40CFR60-AA	Visual determination of fugitive emissions from material sources and smoke emissions from flares
Gasoline	Method 27	40CFR60-AA	Determination of vapor tightness of gasoline delivery tank using pressure-vacuum test
Hydrogen sulfide	Method 15	40CFR60-AA	Determination of hydrogen sulfide, carbonyl sulfide, and carbon disulfide emissions from stationary sources
Hydrogen sulfide	Method 11	40CFR60-AA	Determination of hydrogen sulfide content of fuel gas streams in petroleum refineries
Lead	Method 12	40CFR60-AA	Determination of inorganic lead emissions from stationary sources
Moisture	Method 4	40CFR60-AA	Determination of moisture content in stack gases Method 5 - Determination of particulate emissions from stationary sources
Nitrogen oxide	Method 7D	40CFR60-AA	Determination of nitrogen oxide emissions from stationary sources -- Alkaline-permanganate/ion chromatographic method
Nitrogen oxide	Method 7E	40CFR60-AA	Determination of nitrogen oxides emissions from stationary sources (instrumental analyzer procedure)
Nitrogen oxide	Method 7	40CFR60-AA	Determination of nitrogen oxide emissions from stationary sources
Nitrogen oxide	Method 7C	40CFR60-AA	Determination of nitrogen oxide emissions from stationary sources -- Alkaline-permanganate/colorimetric method
Nitrogen oxide	Method 7A	40CFR60-AA	Determination of nitrogen oxide emissions from stationary sources -- Ion chromatographic method

		TABLE 6.3A. REFERENCE METHODS--ALPHABETICAL ORDER BY METHOD PARAMETER	
Parameter	Method	Source	Description
Nitrogen oxide	Method 7B	40CFR60-AA	Determination of nitrogen oxide emissions from stationary sources (Ultraviolet spectrophotometry)
Nitrogen oxides	Method 20	40CFR60-AA	Determination of nitrogen oxides, sulfur dioxide, and diluent emissions from stationary gas turbines
Opacity	Alternate method 1	40CFR60-AA	Determination of the opacity of emissions from stationary sources remotely by lidar
Opacity	Method 9	40CFR60-AA	Visual determination of the opacity of emissions from stationary sources
Organic	Method 25B	40CFR60-AA	Determination of total gaseous organic concentration using a nondispersive infrared analyzer
Organic compound	Method 18	40CFR60-AA	Measurement of gaseous organic compound emissions by gas chromatography
Organic	Method 25	40CFR60-AA	Determination of total gaseous nonmethane organic emissions as carbon
Organic	Method 25A	40CFR60-AA	Determination of total gaseous organic concentration using a flame ionization analyzer
PM	Method 5F	40CFR60-AA	Determination of nonsulfate particulate matter from stationary sources
PM	Method 5H	40CFR60-AA	Determination of particulate emissions from wood heaters from a stack location
PM	Method 5G	40CFR60-AA	Determination of particulate emissions from wood heaters from a dilution tunnel sampling location
PM	Method 5E	40CFR60-AA	Determination of particulate emissions from the wool fiberglass insulation industry
PM	Method 5A	40CFR60-AA	Determination of particulate emissions from the asphalt processing and asphalt roofing industry
PM	Method 17	40CFR60-AA	Determination of particulate emissions from stationary sources (instack filtration method)
PM	Method 5B	40CFR60-AA	Determination of nonsulfuric acid particulate matter from stationary sources
PM	Method 5D	40CFR60-AA	Determination of particulate emissions from positive pressure fabric filters
Reduced sulfur	Method 16B	40CFR60-AA	Determination of total reduced sulfur emissions from stationary sources
Reduced sulfur	Method 16A	40CFR60-AA	Determination of total reduced sulfur emissions from stationary sources (impinger technique)
Reduced sulfur	Method 15A	40CFR60-AA	Determination of total reduced sulfur emissions from sulfur recovery plants in petroleum refineries
SO_2, CO_2	Method 6B	40CFR60-AA	Determination of sulfur dioxide and carbon dioxide daily average emissions from fossil fuel combustion sources
SO_2, H_2O, CO_2	Method 6A	40CFR60-AA	Determination of sulfur dioxide, moisture, and carbon dioxide emissions from fossil fuel combustion sources
Sulfur	Method 16	40CFR60-AA	Semicontinuous determination of sulfur emissions from stationary sources
Sulfur dioxide	Method 6	40CFR60-AA	Determination of sulfur dioxide emissions from stationary sources
Sulfur dioxide	Method 19	40CFR60-AA	Determination of sulfur dioxide removal efficiency and particulate, sulfur dioxide and nitrogen oxides emission rates

Parameter	Method	Source	Description
			TABLE 6.3A. REFERENCE METHODS--ALPHABETICAL ORDER BY METHOD PARAMETER
Sulfur dioxide	Appendix I	40CFR60.47	Determination of sulfur dioxide in the atmosphere (pararosaniline method).
Sulfur dioxide	Method 6C	40CFR60-AA	Determination of sulfur dioxide emissions from stationary sources (instrumental analyzer procedure)
Sulfuric acid	Method 8	40CFR60-AA	Determination of sulfuric acid mist and sulfur dioxide emissions from stationary sources
Velocity	Method 2C	40CFR60-AA	Determination of stack gas velocity and volumetric flow rate in small stacks or ducts (standard pitot tube)
Velocity	Method 2	40CFR60-AA	Determination of stack gas velocity and volumetric flow rate (Type S pitot tube)
Velocity	Method 1A	40CFR60-AA	Sample and velocity traverses for sources with small stacks or ducts
Velocity	Method 1	40CFR60-AA	Sample and velocity traverses for stationary sources
Volatile matter	Method 24A	40CFR60-AA	Determination of volatile matter content and density of printing inks and related coatings
Volatile matter	Method 24	40CFR60-AA	Determination of volatile matter content, water content, density, volume solids, and weight solids of surface coatings
Volatile organic	Method 21	40CFR60-AA	Determination of volatile organic compound leaks
Volumetric rate	Method 2D	40CFR60-AA	Measurement of gas volumetric flow rates in small pipes and ducts
Volumetric rate	Method 2A	40CFR60-AA	Direct measurement of gas volume through pipes and small ducts
Volumetric rate	Method 2B	40CFR60-AA	Determination of exhaust gas volume flow rate from gasoline vapor incinerators
	Method 5C	40CFR60-AA	Reserved

Parameter	Method	Source	Description
			TABLE 6.3B. REFERENCE METHODS--NUMBER ORDER BY METHOD
Velocity	Method 1	40CFR60-AA	Sample and velocity traverses for stationary sources
Velocity	Method 1A	40CFR60-AA	Sample and velocity traverses for sources with small stacks or ducts
Velocity	Method 2	40CFR60-AA	Determination of stack gas velocity and volumetric flow rate (Type S pitot tube)
Volume rate	Method 2A	40CFR60-AA	Direct measurement of gas volume through pipes and small ducts
Volume rate	Method 2B	40CFR60-AA	Determination of exhaust gas volume flow rate from gasoline vapor incinerators
Velocity	Method 2C	40CFR60-AA	Determination of stack gas velocity and volumetric flow rate in small stacks or ducts (standard pitot tube)
Volume	Method 2D	40CFR60-AA	Measurement of gas volumetric flow rates in small pipes and ducts
CO_2, O_2	Method 3	40CFR60-AA	Gas analysis for carbon dioxide, oxygen, excess air, and dry molecular weight
CO_2, O_2	Method 3A	40CFR60-AA	Determination of Oxygen and Carbon Dioxide Concentrations in Emissions From Stationary Sources (Instrumental Analyzer Procedure)
Moisture	Method 4	40CFR60-AA	Determination of moisture content in stack gases Method 5 - Determination of particulate emissions from stationary sources
PM	Method 5A	40CFR60-AA	Determination of particulate emissions from the asphalt processing and asphalt roofing industry
PM	Method 5B	40CFR60-AA	Determination of nonsulfuric acid particulate matter from stationary sources
	Method 5C	40CFR60-AA	Reserved
PM	Method 5D	40CFR60-AA	Determination of particulate emissions from positive pressure fabric filters
PM	Method 5E	40CFR60-AA	Determination of particulate emissions from the wool fiberglass insulation industry
PM	Method 5F	40CFR60-AA	Determination of nonsulfate particulate matter from stationary sources
PM	Method 5G	40CFR60-AA	Determination of particulate emissions from wood heaters from a dilution tunnel sampling location
PM	Method 5H	40CFR60-AA	Determination of particulate emissions from wood heaters from a stack location
Sulfur dioxide	Method 6	40CFR60-AA	Determination of sulfur dioxide emissions from stationary sources
SO_2, H_2O, CO_2	Method 6A	40CFR60-AA	Determination of sulfur dioxide, moisture, and carbon dioxide emissions from fossil fuel combustion sources
SO_2, CO_2	Method 6B	40CFR60-AA	Determination of sulfur dioxide and carbon dioxide daily average emissions from fossil fuel combustion sources
Sulfur dioxide	Method 6C	40CFR60-AA	Determination of sulfur dioxide emissions from stationary sources (instrumental analyzer procedure)
Nitrogen oxide	Method 7	40CFR60-AA	Determination of nitrogen oxide emissions from stationary sources
Nitrogen oxide	Method 7A	40CFR60-AA	Determination of nitrogen oxide emissions from stationary sources -- Ion chromatographic method
Nitrogen oxide	Method 7B	40CFR60-AA	Determination of nitrogen oxide emissions from stationary sources (Ultraviolet spectrophotometry)
Nitrogen oxide	Method 7C	40CFR60-AA	Determination of nitrogen oxide emissions from stationary sources -- Alkaline-permanganate/colorimetric method

Parameter	Method	Source	Description
			TABLE 6.3B. REFERENCE METHODS--NUMBER ORDER BY METHOD
Nitrogen oxide	Method 7D	40CFR60-AA	Determination of nitrogen oxide emissions from stationary sources -- Alkaline-permanganate/ion chromatographic method
Nitrogen oxide	Method 7E	40CFR60-AA	Determination of nitrogen oxides emissions from stationary sources (instrumental analyzer procedure)
Sulfuric acid	Method 8	40CFR60-AA	Determination of sulfuric acid mist and sulfur dioxide emissions from stationary sources
Opacity	Method 9	40CFR60-AA	Visual determination of the opacity of emissions from stationary sources
Opacity	Alternate method 1	40CFR60-AA	Determination of the opacity of emissions from stationary sources remotely by lidar
Carbon monoxide	Method 10	40CFR60-AA	Determination of carbon monoxide emissions from stationary sources
Carbon monoxide	Method 1OA	40CFR60-AA	Determination of carbon monoxide emissions in certifying continuous emission monitoring systems at petroleum refineries
Carbon monoxide	Method 1OB	40CFR60-AA	Determination of carbon monoxide emissions from stationary sources
Hydrogen sulfide	Method 11	40CFR60-AA	Determination of hydrogen sulfide content of fuel gas streams in petroleum refineries
Lead	Method 12	40CFR60-AA	Determination of inorganic lead emissions from stationary sources
Flouride	Method 13A	40CFR60-AA	Determination of total fluoride emissions from stationary sources -- SPADNS zirconium lake method
Flouride	Method 13B	40CFR60-AA	Determination of total fluoride emissions from stationary sources -- Specific ion electrode method
Flouride	Method 14	40CFR60-AA	Determination of fluoride emissions from potroom roof monitors for primary aluminum plants
Hydrogen sulfide	Method 15	40CFR60-AA	Determination of hydrogen sulfide, carbonyl sulfide, and carbon disulfide emissions from stationary sources
Reduced sulfur	Method 15A	40CFR60-AA	Determination of total reduced sulfur emissions from sulfur recovery plants in petroleum refineries
Sulfur	Method 16	40CFR60-AA	Semicontinuous determination of sulfur emissions from stationary sources
Reduced sulfur	Method 16A	40CFR60-AA	Determination of total reduced sulfur emissions from stationary sources (impinger technique)
Reduced sulfur	Method 16B	40CFR60-AA	Determination of total reduced sulfur emissions from stationary sources
PM	Method 17	40CFR60-AA	Determination of particulate emissions from stationary sources (instack filtration method)
Organic compound	Method 18	40CFR60-AA	Measurement of gaseous organic compound emissions by gas chromatography
Sulfur dioxide	Method 19	40CFR60-AA	Determination of sulfur dioxide removal efficiency and particulate, sulfur dioxide and nitrogen oxides emission rates
Nitrogen oxides	Method 20	40CFR60-AA	Determination of nitrogen oxides, sulfur dioxide, and diluent emissions from stationary gas turbines

		TABLE 6.3B. REFERENCE METHODS--NUMBER ORDER BY METHOD		
Parameter	Method	Source	Description	
Volatile organic	Method 21	40CFR60-AA	Determination of volatile organic compound leaks	
Fugitive emission	Method 22	40CFR60-AA	Visual determination of fugitive emissions from material sources and smoke emissions from flares	
Volatile matter	Method 24	40CFR60-AA	Determination of volatile matter content, water content, density, volume solids, and weight solids of surface coatings	
Volatile matter	Method 24A	40CFR60-AA	Determination of volatile matter content and density of printing inks and related coatings	
Organic	Method 25	40CFR60-AA	Determination of total gaseous nonmethane organic emissions as carbon	
Organic	Method 25A	40CFR60-AA	Determination of total gaseous organic concentration using a flame ionization analyzer	
Organic	Method 25B	40CFR60-AA	Determination of total gaseous organic concentration using a nondispersive infrared analyzer	
Gasoline	Method 27	40CFR60-AA	Determination of vapor tightness of gasoline delivery tank using pressure-vacuum test	
Auditing	Method 28	40CFR60-AA	Certification and auditing of wood heaters	
Air to fuel ratio	Method 28A	40CFR60-AA	Measurement of air to fuel ratio and minimum achievable burn rates for wood-fired appliances	
Sulfur dioxide	Appendix I	40CFR60.47	Determination of sulfur dioxide in the atmosphere (pararosaniline method).	

6.5 NATIONAL EMISSION STANDARDS FOR HAZARDOUS AIR POLLUTANTS (HAPs)(40 CFR 60)

TABLE 6.4. HAZARDOUS AIR POLLUTANTS AND THEIR STANDARDS				hap[1]
CAS RN Order		Alphabetic order		
50-000	Formaldehyde	75-070	Acetaldehyde	hap
51-285	Dinitrophenol(2,4-)	60-355	Acetamide	hap
51-796	Ethyl carbamate	75-058	Acetonitrile	hap
53-963	Acetylaminofluorene(2-)	98-862	Acetophenone	hap
56-235	Carbon tetrachloride	53-963	Acetylaminofluorene(2-)	hap
56-382	Parathion	107-028	Acrolein	hap
57-147	Dimethyl(1,1-) hydrazine	79-061	Acrylamide	hap
57-578	Propiolactone (beta-)	79-107	Acrylic acid	hap
57-749	Chlordane	107-131	Acrylonitrile	hap*
58-899	Lindane (all isomers)	107-051	Allyl chloride	hap
59-892	Nitrosomorpholine(n-)	92-671	Aminobiphenyl(4-)	hap
60-117	Dimethyl aminoazobenzene	62-533	Aniline	hap
60-344	Methyl hydrazine	90-040	Anisidine(o-)	hap
60-355	Acetamide		Antimony compounds	hap
62-533	Aniline		Arsenic compounds (inorganic including arsine)	hap*
62-737	Dichlorvos	1332-214	Asbestos	hap*
62-759	Nitrosodimethylamine(n-)	71-432	Benzene (including benzene from gasoline)	hap*
63-252	Carbaryl	92-875	Benzidine	hap
64-675	Diethyl sulfate	98-077	Benzotrichloride	hap
67-561	Methanol	100-447	Benzyl chloride	hap
67-663	Chloroform		Beryllium compounds	hap*
67-721	Hexachloroethane	92-524	Biphenyl	hap
68-122	Dimethyl formamide	117-817	Bis(2-ethylhexyl)phthalate (DEHP)	hap
71-432	Benzene (including benzene from gasoline)	542-881	Bis(chloromethyl)ether	hap
71-556	Methyl chloroform	75-252	Bromoform	hap
72-435	Methoxychlor	106-990	Butadiene(1,3-)	hap*
74-839	Methyl bromide		Cadmium compounds	hap*
74-873	Methyl chloride	156-627	Calcium cyanamide	hap
74-884	Methyl iodide	105-602	Caprolactam	hap
75-003	Ethyl chloride	133-062	Captan	hap
75-014	Vinyl chloride	63-252	Carbaryl	hap

TABLE 6.4. HAZARDOUS AIR POLLUTANTS AND THEIR STANDARDS				
CAS RN Order		Alphabetic order		hap[1]
75-058	Acetonitrile	75-150	Carbon disulfide	hap
75-070	Acetaldehyde	56-235	Carbon tetrachloride	hap*
75-092	Methylene chloride	463-581	Carbonyl sulfide	hap
75-150	Carbon disulfide	120-809	Catechol	hap
75-218	Ethylene oxide	133-904	Chloramben	hap
75-252	Bromoform	57-749	Chlordane	hap
75-343	Ethylidene dichloride		Chlorinated benzenes	hap**
75-354	Vinylidene chloride	7782-505	Chlorine	hap
75-445	Phosgene	79-118	Chloroacetic acid	hap
75-558	Propylenimine(1,2-)	532-274	Chloroacetophenone(2-)	hap
75-569	Propylene oxide	108-907	Chlorobenzene	hap
76-448	Heptachlor	510-156	Chlorobenzilate	hap
77-474	Hexachlorocyclopentadiene		Chlorofluorocarbon	hap**
77-781	Dimethyl sulfate	67-663	Chloroform	hap*
78-591	Isophorone	107-302	Chloromethyl methyl ether	hap
78-875	Propylene dichloride	126-998	Chloroprene	hap*
78-933	Methyl ethyl ketone		Chromium compounds	hap*
79-005	Trichloroethane(1,1,2-)		Cobalt compounds	hap
79-016	Trichloroethylene		Coke oven emissions	hap*
79-061	Acrylamide		Copper	hap**
79-107	Acrylic acid	108-394	Cresol(m-)	hap
79-118	Chloroacetic acid	95-487	Cresol(o-)	hap
79-345	Tetrachloroethane(1,1,2,2-)	106-445	Cresol(p-)	hap
79-447	Dimethyl carbamoyl chloride	1319-773	Cresols/cresylic acid (isomers and mixture)	hap
79-469	Nitropropane(2-)	98-828	Cumene	hap
80-626	Methyl methacrylate		Cyanide compounds	hap
82-688	Pentachloronitrobenzene	94-757	D(2,4-), salts and esters	hap
84-742	Dibutylphthalate	3547-044	DDE	hap
85-449	Phthalic anhydride	334-883	Diazomethane	hap
87-683	Hexachlorobutadiene	132-649	Dibenzofurans	hap
87-865	Pentachlorophenol	96-128	Dibromo(1,2-)-3-chloropropane	hap
88-062	Trichlorophenol(2,4,6-)	84-742	Dibutylphthalate	hap
90-040	Anisidine(o-)	106-467	Dichlorobenzene(1,4-)(p)	hap

TABLE 6.4. HAZARDOUS AIR POLLUTANTS AND THEIR STANDARDS				
CAS RN Order		Alphabetic order		hap[1]
91-203	Naphthalene	91-941	Dichlorobenzidene(3,3-)	hap
91-225	Quinoline	111-444	Dichloroethyl ether	hap
91-941	Dichlorobenzidene(3,3-)	542-756	Dichloropropene(1,3-)	hap
92-524	Biphenyl	62-737	Dichlorvos	hap
92-671	Aminobiphenyl(4-)	111-422	Diethanolamine	hap
92-875	Benzidine	64-675	Diethyl sulfate	hap
92-933	Nitrobiphenyl(4-)	121-697	Diethyl(n,n-) aniline	hap
94-757	D(2,4-), salts and esters	119-904	Dimethoxybenzidine(3,3-)	hap
95-476	Xylenes(o-)	60-117	Dimethyl aminoazobenzene	hap
95-487	Cresol(o-)	79-447	Dimethyl carbamoyl chloride	hap
95-534	Toluidine(o-)	68-122	Dimethyl formamide	hap
95-807	Toluene(2,4-) diamine	131-113	Dimethyl phthalate	hap
95-954	Trichlorophenol(2,4,5-)	77-781	Dimethyl sulfate	hap
96-093	Styrene oxide	57-147	Dimethyl(1,1-) hydrazine	hap
96-128	Dibromo(1,2-)-3-chloropropane	119-937	Dimethyl(3,3'-) benzidine	hap
96-457	Ethylene thiourea	534-521	Dinitro(4,6-)-o-cresol, and salts	hap
98-077	Benzotrichloride	51-285	Dinitrophenol(2,4-)	hap
98-828	Cumene	121-142	Dinitrotoluene(2,4-)	hap
98-862	Acetophenone	123-911	Dioxane(1,4-)	hap
98-953	Nitrobenzene	122-667	Diphenylhydrazine(1,2-)	hap
100-027	Nitrophenol(4-)	106-898	Epichlorohydrin	hap*
100-414	Ethyl benzene	106-887	Epoxybutane(1,2-)	hap
100-425	Styrene	140-885	Ethyl acrylate	hap
100-447	Benzyl chloride	100-414	Ethyl benzene	hap
101-144	Methylene(4,4-) bis(2-chloroaniline)	51-796	Ethyl carbamate	hap
101-688	Methylene diphenyl diisocyanate	75-003	Ethyl chloride	hap
101-779	Methylenedianiline(4,4'-)	106-934	Ethylene dibromide	hap
105-602	Caprolactam	107-062	Ethylene dichloride	hap*
106-423	Xylenes(p-)	107-211	Ethylene glycol	hap
106-445	Cresol(p-)	151-564	Ethylene imine	hap
106-467	Dichlorobenzene(1,4-)(p)	75-218	Ethylene oxide	hap*
106-503	Phenylenediamine(p-)	96-457	Ethylene thiourea	hap
106-514	Quinone	75-343	Ethylidene dichloride	hap

TABLE 6.4. HAZARDOUS AIR POLLUTANTS AND THEIR STANDARDS				
CAS RN Order		Alphabetic order		hap[1]
106-887	Epoxybutane(1,2-)		Fine mineral fibers	hap
106-898	Epichlorohydrin	50-000	Formaldehyde	hap
106-934	Ethylene dibromide		Glycol ethers	hap
106-990	Butadiene(1,3-)	76-448	Heptachlor	hap
107-028	Acrolein	118-741	Hexachlorobenzene	hap
107-051	Allyl chloride	87-683	Hexachlorobutadiene	hap
107-062	Ethylene dichloride	77-474	Hexachlorocyclopentadiene	hap*
107-131	Acrylonitrile	67-721	Hexachloroethane	hap
107-211	Ethylene glycol	822-060	Hexamethylene-1,6-diisocyanate	hap
107-302	Chloromethyl methyl ether	680-319	Hexamethylphosphoramide	hap
108-054	Vinyl acetate	110-543	Hexane	hap
108-101	Methyl isobutyl ketone	302-012	Hydrazine	hap
108-316	Maleic anhydride	7647-010	Hydrochloric acid	hap
108-383	Xylenes(m-)	7664-393	Hydrogen fluoride	hap
108-394	Cresol(m-)	7783-064	Hydrogen sulfide	hap
108-883	Toluene	123-319	Hydroquinone	hap
108-907	Chlorobenzene	78-591	Isophorone	hap
108-952	Phenol		Lead compounds	hap
110-543	Hexane	58-899	Lindane (all isomers)	hap
111-422	Diethanolamine	108-316	Maleic anhydride	hap
111-444	Dichloroethyl ether		Manganese compounds	hap*
114-261	Propoxur		Mercury compounds	hap
117-817	Bis(2-ethylhexyl)phthalate (DEHP)	67-561	Methanol	hap
118-741	Hexachlorobenzene	72-435	Methoxychlor	hap
119-904	Dimethoxybenzidine(3,3-)	74-839	Methyl bromide	hap
119-937	Dimethyl(3,3'-) benzidine	74-873	Methyl chloride	hap
120-809	Catechol	71-556	Methyl chloroform	hap*
120-821	Trichlorobenzene(1,2,4-)	78-933	Methyl ethyl ketone	hap
121-142	Dinitrotoluene(2,4-)	60-344	Methyl hydrazine	hap
121-448	Triethylamine	74-884	Methyl iodide	hap
121-697	Diethyl(n,n-) aniline	108-101	Methyl isobutyl ketone	hap
122-667	Diphenylhydrazine(1,2-)	624-839	Methyl isocyanate	hap
123-319	Hydroquinone	80-626	Methyl methacrylate	hap

TABLE 6.4. HAZARDOUS AIR POLLUTANTS AND THEIR STANDARDS				
CAS RN Order		Alphabetic order		hap[1]
123-386	Propionaldehyde	1634-044	Methyl tert butyl ether	hap
123-911	Dioxane(1,4-)	75-092	Methylene chloride	hap*
126-998	Chloroprene	101-688	Methylene diphenyl diisocyanate	hap
127-184	Tetrachloroethylene	101-144	Methylene(4,4-) bis(2-chloroaniline)	hap
131-113	Dimethyl phthalate	101-779	Methylenedianiline(4,4'-)	hap
132-649	Dibenzofurans	91-203	Naphthalene	hap
133-062	Captan		Nickel compounds	hap*
133-904	Chloramben	98-953	Nitrobenzene	hap
140-885	Ethyl acrylate	92-933	Nitrobiphenyl(4-)	hap
151-564	Ethylene imine	100-027	Nitrophenol(4-)	hap
156-627	Calcium cyanamide	79-469	Nitropropane(2-)	hap
302-012	Hydrazine	684-935	Nitroso(n-)-n-methylurea	hap
334-883	Diazomethane	62-759	Nitrosodimethylamine(n-)	hap
463-581	Carbonyl sulfide	59-892	Nitrosomorpholine(n-)	hap
510-156	Chlorobenzilate	56-382	Parathion	hap
532-274	Chloroacetophenone(2-)	82-688	Pentachloronitrobenzene	hap
534-521	Dinitro(4,6-)-o-cresol, and salts	87-865	Pentachlorophenol	hap
540-841	Trimethylpentane(2,2,4-)		Perchloroethylene	hap**
542-756	Dichloropropene(1,3-)	108-952	Phenol	hap*
542-881	Bis(chloromethyl)ether	106-503	Phenylenediamine(p-)	hap
584-849	Toluene(2,4-) diisocyanate	75-445	Phosgene	hap
593-602	Vinyl bromide	7803-512	Phosphine	hap
624-839	Methyl isocyanate	7723-140	Phosphorus	hap
680-319	Hexamethylphosphoramide	85-449	Phthalic anhydride	hap
684-935	Nitroso(n-)-n-methylurea	1336-363	Polychlorinated biphenyls	hap
822-060	Hexamethylene-1,6-diisocyanate		Polycylic organic matter	hap*
1120-714	Propane sultone(1,3-)	1120-714	Propane sultone(1,3-)	hap
1319-773	Cresols/cresylic acid (isomers and mixture)	57-578	Propiolactone (beta-)	hap
1330-207	Xylenes (isomers and mixture)	123-386	Propionaldehyde	hap
1332-214	Asbestos	114-261	Propoxur	hap
1336-363	Polychlorinated biphenyls	78-875	Propylene dichloride	hap
1582-098	Trifluralin	75-569	Propylene oxide	hap
1634-044	Methyl tert butyl ether	75-558	Propylenimine(1,2-)	hap

TABLE 6.4. HAZARDOUS AIR POLLUTANTS AND THEIR STANDARDS

CAS RN Order		Alphabetic order		hap[1]
1746-016	Tetrachlorodibenzo(2,3,7,8-)-p-dioxin	91-225	Quinoline	hap
3547-044	DDE	106-514	Quinone	hap
7550-450	Titanium tetrachloride		Radionuclides (including radon)	hap
7647-010	Hydrochloric acid		Selenium compounds	hap
7664-393	Hydrogen fluoride	100-425	Styrene	hap
7723-140	Phosphorus	96-093	Styrene oxide	hap
7782-505	Chlorine	1746-016	Tetrachlorodibenzo(2,3,7,8-)-p-dioxin	hap
7783-064	Hydrogen sulfide	79-345	Tetrachloroethane(1,1,2,2-)	hap
7803-512	Phosphine	127-184	Tetrachloroethylene	hap
8001-352	Toxaphene (chlorinated camphene)	7550-450	Titanium tetrachloride	hap
	Antimony compounds	108-883	Toluene	hap*
	Arsenic compounds (inorganic including arsine)	95-807	Toluene(2,4-) diamine	hap
	Beryllium compounds	584-849	Toluene(2,4-) diisocyanate	hap
	Cadmium compounds	95-534	Toluidine(o-)	hap
	Chlorinated benzenes	8001-352	Toxaphene (chlorinated camphene)	hap
	Chlorofluorocarbon	120-821	Trichlorobenzene(1,2,4-)	hap
	Chromium compounds	79-005	Trichloroethane(1,1,2-)	hap
	Cobalt compounds	79-016	Trichloroethylene	hap*
	Coke oven emissions	95-954	Trichlorophenol(2,4,5-)	hap
	Copper	88-062	Trichlorophenol(2,4,6-)	hap
	Cyanide compounds	121-448	Triethylamine	hap
	Fine mineral fibers	1582-098	Trifluralin	hap
	Glycol ethers	540-841	Trimethylpentane(2,2,4-)	hap
	Lead compounds	108-054	Vinyl acetate	hap
	Manganese compounds	593-602	Vinyl bromide	hap
	Mercury compounds	75-014	Vinyl chloride	hap
	Nickel compounds	75-354	Vinylidene chloride	hap*
	Perchloroethylene	1330-207	Xylenes (isomers and mixture)	hap
	Polycylic organic matter	108-383	Xylenes(m-)	hap
	Radionuclides (including radon)	95-476	Xylenes(o-)	hap
	Selenium compounds	106-423	Xylenes(p-)	hap
	Zinc and zinc oxide		Zinc and zinc oxide	hap**

TABLE 6.4. HAZARDOUS AIR POLLUTANTS AND THEIR STANDARDS		
CAS RN Order	Alphabetic order	hap[1]

[1] hap: Hazardous Air Pollutants listed in Section 112 of the Clean Air Act
  hap*: hap compounds that EPA has developed standards for implementation and has listed them in 40CFR61.01.
  hap**: hap compounds that are not listed in CAA Section 112, but EPA is regulating these compounds under the hazardous air pollutant regulations

Pollutant	Method	Description
		TABLE 6.5A. TEST METHODS FOR HAPs--ALPHABETICAL ORDER BY POLLUTANT (40CFRFR61 APPENDIX B)
Arsenic	Method 108A	Determination of arsenic content in ore samples from nonferrous smelters
Arsenic	Method 108	Determination of particulate and gaseous arsenic emissions
Arsenic	Method 108B	Determination of arsenic content in ore samples from nonferrous smelters
Arsenic	Method 108C	Determination of arsenic content in ore samples from nonferrous smelters
Beryllium	Method 104	Determination of beryllium emissions from stationary sources
Beryllium	Method 103	Beryllium screening method
Mercury	Method 105	Determination of mercury in wastewater treatment plant sewage sludges
Mercury	Method 101A	Determination of particulate and gaseous mercury emissions from sewage sludge incinerators
Mercury	Method 101	Determination of particulate and gaseous mercury emissions from chlor-alkali plants--air streams
Mercury	Method 102	Determination of particulate and gaseous mercury emissions from chlor-alkali plants--hydrogen streams
Polonium-210	Method 111	Determination of Polonium-210 emissions from stationary sources
Radionuclide	Method 114	Test methods for measuring radionuclide emissions from stationary Sources
Radon-222	Method 115	Monitoring for radon-222 emissions
Vinyl chloride	Method 107	Determination of vinyl chloride content of inprocess wastewater samples, and vinyl chloride content of polyvinyl chloride resin, slurry, wet cake, and latex samples
Vinyl chloride	Method 107A	Determination of vinyl chloride content of solvents, resin-solvent solution, polyvinyl chloride resin, resin slurry, wet resin, and latex samples
Vinyl chloride	Method 106	Determination of vinyl chloride from stationary sources

SAMPLING, ANALYSIS, AND MONITORING METHODS: A GUIDE TO EPA REQUIREMENTS

Pollutant	Method	Description
		TABLE 6.5B. TEST METHODS FOR HAPs--NUMBER ORDER BY METHOD (40CFRFR61 APPENDIX B)
Mercury	Method 101	Determination of particulate and gaseous mercury emissions from chlor-alkali plants--air streams
Mercury	Method 101A	Determination of particulate and gaseous mercury emissions from sewage sludge incinerators
Mercury	Method 102	Determination of particulate and gaseous mercury emissions from chlor-alkali plants--hydrogen streams
Beryllium	Method 103	Beryllium screening method
Beryllium	Method 104	Determination of beryllium emissions from stationary sources
Mercury	Method 105	Determination of mercury in wastewater treatment plant sewage sludges
Vinyl chloride	Method 106	Determination of vinyl chloride from stationary sources
Vinyl chloride	Method 107	Determination of vinyl chloride content of inprocess wastewater samples, and vinyl chloride content of polyvinyl chloride resin, slurry, wet cake, and latex samples
Vinyl chloride	Method 107A	Determination of vinyl chloride content of solvents, resin-solvent solution, polyvinyl chloride resin, resin slurry, wet resin, and latex samples
Arsenic	Method 108	Determination of particulate and gaseous arsenic emissions
Arsenic	Method 108A	Determination of arsenic content in ore samples from nonferrous smelters
Arsenic	Method 108B	Determination of arsenic content in ore samples from nonferrous smelters
Arsenic	Method 108C	Determination of arsenic content in ore samples from nonferrous smelters
Polonium-210	Method 111	Determination of Polonium-210 emissions from stationary sources
Radionuclide	Method 114	Test methods for measuring radionuclide emissions from stationary Sources
Radon-222	Method 115	Monitoring for radon-222 emissions

6.6. NATIONAL EMISSION STANDARDS FOR HAPs FOR SOURCE CATEGORIES (40CFR63)

			TABLE 6.6. TEST METHODS FOR HAPs (40CFRFR63 APPENDIX A)
Pollutant	Method	Source	Description
Field Validation	Method 301	40CFR63-AA	Field Validation of Pollutant Measurement Methods from Various Waste Media
Visible emissions	Method 303	40CFR63-AA	Determination of visible emissions from by-product coke oven batteries
Visible emissions	Method 303A	40CFR63-AA	Determination of visible emissions from nonrecovery coke oven batteries
Organic Compounds	Method 304A	40CFR63-AA	Determination of Biodegradation Rates of Organic Compounds (Vent Option)
Organic Compounds	Method 304B	40CFR63-AA	Determination of Biodegradation Rates of Organic Compounds (Scrubber Option)
Volatile Organic Compounds	Method 305	40CFR63-AA	Measurement of Emission Potential of Individual Volatile Organic Compounds in Waste

6.7. TESTS METHODS FOR FUELS AND FUEL ADDITIVES (40CFR80 APPENDIXES A-G)

- Appendix A to Part 80-Test for the Determination of Phosphorus in Gasoline

- Appendix B to Part 80-Test Methods for Lead in Gasoline

- Appendix C to Part 80-[Reserved]

- Appendix D to Part 80-Sampling Procedures for Fuel Volatility

- Appendix E to Part 80-Tests for Determining Reid Vapor Pressure (RVP) of Gasoline and Gasoline-Oxygenate Blends

- Appendix F to Part 80-Test for Determining the Quantity of Alcohol in Gasoline

- Appendix G to Part 80-Sampling Procedures for Diesel Fuel

RISK ASSESSMENT PROGRAMS

7.1. INTRODUCTION 2

7.2. INTEGRATED RISK INFORMATION SYSTEM (IRIS) 2

    TABLE 7.1. INTEGRATED RISK INFORMATION SYSTEM COMPOUNDS 2

    TABLE 7.2. REFERENCE AIR CONCENTRATIONS (RAC) (40CFR266, APPENDIX IV) 23

    TABLE 7.3. RISK SPECIFIC DOSES (RsD) (10-5) (40CFR266, APPENDIX V) 26

    TABLE 7.4. PRODUCTS OF INCOMPLETE COMBUSTION (PICs) FOUND IN

    STACK EFFLUENTS 29

7.1. INTRODUCTION

Risk analysis has recently been increasingly receiving attention in making environmental decisions. For example, on its May 18, 1993 Combustion Strategy announcement, EPA requires that any issuance of a new hazardous waste combustion permit be preceded by a complete risk assessment (both direct and indirect risk assessment). This section provides chemical lists that would allow professionals to obtain more information needed to perform risk assessment. Four chemical Tables are provided in this section as follows.

- TABLE 7.1. INTEGRATED RISK INFORMATION SYSTEM COMPOUNDS
- TABLE 7.2. REFERENCE AIR CONCENTRATIONS (RAC)[1] (40CFR266 APPENDIX IV)
- TABLE 7.3. RISK SPECIFIC DOSES (RsD) (10^{-5}) (40CFR266 APPENDIX V)
- TABLE 7.4. PRODUCTS OF INCOMPLETE COMBUSTION (PICs) FOUND IN STACK EFFLUENTS (40CFR266 APPENDIX VIII)

7.2. INTEGRATED RISK INFORMATION SYSTEM (IRIS)

As part of an EPA-wide effort to improve the quality of science used to evaluate and manage risks, EPA has initiated and developed the Integrated Risk Information System (IRIS) since 1986. IRIS currently contains summaries of EPA human health hazard information that support two of the four steps, hazard identification and dose-response evaluation, of the risk assessment process. It is a data base that accommodates EPA consensus scientific positions on potential adverse human health effects that may result from exposure to environmental pollutants (EPA-93/1). Chemicals contained in IRIS are provided in Table 7.1. IRIS information can be available from the EPA National Risk Management Research Laboratory in Cincinnati, Ohio.

TABLE 7.1. INTEGRATED RISK INFORMATION SYSTEM COMPOUNDS			
CAS RN Order		IRIS Compounds--Alphabetical Order	
50-000	Formaldehyde	83-329	Acenaphthene
50-293	p,p'-Dichlorodiphenyltrichloroethane	208-968	Acenaphthylene
50-328	Benzo(a)pyrene	30560-191	Acephate
51-285	Dinitrophenol(2,4-)	75-070	Acetaldehyde
51-796	Ethyl carbamate	60-355	Acetamide
52-857	Famphur	34256-821	Acetochlor
53-703	Dibenz(ah)anthracene	67-641	Acetone
55-185	Nitrosodiethylamine(n-)	75-058	Acetonitrile
56-235	Carbon tetrachloride	98-862	Acetophenone
56-359	Tributyltin oxide	75-365	Acetyl chloride
56-382	Parathion	62476-599	Acifluorfen, sodium
56-553	Benz(a)anthracene	107-028	Acrolein
56-724	Coumaphos	79-061	Acrylamide
57-125	Cyanide, free	79-107	Acrylic acid
57-147	Dimethylhydrazine(1,1-)	107-131	Acrylonitrile
57-249	Strychnine	111-693	Adiponitrile
57-556	Propylene glycol	15972-608	Alachlor
57-578	Propiolactone(beta-)	1596-845	Alar
57-749	Chlordane	116-063	Aldicarb
58-899	gamma-Hexachlorocyclohexane	1646-884	Aldicarb sulfone

TABLE 7.1. INTEGRATED RISK INFORMATION SYSTEM COMPOUNDS			
CAS RN Order		IRIS Compounds--Alphabetical Order	
58-902	Tetrachlorophenol(2,3,4,6-)	309-002	Aldrin
59-507	Chloro-m-cresol(p-)	74223-646	Ally
60-117	Dimethyl aminoazobenzene	107-186	Allyl alcohol
60-297	Ethyl ether	107-051	Allyl chloride
60-344	Monomethylhydrazine		alpha emitters
60-355	Acetamide	7429-905	Aluminum
60-515	Dimethoate	20859-738	Aluminum phosphide
60-571	Dieldrin	67485-294	Amdro
62-384	Phenylmercuric acetate	834-128	Ametryn
62-533	Aniline	504-245	Aminopyridine(4-)
62-737	Dichlorvos	33089-611	Amitraz
62-748	Sodium fluoroacetate	7664-417	Ammonia
62-759	Nitrosodimethylamine(n-)	631-618	Ammonium acetate
63-252	Carbaryl	16325-476	Ammonium methacrylate
64-186	Formic acid	7773-060	Ammonium sulfamate
64-675	Diethyl sulfate	62-533	Aniline
65-850	Benzoic acid	90-040	Anisidine(ortho-)
67-561	Methanol	120-127	Anthracene
67-641	Acetone	7440-360	Antimony
67-663	Chloroform	1309-644	Antimony trioxide
67-721	Hexachloroethane	74115-245	Apollo
68-122	N,N-Dimethylformamide	140-578	Aramite
70-304	Hexachlorophene	12674-112	Aroclor 1016
70-382	Dimethrin	12672-296	Aroclor 1248
71-363	Butanol(n-)	11097-691	Aroclor 1254
71-432	Benzene	7440-382	Arsenic, inorganic
71-556	Trichloroethane(1,1,1-)	7784-421	Arsine
72-208	Endrin	1332-214	Asbestos
72-435	Methoxychlor	81405-858	Assert
72-548	p,p'-Dichlorodiphenyl dichloroethane	76578-148	Assure
72-559	p,p'-Dichlorodiphenyldichloroethylene	3337-711	Asulam
74-839	Bromomethane	1912-249	Atrazine
74-873	Chloromethane	65195-553	Avermectin B1

TABLE 7.1. INTEGRATED RISK INFORMATION SYSTEM COMPOUNDS			
CAS RN Order		IRIS Compounds--Alphabetical Order	
74-884	Methyl iodide	103-333	Azobenzene
74-908	Hydrogen cyanide	7440-393	Barium
74-931	Methanethiol	542-621	Barium cyanide
74-953	Dibromomethane	55179-312	Baycor
74-964	Bromoethane	114-261	Baygon
74-975	Bromochloromethane	43121-433	Bayleton
75-003	Ethyl chloride	55219-653	Baytan
75-014	Vinyl chloride	68359-375	Baythroid
75-058	Acetonitrile	1861-401	Benefin
75-070	Acetaldehyde	17804-352	Benomyl
75-092	Dichloromethane	25057-890	Bentazon
75-150	Carbon disulfide	100-527	Benzaldehyde
75-252	Bromoform	71-432	Benzene
75-274	Bromodichloromethane	108-985	Benzenethiol
75-343	Dichloroethane(1,1-)	92-875	Benzidine
75-354	Dichloroethylene(1,1-)	65-850	Benzoic acid
75-365	Acetyl chloride	98-077	Benzotrichloride
75-376	Difluoroethane(1,1-)	50-328	Benzo(a)pyrene
75-445	Phosgene	205-992	Benzo(b)fluoranthene
75-456	Chlorodifluoromethane	192-972	Benzo(e)pyrene
75-467	Trifluoromethane	191-242	Benzo(ghi)perylene
75-558	Propyleneimine	205-823	Benzo(j)fluoranthene
75-569	Propylene oxide	207-089	Benzo(k)fluoranthene
75-605	Cacodylic acid	100-447	Benzyl chloride
75-627	Bromotrichloromethane	56-553	Benz(a)anthracene
75-683	Chloro-1,1-difluoroethane(1-)	7440-417	Beryllium
75-694	Trichlorofluoromethane	13510-491	Beryllium sulfate
75-718	Dichlorodifluoromethane	141-662	Bidrin
75-865	Methyllactonitrile(2-)	82657-043	Biphenthrin
75-876	Chloral	92-524	Biphenyl(1,1-)
75-990	Dalapon, sodium salt	111-911	Bis(2-chloroethoxy)methane
76-017	Pentachloroethane	39638-329	Bis(2-chloroisopropyl) ether
76-039	Trichloroacetic acid	111-444	Bis(chloroethyl)ether

TABLE 7.1. INTEGRATED RISK INFORMATION SYSTEM COMPOUNDS

CAS RN Order		IRIS Compounds--Alphabetical Order	
76-131	Trichloro-1,2,2-trifluoroethane(1,1,2-)	542-881	Bis(chloromethyl)ether
76-448	Heptachlor	80-057	Bisphenol A.
76-879	Triphenyltin hydroxide	35400-432	Bolstar
77-474	Hexachlorocyclopentadiene	7440-428	Boron (Boron and Borates only)
77-781	Dimethyl sulfate		Brominated dibenzofurans
78-002	Tetraethyl lead	74-975	Bromochloromethane
78-488	Merphos oxide	75-274	Bromodichloromethane
78-591	Isophorone	1511-622	Bromodifluoromethane
78-831	Isobutyl alcohol	101-553	Bromodiphenyl ether(p-)
78-864	Chlorobutane(2-)	74-964	Bromoethane
78-875	Dichloropropane(1,2-)	75-252	Bromoform
78-933	Methyl ethyl ketone	74-839	Bromomethane
79-005	Trichloroethane(1,1,2-)	75-627	Bromotrichloromethane
79-016	Trichloroethylene	1689-845	Bromoxynil
79-061	Acrylamide	1689-992	Bromoxynil octanoate
79-094	Propionic acid	357-573	Brucine
79-107	Acrylic acid	106-990	Butadiene(1,3-)
79-196	Thiosemicarbazide	71-363	Butanol(n-)
79-209	Methyl acetate	85-687	Butyl benzyl phthalate
79-221	Methyl chlorocarbonate	2008-415	Butylate
79-345	Tetrachloroethane(1,1,2,2-)	507-200	Butylchloride(t-)
79-436	Dichloroacetic acid	85-701	Butylphthalyl butylglycolate
79-447	Dimethylcarbamoyl chloride	75-605	Cacodylic acid
79-469	Nitropropane(2-)	7440-439	Cadmium
80-057	Bisphenol A.	104098-499	Cadre
80-568	Pinene(alpha-)	592-018	Calcium cyanide
80-626	Methyl methacrylate	105-602	Caprolactam
81-812	Warfarin	2425-061	Captafol
82-688	Pentachloronitrobenzene	133-062	Captan
83-329	Acenaphthene	63-252	Carbaryl
83-794	Rotenone	1563-662	Carbofuran
84-662	Diethyl phthalate	75-150	Carbon disulfide
84-720	Ethylphthalyl ethylglycolate	56-235	Carbon tetrachloride

TABLE 7.1. INTEGRATED RISK INFORMATION SYSTEM COMPOUNDS			
CAS RN Order		**IRIS Compounds--Alphabetical Order**	
84-742	Dibutyl phthalate	353-504	Carbonyl fluoride
85-007	Diquat	463-581	Carbonyl sulfide
85-018	Phenanthrene	55285-148	Carbosulfan
85-449	Phthalic anhydride	5234-684	Carboxin
85-687	Butyl benzyl phthalate	120-809	Catechol
85-701	Butylphthalyl butylglycolate	75-876	Chloral
86-306	Nitrosodiphenylamine(n-)	302-170	Chloral hydrate
86-500	Guthion	133-904	Chloramben
86-737	Fluorene	57-749	Chlordane
86-884	Naphthylthiourea(alpha-)	90982-324	Chlorimuron-ethyl
87-683	Hexachlorobutadiene	7782-505	Chlorine
87-821	Hexabromobenzene	506-774	Chlorine cyanide
87-865	Pentachlorophenol	10049-044	Chlorine dioxide
88-062	Trichlorophenol(2,4,6-)	7758-192	Chlorite
88-744	Nitroaniline(2-)	75-683	Chloro-1,1-difluoroethane(1-)
88-857	Dinoseb	2837-890	Chloro-1,1,1,2-tetrafluoroethane(2-)
90-040	Anisidine(ortho-)	59-507	Chloro-m-cresol(p-)
91-203	Naphthalene	532-274	Chloroacetophenone(2-)
91-225	Quinoline	106-478	Chloroaniline(p-)
91-587	Chloronaphthalene(beta-)	108-907	Chlorobenzene
91-598	Naphthylamine(2-)	510-156	Chlorobenzilate
91-941	Dichlorobenzidine(3,3'-)	109-693	Chlorobutane(1-)
92-524	Biphenyl(1,1-)	78-864	Chlorobutane(2-)
92-875	Benzidine	41851-507	Chlorocyclopentadiene
92-933	Nitrobiphenyl(4-)	75-456	Chlorodifluoromethane
93-652	Methyl-4-chlorophenoxy(2-(2-))propionic acid	67-663	Chloroform
93-721	Trichlorophenoxy(2-(2,4,5-)) propionic acid	74-873	Chloromethane
93-765	Trichlorophenoxyacetic acid(2,4,5-)	107-302	Chloromethyl methyl ether
94-746	Methyl-4-chlorophenoxyacetic acid(2-)	91-587	Chloronaphthalene(beta-)
94-757	Dichlorophenoxyacetic acid(2,4-)	95-578	Chlorophenol(2-)
94-815	Methyl-4-chlorophenoxy(4-(2-)) butyric acid	122-883	Chlorophenoxyacetic acid(4-)
94-826	Dichlorophenoxy(4-(2,4-))butyric acid	123-091	Chlorophenyl methyl sulfide(p-)
95-476	Xylene(o-)	98-571	Chlorophenyl methyl sulfone(p-)

TABLE 7.1. INTEGRATED RISK INFORMATION SYSTEM COMPOUNDS			
CAS RN Order		IRIS Compounds--Alphabetical Order	
95-487	Methylphenol(2-)	934-736	Chlorophenyl methyl sulfoxide(p-)
95-498	Chlorotoluene(o-)	126-998	Chloroprene
95-501	Dichlorobenzene(1,2-)	1897-456	Chlorothalonil
95-578	Chlorophenol(2-)	95-498	Chlorotoluene(o-)
95-658	Dimethylphenol(3,4-)	101-213	Chlorpropham
95-807	Diaminotoluene(2,4-)	2921-882	Chlorpyrifos
95-943	Tetrachlorobenzene(1,2,4,5-)	5598-130	Chlorpyrifos-methyl
95-954	Trichlorophenol(2,4,5-)	64902-723	Chlorsulfuron
96-128	Dibromo-3-chloropropane(1,2-)	16065-831	Chromium(III), insoluble salts
96-184	Trichloropropane(1,2,3-)		Chromium(III), soluble salts
96-333	Methyl acrylate	18540-299	Chromium(VI)
96-457	Ethylene thiourea	218-019	Chrysene
97-632	Ethyl methacrylate	7440-484	Cobalt
98-011	Furfural	8007-452	Coke oven emissions
98-077	Benzotrichloride	7440-508	Copper
98-571	Chlorophenyl methyl sulfone(p-)	544-923	Copper cyanide
98-828	Cumene	56-724	Coumaphos
98-862	Acetophenone	8001-589	Creosote
98-953	Nitrobenzene	123-739	Crotonaldehyde
99-354	Trinitrobenzene(1,3,5-)	98-828	Cumene
99-650	Dinitrobenzene(m-)	21725-462	Cyanazine
100-027	Nitrophenol(p-)	57-125	Cyanide, free
100-414	Ethylbenzene	460-195	Cyanogen
100-425	Styrene	506-683	Cyanogen bromide
100-447	Benzyl chloride	1134-232	Cycloate
100-527	Benzaldehyde	108-941	Cyclohexanone
101-144	Methylene bis (2-chloroaniline)(4,4'-)	108-918	Cyclohexylamine
101-213	Chlorpropham	68085-858	Cyhalothrin/Karate
101-553	Bromodiphenyl ether(p-)	52315-078	Cypermethrin
101-611	Methylene bis(N,N'-dimethyl)aniline(4,4'-)	66215-278	Cyromazine
101-688	Methylene diphenyl isocyanate	1861-321	Dacthal
101-779	Methylene dianiline(4,4'-)	75-990	Dalapon, sodium salt
103-231	Di(2-ethylhexyl)adipate	39515-418	Danitol

TABLE 7.1. INTEGRATED RISK INFORMATION SYSTEM COMPOUNDS			
CAS RN Order		IRIS Compounds--Alphabetical Order	
103-333	Azobenzene	1163-195	Decabromodiphenyl ether
103-855	phenylthiourea(n-)	52918-635	Deltamethrin
105-602	Caprolactam	8065-483	Demeton
105-679	Dimethylphenol(2,4-)	103-231	Di(2-ethylhexyl)adipate
106-376	Dibromobenzene(1,4-)	117-817	Di(2-ethylhexyl)phthalate
106-423	Xylene(p-)	117-840	Di-n-octyl phthalate
106-445	Methylphenol(4-)	95-807	Diaminotoluene(2,4-)
106-467	Dichlorobenzene(1,4-)	333-415	Diazinon
106-478	Chloroaniline(p-)	334-883	Diazomethane
106-514	Quinone	132-649	Dibenzofuran
106-887	Epoxybutane(1,2-)	5385-751	Dibenzo(ae)fluoranthene
106-898	Epichlorohydrin	53-703	Dibenz(ah)anthracene
106-934	Dibromoethane(1,2-)	96-128	Dibromo-3-chloropropane(1,2-)
106-990	Butadiene(1,3-)	106-376	Dibromobenzene(1,4-)
107-028	Acrolein	124-481	Dibromochloromethane
107-051	Allyl chloride	594-183	Dibromodichloromethane
107-062	Dichloroethane(1,2-)	106-934	Dibromoethane(1,2-)
107-131	Acrylonitrile	74-953	Dibromomethane
107-153	Ethylene diamine	84-742	Dibutyl phthalate
107-186	Allyl alcohol	1918-009	Dicamba
107-197	Propargyl alcohol	1717-006	Dichloro-1-fluoroethane(1,1-)
107-211	Ethylene glycol	306-832	Dichloro-2,2,2-trifluoroethane(1,1-)
107-302	Chloromethyl methyl ether	79-436	Dichloroacetic acid
107-493	Tetraethylpyrophosphate	95-501	Dichlorobenzene(1,2-)
107-982	Propylene glycol monomethyl ether	541-731	Dichlorobenzene(1,3-)
108-054	Vinyl acetate	106-467	Dichlorobenzene(1,4-)
108-101	Methyl isobutyl ketone	91-941	Dichlorobenzidine(3,3'-)
108-316	Maleic anhydride	75-718	Dichlorodifluoromethane
108-383	Xylene(m-)	75-343	Dichloroethane(1,1-)
108-394	Methylphenol(3-)	107-062	Dichloroethane(1,2-)
108-452	Phenylenediamine(m-)	75-354	Dichloroethylene(1,1-)
108-883	Toluene	156-592	Dichloroethylene(cis-1,2-)
108-907	Chlorobenzene	156-605	Dichloroethylene(trans-1,2-)

TABLE 7.1. INTEGRATED RISK INFORMATION SYSTEM COMPOUNDS			
CAS RN Order		IRIS Compounds--Alphabetical Order	
108-918	Cyclohexylamine	75-092	Dichloromethane
108-941	Cyclohexanone	120-832	Dichlorophenol(2,4-)
108-952	Phenol	94-826	Dichlorophenoxy(4-(2,4-))butyric acid
108-985	Benzenethiol	94-757	Dichlorophenoxyacetic acid(2,4-)
109-693	Chlorobutane(1-)	120-365	Dichloroprop
109-864	Methoxyethanol(2-)	78-875	Dichloropropane(1,2-)
109-999	Tetrahydrofuran	616-239	Dichloropropanol(2,3-)
110-009	Furan	542-756	Dichloropropene(1,3-)
110-543	Hexane(n-)	62-737	Dichlorvos
110-805	Ethoxyethanol(2-)	115-322	Dicofol
110-861	Pyridine	60-571	Dieldrin
111-159	Ethylene glycol monoethyl ether acetate	Not found	Diesel engine emissions
111-444	Bis(chloroethyl)ether	84-662	Diethyl phthalate
111-693	Adiponitrile	64-675	Diethyl sulfate
111-762	Ethylene glycol monobutyl ether	297-972	Diethyl-o-pyrazinyl(o,o-)
111-911	Bis(2-chloroethoxy)methane	311-455	Diethyl-p-nitrophenylphosphate
112-345	Diethylene glycol monobutyl ether	693-210	Diethylene glycol dinitrate
112-356	Triethylene glycol monomethyl ether	124-174	Diethylene glycol monobutyl ether acetate
112-505	Triethylene glycol monoethyl ether	112-345	Diethylene glycol monobutyl ether
114-261	Baygon	1615-801	Diethylhydrazine(1,2-)
115-297	Endosulfan	43222-486	Difenzoquat
115-322	Dicofol	35367-385	Diflubenzuron
116-063	Aldicarb	75-376	Difluoroethane(1,1-)
117-817	Di(2-ethylhexyl)phthalate	1445-756	Diisopropyl methylphosphonate
117-840	Di-n-octyl phthalate	81777-891	Dimethazone
118-741	Hexachlorobenzene	55290-647	Dimethipin
118-967	Trinitrotoluene(2,4,6-)	60-515	Dimethoate
119-937	Dimethylbenzidine(3,3-)	70-382	Dimethrin
120-127	Anthracene	60-117	Dimethyl aminoazobenzene
120-365	Dichloroprop	756-796	Dimethyl methylphosphonate
120-616	Dimethyl terephthalate	131-113	Dimethyl phthalate
120-809	Catechol	77-781	Dimethyl sulfate
120-821	Trichlorobenzene(1,2,4-)	120-616	Dimethyl terephthalate

TABLE 7.1. INTEGRATED RISK INFORMATION SYSTEM COMPOUNDS			
CAS RN Order		IRIS Compounds--Alphabetical Order	
120-832	Dichlorophenol(2,4-)	124-403	Dimethylamine
121-142	Dinitrotoluene(2,4-)	119-937	Dimethylbenzidine(3,3-)
121-448	Triethylamine	79-447	Dimethylcarbamoyl chloride
121-697	N-Dimethylaniline(N-)	57-147	Dimethylhydrazine(1,1-)
121-755	Malathion	540-738	Dimethylhydrazine(1,2-)
121-824	Hexahydro-1,3,5-trinitro-1,3,5-triazine	105-679	Dimethylphenol(2,4-)
122-145	Fenitrothion	576-261	Dimethylphenol(2,6-)
122-349	Simazine	95-658	Dimethylphenol(3,4-)
122-394	Diphenylamine	534-521	Dinitro-o-cresol(4,6-)
122-429	Propham	131-895	Dinitro-o-cyclohexyl phenol(4,6-)
122-667	Diphenylhydrazine(1,2-)	99-650	Dinitrobenzene(m-)
122-883	Chlorophenoxyacetic acid(4-)	528-290	Dinitrobenzene(o-)
123-091	Chlorophenyl methyl sulfide(p-)	51-285	Dinitrophenol(2,4-)
123-319	Hydroquinone		Dinitrotoluene mixture, 2,4-/2,6-
123-331	Maleic hydrazide	121-142	Dinitrotoluene(2,4-)
123-386	Propionaldehyde	606-202	Dinitrotoluene(2,6-)
123-637	Paraldehyde	88-857	Dinoseb
123-739	Crotonaldehyde	123-911	Dioxane(1,4-)
123-911	Dioxane(1,4-)	957-517	Diphenamid
124-174	Diethylene glycol monobutyl ether acetate	122-394	Diphenylamine
124-403	Dimethylamine	122-667	Diphenylhydrazine(1,2-)
124-481	Dibromochloromethane	85-007	Diquat
126-987	Methacrylonitrile	1937-377	Direct black 38
126-998	Chloroprene	2602-462	Direct blue 6
127-184	Tetrachloroethylene	16071-866	Direct brown 95
127-913	Pinene(beta-)	298-044	Disulfoton
129-000	Pyrene	505-293	Dithiane(1,4-)
130-154	Naphthoquinone(1,4-)	330-541	Diuron
131-113	Dimethyl phthalate	2439-103	Dodine
131-895	Dinitro-o-cyclohexyl phenol(4,6-)	95465-999	Ebufos
132-649	Dibenzofuran	115-297	Endosulfan
133-062	Captan	145-733	Endothall
133-073	Folpet	72-208	Endrin

TABLE 7.1. INTEGRATED RISK INFORMATION SYSTEM COMPOUNDS			
CAS RN Order		IRIS Compounds--Alphabetical Order	
133-904	Chloramben	Not found	Environmental Tobacco Smoke
137-268	Thiram	106-898	Epichlorohydrin
139-402	Propazine	106-887	Epoxybutane(1,2-)
140-578	Aramite	6108-107	Epsilon-Hexachlorocyclohexane
140-885	Ethyl acrylate	16672-870	Ethephon
141-662	Bidrin	563-122	Ethion
141-786	Ethyl acetate	64529-562	Ethiozin
142-596	Nabam	80844-071	Ethofenprox
142-825	Heptane(n-)	110-805	Ethoxyethanol(2-)
143-226	Triethylene glycol monobutyl ether	141-786	Ethyl acetate
143-339	Sodium cyanide	140-885	Ethyl acrylate
145-733	Endothall	51-796	Ethyl carbamate
148-185	Sodium diethyldithiocarbamate	75-003	Ethyl chloride
150-505	Merphos	759-944	Ethyl dipropylthiocarbamate(s-)
151-508	Potassium cyanide	60-297	Ethyl ether
151-564	Ethyleneimine	97-632	Ethyl methacrylate
152-169	Octamethylpyrophosphoramide	2104-645	Ethyl p-nitrophenyl phenylphosphorothioate
156-592	Dichloroethylene(cis-1,2-)	100-414	Ethylbenzene
156-605	Dichloroethylene(trans-1,2-)	107-153	Ethylene diamine
191-242	Benzo(ghi)perylene	107-211	Ethylene glycol
192-972	Benzo(e)pyrene	111-762	Ethylene glycol monobutyl ether
193-395	Indeno(1,2,3-cd)pyrene	111-159	Ethylene glycol monoethyl ether acetate
205-823	Benzo(j)fluoranthene	96-457	Ethylene thiourea
205-992	Benzo(b)fluoranthene	151-564	Ethyleneimine
206-440	Fluoranthene	84-720	Ethylphthalyl ethylglycolate
207-089	Benzo(k)fluoranthene	101200-480	Express
208-968	Acenaphthylene	52-857	Famphur
218-019	Chrysene	22224-926	Fenamiphos
297-972	Diethyl-o-pyrazinyl(o,o-)	122-145	Fenitrothion
298-000	Methyl parathion	2164-172	Fluometuron
298-022	Phorate	206-440	Fluoranthene
298-044	Disulfoton	86-737	Fluorene
299-843	Ronnel	7782-414	Fluorine (soluble fluoride)

TABLE 7.1. INTEGRATED RISK INFORMATION SYSTEM COMPOUNDS			
CAS RN Order		IRIS Compounds--Alphabetical Order	
300-765	Naled	640-197	Fluoroacetamide(2-)
302-012	Hydrazine/Hydrazine sulfate	59756-604	Fluridone
302-170	Chloral hydrate	56425-913	Flurprimidol
306-832	Dichloro-2,2,2-trifluoroethane(1,1-)	66332-965	Flutolanil
309-002	Aldrin	69409-945	Fluvalinate
311-455	Diethyl-p-nitrophenylphosphate	133-073	Folpet
319-846	Hexachlorocyclohexane(alpha-)	72178-020	Fomesafen
319-857	Hexachlorocyclohexane(beta-)	944-229	Fonofos
319-868	Hexachlorocyclohexane(delta-)	50-000	Formaldehyde
330-541	Diuron	23422-539	Formetanate hydrochloride
330-552	Linuron	64-186	Formic acid
333-415	Diazinon	39148-248	Fosetyl-al
334-883	Diazomethane	110-009	Furan
353-504	Carbonyl fluoride	98-011	Furfural
354-336	Pentafluoroethane	60568-050	Furmecyclox
355-259	Perfluorobutane	69806-504	Fusilade
355-420	Perfluorohexane	58-899	gamma-Hexachlorocyclohexane
357-573	Brucine	77182-822	Glufosinate-ammonium
420-462	Trifluoroethane(1,1,1-)	765-344	Glycidaldehyde
431-890	Heptafluoropropane(1,1,1,2,3,3,3-)	1071-836	Glyphosate
460-195	Cyanogen	86-500	Guthion
463-581	Carbonyl sulfide	100784-201	Halosulfuron methyl
465-736	Isodrin	69806-402	Haloxyfop-methyl
504-245	Aminopyridine(4-)	79277-273	Harmony
505-293	Dithiane(1,4-)	76-448	Heptachlor
506-616	Potassium silver cyanide	1024-573	Heptachlor epoxide
506-649	Silver cyanide	431-890	Heptafluoropropane(1,1,1,2,3,3,3-)
506-683	Cyanogen bromide	142-825	Heptane(n-)
506-774	Chlorine cyanide	87-821	Hexabromobenzene
507-200	Butylchloride(t-)	36483-600	Hexabromodiphenyl ether
509-148	Tetranitromethane	118-741	Hexachlorobenzene
510-156	Chlorobenzilate	87-683	Hexachlorobutadiene
528-290	Dinitrobenzene(o-)	319-846	Hexachlorocyclohexane(alpha-)

TABLE 7.1. INTEGRATED RISK INFORMATION SYSTEM COMPOUNDS

CAS RN Order		IRIS Compounds--Alphabetical Order	
532-274	Chloroacetophenone(2-)	319-857	Hexachlorocyclohexane(beta-)
534-521	Dinitro-o-cresol(4,6-)	319-868	Hexachlorocyclohexane(delta-)
540-738	Dimethylhydrazine(1,2-)	77-474	Hexachlorocyclopentadiene
540-841	Trimethylpentane(2,2,4-)	19408-743	Hexachlorodibenzo-p-dioxin, mixture
541-731	Dichlorobenzene(1,3-)	67-721	Hexachloroethane
542-621	Barium cyanide	70-304	Hexachlorophene
542-756	Dichloropropene(1,3-)	121-824	Hexahydro-1,3,5-trinitro-1,3,5-triazine
542-881	Bis(chloromethyl)ether	822-060	Hexamethylene diisocyanate(1,6-)
544-923	Copper cyanide	680-319	Hexamethylphosphoramide
556-887	Nitroguanidine	110-543	Hexane(n-)
557-197	Nickel cyanide	591-786	Hexanone(2-)
557-211	Zinc cyanide	51235-042	Hexazinone
563-122	Ethion	302-012	Hydrazine/Hydrazine sulfate
563-688	Thallium acetate	7647-010	Hydrogen chloride
576-261	Dimethylphenol(2,6-)	74-908	Hydrogen cyanide
591-786	Hexanone(2-)	7664-393	Hydrogen fluoride
592-018	Calcium cyanide	7783-064	Hydrogen sulfide
593-602	Vinyl bromide	123-319	Hydroquinone
594-150	Tribromochloromethane	35554-440	Imazalil
594-183	Dibromodichloromethane	81335-377	Imazaquin
598-776	Trichloropropane(1,1,2-)	193-395	Indeno(1,2,3-cd)pyrene
606-202	Dinitrotoluene(2,6-)		Ionizing radiation
608-731	technical Hexachlorocyclohexane	36734-197	Iprodione
608-935	Pentachlorobenzene	42509-808	Isazophos
615-543	Tribromobenzene(1,2,4-)	78-831	Isobutyl alcohol
616-239	Dichloropropanol(2,3-)	465-736	Isodrin
621-647	Nitrosodi-n-propylamine(n-)	78-591	Isophorone
624-839	Methyl isocyanate	33820-530	Isopropalin
628-864	Mercury fulminate	1832-548	Isopropyl methyl phosphonic acid
630-104	Selenourea	82558-507	Isoxaben
630-206	Tetrachloroethane(1,1,1,2-)	77501-634	Lactofen
631-618	Ammonium acetate	7439-921	Lead and compounds (inorganic)
640-197	Fluoroacetamide(2-)	5989-275	Limonene(d-)

TABLE 7.1. INTEGRATED RISK INFORMATION SYSTEM COMPOUNDS			
CAS RN Order		IRIS Compounds--Alphabetical Order	
680-319	Hexamethylphosphoramide	330-552	Linuron
684-935	Nitroso-n-methylurea(n-)	83055-996	Londax
693-210	Diethylene glycol dinitrate	1702-176	Lontrel
695-772	Tetrachlorocyclopentadiene	121-755	Malathion
709-988	Propanil	108-316	Maleic anhydride
732-116	Phosmet	123-331	Maleic hydrazide
756-796	Dimethyl methylphosphonate	12427-382	Maneb
759-739	Nitroso-n-ethylurea(n-)	7439-965	Manganese
759-944	Ethyl dipropylthiocarbamate(s-)	24307-264	Mepiquat chloride
765-344	Glycidaldehyde	7487-947	Mercuric chloride
811-972	Tetrafluoroethane(1,1,1,2-)	7439-976	Mercury, elemental
822-060	Hexamethylene diisocyanate(1,6-)	628-864	Mercury fulminate
834-128	Ametryn	150-505	Merphos
886-500	Terbutryn	78-488	Merphos oxide
924-163	Nitroso-di-n-butylamine(n-)	2032-657	Mesurol
930-552	Nitrosopyrrolidine(n-)	57837-191	Metalaxyl
934-736	Chlorophenyl methyl sulfoxide(p-)	126-987	Methacrylonitrile
944-229	Fonofos	10265-926	Methamidophos
950-378	Methidathion	74-931	Methanethiol
957-517	Diphenamid	67-561	Methanol
961-115	Tetrachlorovinphos	950-378	Methidathion
1024-573	Heptachlor epoxide	16752-775	Methomyl
1071-836	Glyphosate	72-435	Methoxychlor
1116-547	Nitrosodiethanolamine(n-)	109-864	Methoxyethanol(2-)
1120-714	Propane sultone(1,3-)	79-209	Methyl acetate
1134-232	Cycloate	96-333	Methyl acrylate
1163-195	Decabromodiphenyl ether	79-221	Methyl chlorocarbonate
1309-644	Antimony trioxide	78-933	Methyl ethyl ketone
1314-325	Thallic oxide	1338-234	Methyl Ethyl Ketone Peroxide
1314-621	Vanadium pentoxide	74-884	Methyl iodide
1314-847	Zinc phosphide	108-101	Methyl isobutyl ketone
1319-773	Tricresol	624-839	Methyl isocyanate
1330-207	Xylenes	80-626	Methyl methacrylate

TABLE 7.1. INTEGRATED RISK INFORMATION SYSTEM COMPOUNDS			
CAS RN Order		IRIS Compounds--Alphabetical Order	
1332-214	Asbestos	298-000	Methyl parathion
1336-363	Polychlorinated biphenyls	25013-154	Methyl styrene
1338-234	Methyl Ethyl Ketone Peroxide	1634-044	Methyl tert-butyl ether
1445-756	Diisopropyl methylphosphonate	93-652	Methyl-4-chlorophenoxy(2-(2-))propionic acid
1511-622	Bromodifluoromethane	94-746	Methyl-4-chlorophenoxyacetic acid(2-)
1563-662	Carbofuran	94-815	Methyl-4-chlorophenoxy(4-(2-)) butyric acid
1582-098	Trifluralin	101-144	Methylene bis (2-chloroaniline)(4,4'-)
1596-845	Alar	101-611	Methylene bis(N,N'-dimethyl)aniline(4,4'-)
1610-180	Prometon	101-779	Methylene dianiline(4,4'-)
1615-801	Diethylhydrazine(1,2-)	101-688	Methylene diphenyl isocyanate
1634-044	Methyl tert-butyl ether	75-865	Methyllactonitrile(2-)
1646-884	Aldicarb sulfone	22967-926	Methylmercury
1689-845	Bromoxynil	95-487	Methylphenol(2-)
1689-992	Bromoxynil octanoate	108-394	Methylphenol(3-)
1702-176	Lontrel	106-445	Methylphenol(4-)
1717-006	Dichloro-1-fluoroethane(1,1-)	51218-452	Metolachlor
1832-548	Isopropyl methyl phosphonic acid	21087-649	Metribuzin
1861-321	Dacthal	2385-855	Mirex
1861-401	Benefin	2212-671	Molinate
1897-456	Chlorothalonil	7439-987	Molybdenum
1910-425	Paraquat	10599-903	Monochloramine
1912-249	Atrazine	6923-224	Monocrotophos
1918-009	Dicamba	60-344	Monomethylhydrazine
1918-021	Picloram	121-697	N-Dimethylaniline(N-)
1918-167	Propachlor	142-596	Nabam
1929-777	Vernam	300-765	Naled
1929-824	Nitrapyrin	91-203	Naphthalene
1937-377	Direct black 38	130-154	Naphthoquinone(1,4-)
2008-415	Butylate	91-598	Naphthylamine(2-)
2032-657	Mesurol	86-884	Naphthylthiourea(alpha-)
2050-477	p,p'-Dibromodiphenyl ether	15299-997	Napropamide
2104-645	Ethyl p-nitrophenyl phenylphosphorothioate	13463-393	Nickel carbonyl
2164-172	Fluometuron	557-197	Nickel cyanide

TABLE 7.1. INTEGRATED RISK INFORMATION SYSTEM COMPOUNDS			
CAS RN Order		IRIS Compounds--Alphabetical Order	
2212-671	Molinate		Nickel refinery dust
2303-175	Triallate	Various	Nickel, soluble salts
2310-170	Phosalone	12035-722	Nickel subsulfide
2312-358	Propargite	1929-824	Nitrapyrin
2385-855	Mirex	14797-558	Nitrate
2425-061	Captafol	10102-439	Nitric oxide
2439-103	Dodine	14797-650	Nitrite
2602-462	Direct blue 6	88-744	Nitroaniline(2-)
2691-410	Octahydro-1,3,5,7-tetranitro-1,3,5,7-tetrazocine	98-953	Nitrobenzene
2837-890	Chloro-1,1,1,2-tetrafluoroethane(2-)	92-933	Nitrobiphenyl(4-)
2921-882	Chlorpyrifos	10102-440	Nitrogen dioxide
3337-711	Asulam	556-887	Nitroguanidine
3689-245	Tetraethyldithiopyrophosphate	100-027	Nitrophenol(p-)
5234-684	Carboxin	79-469	Nitropropane(2-)
5385-751	Dibenzo(ae)fluoranthene	924-163	Nitroso-di-n-butylamine(n-)
5598-130	Chlorpyrifos-methyl	759-739	Nitroso-n-ethylurea(n-)
5902-512	Terbacil	10595-956	Nitroso-n-methylethylamine(n-)
5989-275	Limonene(d-)	684-935	Nitroso-n-methylurea(n-)
6108-107	Epsilon-Hexachlorocyclohexane	621-647	Nitrosodi-n-propylamine(n-)
6533-739	Thallium carbonate	1116-547	Nitrosodiethanolamine(n-)
6923-224	Monocrotophos	55-185	Nitrosodiethylamine(n-)
7287-196	Prometryn	62-759	Nitrosodimethylamine(n-)
7429-905	Aluminum	86-306	Nitrosodiphenylamine(n-)
7439-921	Lead and compounds (inorganic)	930-552	Nitrosopyrrolidine(n-)
7439-965	Manganese	68-122	N,N-Dimethylformamide
7439-976	Mercury, elemental	63936-561	Nonabromodiphenyl ether
7439-987	Molybdenum	27314-132	Norflurazon
7440-144	Radium 226,228	85509-199	NuStar
7440-224	Silver	32536-520	Octabromodiphenyl ether
7440-246	Strontium	2691-410	Octahydro-1,3,5,7-tetranitro-1,3,5,7-tetrazocine
7440-360	Antimony	152-169	Octamethylpyrophosphoramide
7440-382	Arsenic, inorganic	19044-883	Oryzalin
7440-393	Barium	20816-120	Osmium tetroxide

TABLE 7.1. INTEGRATED RISK INFORMATION SYSTEM COMPOUNDS			
CAS RN Order		IRIS Compounds--Alphabetical Order	
7440-417	Beryllium	19666-309	Oxadiazon
7440-428	Boron (Boron and Borates only)	77732-093	Oxadixyl
7440-439	Cadmium	23135-220	Oxamyl
7440-484	Cobalt	42874-033	Oxyfluorfen
7440-508	Copper	76738-620	Paclobutrazol
7440-611	Uranium, natural	123-637	Paraldehyde
7440-622	Vanadium	1910-425	Paraquat
7440-666	Zinc and Compounds	56-382	Parathion
7446-186	Thallium(I) sulfate	40487-421	Pendimethalin
7446-346	Selenium sulfide	32534-819	Pentabromodiphenyl ether
7487-947	Mercuric chloride	608-935	Pentachlorobenzene
7647-010	Hydrogen chloride	25329-355	Pentachlorocyclopentadiene
7664-382	Phosphoric acid	76-017	Pentachloroethane
7664-393	Hydrogen fluoride	82-688	Pentachloronitrobenzene
7664-417	Ammonia	87-865	Pentachlorophenol
7723-140	White phosphorus	354-336	Pentafluoroethane
7758-012	Potassium bromate	355-259	Perfluorobutane
7758-192	Chlorite	355-420	Perfluorohexane
7773-060	Ammonium sulfamate	52645-531	Permethrin
7782-414	Fluorine (soluble fluoride)	85-018	Phenanthrene
7782-492	Selenium and Compounds	13684-634	Phenmedipham
7782-505	Chlorine	108-952	Phenol
7783-008	Selenious acid	108-452	Phenylenediamine(m-)
7783-064	Hydrogen sulfide	62-384	Phenylmercuric acetate
7784-421	Arsine	103-855	phenylthiourea(n-)
7791-120	Thallium chloride	298-022	Phorate
7803-512	Phosphine	2310-170	Phosalone
8001-352	Toxaphene	75-445	Phosgene
8001-589	Creosote	732-116	Phosmet
8007-452	Coke oven emissions	7803-512	Phosphine
8065-483	Demeton	7664-382	Phosphoric acid
10049-044	Chlorine dioxide		photon emitters(beta-)
10102-439	Nitric oxide	85-449	Phthalic anhydride

TABLE 7.1. INTEGRATED RISK INFORMATION SYSTEM COMPOUNDS			
CAS RN Order		IRIS Compounds--Alphabetical Order	
10102-440	Nitrogen dioxide	1918-021	Picloram
10102-451	Thallium nitrate	80-568	Pinene(alpha-)
10265-926	Methamidophos	127-913	Pinene(beta-)
10453-868	Resmethrin	29232-937	Pirimiphos-methyl
10595-956	Nitroso-n-methylethylamine(n-)	1336-363	Polychlorinated biphenyls
10599-903	Monochloramine		Polycyclic organic matter
11097-691	Aroclor 1254	7758-012	Potassium bromate
12035-722	Nickel subsulfide	151-508	Potassium cyanide
12039-520	Thallium selenite	506-616	Potassium silver cyanide
12122-677	Zineb	2050-477	p,p'-Dibromodiphenyl ether
12427-382	Maneb	72-559	p,p'-Dichlorodiphenyldichloroethylene
12672-296	Aroclor 1248	50-293	p,p'-Dichlorodiphenyltrichloroethane
12674-112	Aroclor 1016	72-548	p,p'-Dichlorodiphenyl dichloroethane
13071-799	Terbufos	67747-095	Prochloraz
13463-393	Nickel carbonyl	1610-180	Prometon
13510-491	Beryllium sulfate	7287-196	Prometryn
13593-038	Quinalphos	23950-585	Pronamide
13684-634	Phenmedipham	1918-167	Propachlor
14797-558	Nitrate	1120-714	Propane sultone(1,3-)
14797-650	Nitrite	709-988	Propanil
14859677	Radon 222	2312-358	Propargite
15299-997	Napropamide	107-197	Propargyl alcohol
15972-608	Alachlor	139-402	Propazine
16065-831	Chromium(III), insoluble salts	31218-834	Propetamphos
16071-866	Direct brown 95	122-429	Propham
16325-476	Ammonium methacrylate	60207-901	Propiconazole
16672-870	Ethephon	57-578	Propiolactone(beta-)
16752-775	Methomyl	123-386	Propionaldehyde
17804-352	Benomyl	79-094	Propionic acid
18540-299	Chromium(VI)	57-556	Propylene glycol
19044-883	Oryzalin	52125-538	Propylene glycol monoethyl ether
19408-743	Hexachlorodibenzo-p-dioxin, mixture	107-982	Propylene glycol monomethyl ether
19666-309	Oxadiazon	75-569	Propylene oxide

TABLE 7.1. INTEGRATED RISK INFORMATION SYSTEM COMPOUNDS

CAS RN Order		IRIS Compounds--Alphabetical Order	
20816-120	Osmium tetroxide	75-558	Propyleneimine
20859-738	Aluminum phosphide	81335-775	Pursuit
21087-649	Metribuzin	51630-581	Pydrin
21725-462	Cyanazine	129-000	Pyrene
22224-926	Fenamiphos	110-861	Pyridine
22967-926	Methylmercury	13593-038	Quinalphos
23135-220	Oxamyl	84087-014	Quinclorac
23422-539	Formetanate hydrochloride	91-225	Quinoline
23564-058	Thiophanate-methyl	106-514	Quinone
23950-585	Pronamide	7440-144	Radium 226,228
24307-264	Mepiquat chloride		Radon (Inert gas only)
25013-154	Methyl styrene	14859677	Radon 222
25057-890	Bentazon	Not found	Refractory ceramic fibers
25329-355	Pentachlorocyclopentadiene	10453-868	Resmethrin
26471-625	Toluene diisocyanate mixture(2,4-/2,6)	299-843	Ronnel
26628-228	Sodium azide	83-794	Rotenone
27314-132	Norflurazon	78587-050	Savey
28249-776	Thiobencarb	7783-008	Selenious acid
29232-937	Pirimiphos-methyl	7782-492	Selenium and Compounds
30560-191	Acephate	7446-346	Selenium sulfide
31218-834	Propetamphos	630-104	Selenourea
32534-819	Pentabromodiphenyl ether	74051-802	Sethoxydim
32536-520	Octabromodiphenyl ether	7440-224	Silver
33089-611	Amitraz	506-649	Silver cyanide
33820-530	Isopropalin	122-349	Simazine
34014-181	Tebuthiuron	26628-228	Sodium azide
34256-821	Acetochlor	143-339	Sodium cyanide
35367-385	Diflubenzuron	148-185	Sodium diethyldithiocarbamate
35400-432	Bolstar	62-748	Sodium fluoroacetate
35554-440	Imazalil	7440-246	Strontium
36483-600	Hexabromodiphenyl ether	57-249	Strychnine
36734-197	Iprodione	100-425	Styrene
39148-248	Fosetyl-al	88671-890	Systhane

TABLE 7.1. INTEGRATED RISK INFORMATION SYSTEM COMPOUNDS			
CAS RN Order		IRIS Compounds--Alphabetical Order	
39196-184	Thiofanox	107534-963	Tebuconazole
39515-418	Danitol	34014-181	Tebuthiuron
39638-329	Bis(2-chloroisopropyl) ether	608-731	technical Hexachlorocyclohexane
40088-479	Tetrabromodiphenyl ether	5902-512	Terbacil
40487-421	Pendimethalin	13071-799	Terbufos
41851-507	Chlorocyclopentadiene	886-500	Terbutryn
42509-808	Isazophos	40088-479	Tetrabromodiphenyl ether
42874-033	Oxyfluorfen	95-943	Tetrachlorobenzene(1,2,4,5-)
43121-433	Bayleton	695-772	Tetrachlorocyclopentadiene
43222-486	Difenzoquat	79-345	Tetrachloroethane(1,1,2,2-)
49690-940	Tribromodiphenyl ether	630-206	Tetrachloroethane(1,1,1,2-)
50471-448	Vinclozolin	127-184	Tetrachloroethylene
51218-452	Metolachlor	58-902	Tetrachlorophenol(2,3,4,6-)
51235-042	Hexazinone	961-115	Tetrachlorovinphos
51630-581	Pydrin	78-002	Tetraethyl lead
52125-538	Propylene glycol monoethyl ether	3689-245	Tetraethyldithiopyrophosphate
52315-078	Cypermethrin	107-493	Tetraethylpyrophosphate
52645-531	Permethrin	811-972	Tetrafluoroethane(1,1,1,2-)
52918-635	Deltamethrin	109-999	Tetrahydrofuran
55179-312	Baycor	509-148	Tetranitromethane
55219-653	Baytan	1314-325	Thallic oxide
55285-148	Carbosulfan	563-688	Thallium acetate
55290-647	Dimethipin	6533-739	Thallium carbonate
56425-913	Flurprimidol	7791-120	Thallium chloride
57837-191	Metalaxyl	10102-451	Thallium nitrate
58138-082	Tridiphane	12039-520	Thallium selenite
59756-604	Fluridone	7446-186	Thallium(I) sulfate
60207-901	Propiconazole	28249-776	Thiobencarb
60568-050	Furmecyclox	39196-184	Thiofanox
62476-599	Acifluorfen, sodium	23564-058	Thiophanate-methyl
63936-561	Nonabromodiphenyl ether	79-196	Thiosemicarbazide
64529-562	Ethiozin	137-268	Thiram
64902-723	Chlorsulfuron	108-883	Toluene

TABLE 7.1. INTEGRATED RISK INFORMATION SYSTEM COMPOUNDS			
CAS RN Order		IRIS Compounds--Alphabetical Order	
65195-553	Avermectin B1	26471-625	Toluene diisocyanate mixture(2,4-/2,6)
66215-278	Cyromazine	8001-352	Toxaphene
66332-965	Flutolanil	66841-256	Tralomethrin
66841-256	Tralomethrin	2303-175	Triallate
67485-294	Amdro	82097-505	Triasulfuron
67747-095	Prochloraz	615-543	Tribromobenzene(1,2,4-)
68085-858	Cyhalothrin/Karate	594-150	Tribromochloromethane
68359-375	Baythroid	49690-940	Tribromodiphenyl ether
69409-945	Fluvalinate	56-359	Tributyltin oxide
69806-402	Haloxyfop-methyl	76-131	Trichloro-1,2,2-trifluoroethane(1,1,2-)
69806-504	Fusilade	76-039	Trichloroacetic acid
72178-020	Fomesafen	120-821	Trichlorobenzene(1,2,4-)
74051-802	Sethoxydim	77323-843	Trichlorocyclopentadiene
74115-245	Apollo	71-556	Trichloroethane(1,1,1-)
74223-646	Ally	79-005	Trichloroethane(1,1,2-)
76578-148	Assure	79-016	Trichloroethylene
76738-620	Paclobutrazol	75-694	Trichlorofluoromethane
77182-822	Glufosinate-ammonium	95-954	Trichlorophenol(2,4,5-)
77323-843	Trichlorocyclopentadiene	88-062	Trichlorophenol(2,4,6-)
77501-634	Lactofen	93-721	Trichlorophenoxy(2-(2,4,5-)) propionic acid
77732-093	Oxadixyl	93-765	Trichlorophenoxyacetic acid(2,4,5-)
78587-050	Savey	598-776	Trichloropropane(1,1,2-)
79277-273	Harmony	96-184	Trichloropropane(1,2,3-)
80844-071	Ethofenprox	1319-773	Tricresol
81335-377	Imazaquin	58138-082	Tridiphane
81335-775	Pursuit	121-448	Triethylamine
81405-858	Assert	143-226	Triethylene glycol monobutyl ether
81777-891	Dimethazone	112-505	Triethylene glycol monoethyl ether
82097-505	Triasulfuron	112-356	Triethylene glycol monomethyl ether
82558-507	Isoxaben	420-462	Trifluoroethane(1,1,1-)
82657-043	Biphenthrin	75-467	Trifluoromethane
83055-996	Londax	1582-098	Trifluralin
84087-014	Quinclorac	540-841	Trimethylpentane(2,2,4-)

TABLE 7.1. INTEGRATED RISK INFORMATION SYSTEM COMPOUNDS			
CAS RN Order		IRIS Compounds--Alphabetical Order	
85509-199	NuStar	99-354	Trinitrobenzene(1,3,5-)
88671-890	Systhane	118-967	Trinitrotoluene(2,4,6-)
90982-324	Chlorimuron-ethyl	76-879	Triphenyltin hydroxide
95465-999	Ebufos	7440-611	Uranium, natural
100784-201	Halosulfuron methyl		Uranium, soluble salts
101200-480	Express	7440-622	Vanadium
104098-499	Cadre	1314-621	Vanadium pentoxide
107534-963	Tebuconazole	1929-777	Vernam
	alpha emitters	50471-448	Vinclozolin
	Brominated dibenzofurans	108-054	Vinyl acetate
	Chromium(III), soluble salts	593-602	Vinyl bromide
	Diesel engine emissions	75-014	Vinyl chloride
	Dinitrotoluene mixture, 2,4-/2,6-	81-812	Warfarin
	Environmental Tobacco Smoke	7723-140	White phosphorus
	Ionizing radiation	108-383	Xylene(m-)
	Nickel refinery dust	95-476	Xylene(o-)
	Nickel, soluble salts	106-423	Xylene(p-)
	photon emitters(beta-)	1330-207	Xylenes
	Polycyclic organic matter	7440-666	Zinc and Compounds
	Radon (Inert gas only)	557-211	Zinc cyanide
	Refractory ceramic fibers	1314-847	Zinc phosphide
	Uranium, soluble salts	12122-677	Zineb

TABLE 7.2. REFERENCE AIR CONCENTRATIONS (RAC)[1] (40CFR266 APPENDIX IV)

CAS RN order		Alphabetical order		RAC ($\mu g/m^3$)
51-285	Dinitrophenol(2,4-)	75-070	Acetaldehyde	10
57-125	Cyanide (free)	75-058	Acetonitrile	10
57-249	Strychnine	98-862	Acetophenone	100
58-902	Tetrachlorophenol(2,3,4,6-)	107-028	Acrolein	20
60-515	Dimethoate	116-063	Aldicarb	1
62-384	Phenylmercuric Acetate	20859-738	Aluminum Phosphide	0.3
64-186	Formic Acid	107-186	Allyl Alcohol	5
70-304	Hexachlorophene	7440-360	Antimony	0.3
72-208	Endrin	7440-393	Barium	50
72-435	Methoxychlor	542-621	Barium Cyanide	50
74-839	Bromomethane	74-839	Bromomethane	0.8
74-908	Hydrocyanic Acid	592-018	Calcium Cyanide	30
75-058	Acetonitrile	75-150	Carbon Disulfide	200
75-070	Acetaldehyde	75-876	Chloral	2
75-150	Carbon Disulfide		Chlorine (free)	0.4
75-694	Trichloromonofluoromethane	126-998	Chloro-1,3-butadiene(2-)	3
75-718	Dichlorodifluoromethane	16065-831	Chromium III	1000
75-876	Chloral	544-923	Copper Cyanide	5
77-474	Hexachlorocyclopentadiene	1319-773	Cresols	50
78-002	Tetraethyl Lead	98-828	Cumene	1
78-831	Isobutyl Alcohol	57-125	Cyanide (free)	20
78-933	Methyl Ethyl Ketone	460-195	Cyanogen	30
79-221	Methyl Chlorocarbonate	506-683	Cyanogen Bromide	80
81-812	Warfarin	84-742	Di-n-butyl Phthalate	100
84-662	Diethyl Phthalate	95-501	Dichlorobenzene(o-)	10
84-742	Di-n-butyl Phthalate	106-467	Dichlorobenzene(p-)	10
85-449	Phthalic Anhydride	75-718	Dichlorodifluoromethane	200
87-865	Pentachlorophenol	120-832	Dichlorophenol(2,4-)	3
88-857	Dinoseb	84-662	Diethyl Phthalate	800
95-501	Dichlorobenzene(o-)	60-515	Dimethoate	0.8
95-943	Tetrachlorobenzene(1,2,4,5-)	51-285	Dinitrophenol(2,4-)	2
95-954	Trichlorophenol(2.4.5-)	88-857	Dinoseb	0.9
98-828	Cumene	122-394	Diphenylamine	20

TABLE 7.2. REFERENCE AIR CONCENTRATIONS (RAC)[1] (40CFR266 APPENDIX IV)					
CAS RN order		Alphabetical order			RAC ($\mu g/m^3$)
98-862	Acetophenone	115-297	Endosulfan		0.05
98-953	Nitrobenzene	72-208	Endrin		0.3
106-467	Dichlorobenzene(p-)	7782-414	Fluorine		50
107-028	Acrolein	64-186	Formic Acid		2000
107-186	Allyl Alcohol	765-344	Glycidyaldehyde		0.3
108-316	Maleic Anyhdride	77-474	Hexachlorocyclopentadiene		5
108-452	Phenylenediamine(m-)	70-304	Hexachlorophene		0.3
108-883	Toluene	74-908	Hydrocyanic Acid		20
108-952	Phenol	7647-010	Hydrogen Chloride		7
109-999	Tetrahydrofuran	7783-064	Hydrogen Sulfide		3
110-861	Pyridine	78-831	Isobutyl Alcohol		300
115-297	Endosulfan	7439-921	Lead		0.09
116-063	Aldicarb	108-316	Maleic Anyhdride		100
120-821	Trichlorobenzene(1,2,4-)	7439-976	Mercury		0.3
120-832	Dichlorophenol(2,4-)	126-987	Methacrylonitrile		0.1
122-394	Diphenylamine	16752-775	Methomyl		20
126-987	Methacrylonitrile	72-435	Methoxychlor		50
126-998	Chloro-1,3-butadiene(2-)	79-221	Methyl Chlorocarbonate		1000
137-268	Thiram	78-933	Methyl Ethyl Ketone		80
143-339	Sodium Cyanide	298-000	Methyl Parathion		0.3
151-508	Potassium Cyanide	557-197	Nickel Cyanide		20
298-000	Methyl Parathion	10102-439	Nitric Oxide		100
460-195	Cyanogen	98-953	Nitrobenzene		0.8
506-616	Potassium Silver Cyanide	608-935	Pentachlorobenzene		0.8
506-649	Silver Cyanide	87-865	Pentachlorophenol		30
506-683	Cyanogen Bromide	108-952	Phenol		30
542-621	Barium Cyanide	108-452	Phenylenediamine(m-)		5
544-923	Copper Cyanide	62-384	Phenylmercuric Acetate		0.075
557-197	Nickel Cyanide	7803-512	Phosphine		0.3
557-211	Zinc Cyanide	85-449	Phthalic Anhydride		2000
563-688	Thallium (I) Acetate	151-508	Potassium Cyanide		50
592-018	Calcium Cyanide	506-616	Potassium Silver Cyanide		200
608-935	Pentachlorobenzene	110-861	Pyridine		1

TABLE 7.2. REFERENCE AIR CONCENTRATIONS (RAC)[1] (40CFR266 APPENDIX IV)				
CAS RN order		Alphabetical order		RAC (μg/m^3)
630-104	Selenourea	7783-608	Selenious Acid	3
765-344	Glycidyaldehyde	630-104	Selenourea	5
1314-325	Thallic Oxide	7440-224	Silver	3
1314-621	Vanadium Pentoxide	506-649	Silver Cyanide	100
1314-847	Zinc Phosphide	143-339	Sodium Cyanide	30
1319-773	Cresols	57-249	Strychnine	0.3
1330-207	Xylenes	95-943	Tetrachlorobenzene(1,2,4,5-)	0.3
6533-739	Thallium (I) Carbonate	58-902	Tetrachlorophenol(2,3,4,6-)	30
7439-921	Lead	78-002	Tetraethyl Lead	0.0001
7439-976	Mercury	109-999	Tetrahydrofuran	10
7440-224	Silver	1314-325	Thallic Oxide	0.3
7440-280	Thallium	7440-280	Thallium	0.5
7440-360	Antimony	563-688	Thallium (I) Acetate	0.5
7440-393	Barium	6533-739	Thallium (I) Carbonate	0.3
7446-186	Thallium (I) Sulfate	7791-120	Thallium (I) Chloride	0.3
7647-010	Hydrogen Chloride	10102-451	Thallium (I) Nitrate	0.5
7782-414	Fluorine	12039-520	Thallium Selenite	0.5
7783-064	Hydrogen Sulfide	7446-186	Thallium (I) Sulfate	0.075
7783-608	Selenious Acid	137-268	Thiram	5
7791-120	Thallium (I) Chloride	108-883	Toluene	300
7803-512	Phosphine	120-821	Trichlorobenzene(1,2,4-)	20
10102-439	Nitric Oxide	75-694	Trichloromonofluoromethane	300
10102-451	Thallium (I) Nitrate	95-954	Trichlorophenol(2.4.5-)	100
12039-520	Thallium Selenite	1314-621	Vanadium Pentoxide	20
16065-831	Chromium III	81-812	Warfarin	0.3
16752-775	Methomyl	1330-207	Xylenes	80
20859-738	Aluminum Phosphide	557-211	Zinc Cyanide	50
	Chlorine (free)	1314-847	Zinc Phosphide	0.3

[1] The RAC for other 40CFR261 Appendix VIII Consitituents not listed herein or in Appendix V of this part is 0.1 μg/m^3.

TABLE 7.3. RISK SPECIFIC DOSES (RsD) (10⁻⁵) (40CFR266 APPENDIX V)					
CAS RN order		Alphabetical order		Unit risk ($m^3/\mu g$)	RsD ($\mu g/m^3$)
50-000	Formaldehyde	79-061	Acrylamide	1.30E-03	7.7E-03
50-293	DDT	107-131	Acrylonitrile	6.80E-05	1.5E-01
50-328	Benzo(a)pyrene	309-002	Aldrin	4.90E-03	2.0E-03
50-555	Reserpine	62-533	Aniline	7.40E-06	1.4E+00
53-703	Dibenz(ah)anthracene	7440-382	Arsenic	4.30E-03	2.3E-03
55-185	Nitrosodiethylamine(n-)	56-553	Benz(a)anthracene	8.90E-04	1.1E-02
56-235	Carbon Tetrachloride	71-432	Benxene	8.30E-06	1.2E+00
56-495	Methylcholanthrene(3-)	92-875	Benzidine	6.70E-02	1.5E-04
56-531	Diethylstilbestrol	50-328	Benzo(a)pyrene	3.30E-03	3.0E-03
56-553	Benz(a)anthracene	7440-417	Beryllium	2.40E-03	4.2E-03
57-749	Chlordane	111-444	Bis(2-chloroethyl)ether	3.30E-04	3.0E-02
58-899	Hexachloro-cyclohexane(gamma-)	542-881	Bis(chloromethyl)ether	6.20E-02	1.6E-04
60-344	Methyl Hydrazine	117-817	Bis(2-ethylhexyl)-phthalate	2.40E-07	4.2E+01
60-571	Dieldrin	106-990	Butadiene(1,3-)	2.80E-04	3.6E-02
62-533	Aniline	7440-439	Cadmium	1.80E-03	5.6E-03
62-566	Thiourea	56-235	Carbon Tetrachloride	1.50E-05	6.7E-01
62-759	Dimethylnitrosamine	57-749	Chlordane	3.70E-04	2.7E-02
67-663	Chloroform	67-663	Chloroform	2.30E-05	4.3E-01
67-721	Hexachloroethane	74-873	Chloromethane	3.60E-06	2.8E+00
71-432	Benxene	7440-473	Chromium VI	1.20E-02	8.3E-04
74-873	Chloromethane	50-293	DDT	9.70E-05	1.0E-01
75-014	Vinyl Chloride	53-703	Dibenz(ah)anthracene	1.40E-02	7.1E-04
75-092	Methylene Chloride	96-128	Dibromo-3-chloropropane(1,2-)	6.30E-03	1.6E-03
75-218	Ethylene Oxide	106-934	Dibromoethane(1,2-)	2.20E-04	4.5E-02
75-343	Dichloroethane(1,1-)	75-343	Dichloroethane(1,1-)	2.60E-05	3.8E-01
75-354	Dichloroethylene(1,1-)	107-062	Dichloroethane(1,2-)	2.60E-05	3.8E-01
76-448	Heptachlor	75-354	Dichloroethylene(1,1-)	5.00E-05	2.0E-01
79-005	Trichloroethane(1,1,2-)	542-756	Dichloropropene(1,3-)	3.50E-01	2.9E-05
79-016	Trichloroethylene	60-571	Dieldrin	4.60E-03	2.2E-03
79-061	Acrylamide	56-531	Diethylstilbestrol	1.40E-01	7.1E-05
79-345	Tetrachloroethane(1,1,2,2-)	62-759	Dimethylnitrosamine	1.40E-02	7.1E-04
79-469	Nitropropane(2-)	121-142	Dinitrotoluene(2,4-)	8.80E-05	1.1E-01
82-688	Pentachloronitrobenzene	122-667	Diphenylhydrazine(1,2-)	2.20E-04	4.5E-02

TABLE 7.3. RISK SPECIFIC DOSES (RsD) (10^{-5}) (40CFR266 APPENDIX V)					
CAS RN order		Alphabetical order		Unit risk ($m^3/\mu g$)	RsD ($\mu g/m^3$)
87-683	Hexachlorobutadiene	123-911	Dioxane(1,4-)	1.40E-06	7.1E+00
88-062	Trichlorophenol(2,4,6-)	106-898	Epichlorohydrin	1.20E-06	8.3E+00
92-875	Benzidine	75-218	Ethylene Oxide	1.00E-04	1.0E-01
96-128	Dibromo-3-chloropropane(1,2-)	106-934	Ethylene Dibromide	2.20E-04	4.5E-02
101-144	Methylene-bis(2-chloroaniline(4,4'-)	50-000	Formaldehyde	1.30E-05	7.7E-01
106-898	Epichlorohydrin	76-448	Heptachlor	1.30E-03	7.7E-03
106-934	Dibromoethane(1,2-)	1024-573	Heptachlor Epoxide	2.60E-03	3.8E-03
106-934	Ethylene Dibromide	118-741	Hexachlorobenzene	4.90E-04	2.0E-02
106-990	Butadiene(1,3-)	87-683	Hexachlorobutadiene	2.00E-05	5.0E-01
107-062	Dichloroethane(1,2-)	319-846	Hexachloro-cyclohexane(alpha-)	1.80E-03	5.6E-03
107-131	Acrylonitrile	319-857	Hexachloro-cyclohexane(beta-)	5.30E-04	1.9E-02
111-444	Bis(2-chloroethyl)ether	58-899	Hexachloro-cyclohexane(gamma-)	3.80E-04	2.6E-02
117-817	Bis(2-ethylhexyl)-phthalate		Hexachlorocyclo-hexane, Technical	5.10E-04	2.0E-02
118-741	Hexachlorobenzene		Hexachlorodibenxo-p-dioxin(1,2-Mixture)	1.30e+00	7.7E-06
121-142	Dinitrotoluene(2,4-)	67-721	Hexachloroethane	4.00E-06	2.5E+00
122-667	Diphenylhydrazine(1,2-)	302-012	Hydrazine	2.90E-03	3.4E-03
123-911	Dioxane(1,4-)	302-012	Hydrazine Sulfate	2.90E-03	3.4E-03
127-184	Tetrachloroethylene	56-495	Methylcholanthrene(3-)	2.70E-03	3.7E-03
302-012	Hydrazine	60-344	Methyl Hydrazine	3.10E-04	3.2E-02
302-012	Hydrazine Sulfate	75-092	Methylene Chloride	4.10E-06	2.4E+00
309-002	Aldrin	101-144	Methylene-bis(2-chloroaniline(4,4'-)	4.70E-05	2.1E-01
319-846	Hexachloro-cyclohexane(alpha-)	7440-020	Nickel	2.40E-04	4.2E-02
319-857	Hexachloro-cyclohexane(beta-)	7440-020	Nickel Refinery Dust	2.40E-04	4.2E-02
542-756	Dichloropropene(1,3-)	12035-722	Nickel Subsulfide	4.80E-04	2.1E-02
542-881	Bis(chloromethyl)ether	79-469	Nitropropane(2-)	2.70E-02	3.7E-04
684-935	Nitroso-n-methylurea(n-)	924-163	Nitroso-n-butylamine(n-)	1.60E-03	6.3E-03
924-163	Nitroso-n-butylamine(n-)	684-935	Nitroso-n-methylurea(n-)	8.60E-02	1.2E-04
930-552	Nitrosopyrrolidine(n-)	55-185	Nitrosodiethylamine(n-)	4.30E-02	2.3E-04
1024-573	Heptachlor Epoxide	930-552	Nitrosopyrrolidine(n-)	6.10E-04	1.6E-02
1336-363	PCBs	82-688	Pentachloronitrobenzene	7.30E-05	1.4E-01

TABLE 7.3. RISK SPECIFIC DOSES (RsD) (10⁻⁵) (40CFR266 APPENDIX V)					
CAS RN order		Alphabetical order		Unit risk (m³/μg)	RsD (μg/m³)
1746-016	Tetrachloro-dibenzo-p-dioxin(2,3,7,8-)	1336-363	PCBs	1.20E-03	8.3E-03
7440-020	Nickel	23950-585	Pronamide	4.60E-06	2.2E+00
7440-020	Nickel Refinery Dust	50-555	Reserpine	3.00E-03	3.3E-03
7440-382	Arsenic	1746-016	Tetrachloro-dibenzo-p-dioxin(2,3,7,8-)	4.50e+01	2.2E-07
7440-417	Beryllium	79-345	Tetrachloroethane(1,1,2,2-)	5.80E-05	1.7E-01
7440-439	Cadmium	127-184	Tetrachloroethylene	4.80E-07	2.1E+01
7440-473	Chromium VI	62-566	Thiourea	5.50E-04	1.8E-02
8001-352	Toxaphene	79-005	Trichloroethane(1,1,2-)	1.60E-05	6.3E-01
12035-722	Nickel Subsulfide	79-016	Trichloroethylene	1.30E-06	7.7E+00
23950-585	Pronamide	88-062	Trichlorophenol(2,4,6-)	5.70E-06	1.8E+00
		8001-352	Toxaphene	3.20E-04	3.1E-02
		75-014	Vinyl Chloride	7.10E-06	1.4E+00

TABLE 7.4. PRODUCTS OF INCOMPLETE COMBUSTION (PICs) FOUND IN STACK EFFLUENTS (40CFR266 APPENDIX VIII)	
Constituents	Compound type[1]
Benzene	v
Bis(2-ethylhexyl)phthalate	sv
Bromochloromethane	v
Bromodichloromethane	v
Bromoform	v
Bromomethane	v
Butyl benzyl phthalate	sv
Carbon tetrachloride	v
Chlorobenzene	v
Chloroform	v
Chlorophenol(o-)	sv
Dichloro-2-butene(cis-1,4-)	v
Dichlorobenzene(m-)	sv
Dichlorobenzene(o-)	sv
Dichlorobenzene(p-)	sv
Diethyl phthalate	sv
Dimethyl phthalate	sv
Dimethylphenol(2,4-)	sv
Fluoranthene	sv
Hexachlorobenzene	sv
Methyl ethyl ketone	v
Methylene bromide	v
Methylene chloride	v
Mononitrobenzene	sv
Naphthalene	sv
Nitrophenol(o-)	sv
Pentachlorophenol	sv
Phenol	sv
Pyrene	sv
Tetrachloroethylene	v
Toluene	v
Toluene diisocyanate(2,6-)	sv

TABLE 7.4. PRODUCTS OF INCOMPLETE COMBUSTION (PICs) FOUND IN STACK EFFLUENTS (40CFR266 APPENDIX VIII)	
Constituents	Compound type[1]
Trichlorobenzene(1,2,4-)	sv
Trichloroethane(1,1,1-)	v
Trichloroethylene	v
Trichlorophenol(2,4,6-)	sv
[1] sv: Semivolatiles v: Volatiles	

ACRONYMS AND REFERENCES

ACRONYMS 8-2

REFERENCES 8-4

ACRONYMS AND REFERENCES

ACRONYMS

40CFR51-AM:	40CFR51-Appendix M
40CFR60-AA:	40CFR60-Appendix A
40CFR61-AB:	40CFR61-Appendix B
AAS:	Atomic Absorption Spectroscopy
AOAC:	Association of Official Analytical Chemists
API:	American Petroleum Institute
ASME:	American Society of Mechanical Engineers
ASTM:	American Society for Testing and Materials
BOD:	Biochemical Oxygen Demand (40CFR133)
C:	Colorimetry
CAA:	Clean Air Act
CAS RN:	Chemical Abstract Service Registry Number
CERCLA:	Comprehensive Environmental Response, Compensation, and Liability Act
CFR:	Code of Regulations
COD:	Chemical Oxygen Demand
CWA:	Clean Water Act
CV/AAS:	Cold Vapor/Atomic Absorption Spectroscopy
DCP:	Direct Current Plasma
EP:	Extraction Procedure
FID:	Flame Ionization Detector
FIFRA:	Federal Insecticide, Fungicide, and Rodenticide Act
FPD:	Flame Photometric Detector
G:	Glass
GC:	Gas Chromatography
GC/FID:	Gas Chromatography with Flame Ionization Detector
GC/FPD:	Gas Chromatography with Flame Photometric Detector
GC/MS:	Gas Chromatography/Mass Spectrometry
GC/TCD:	Gas Chromatography with Thermal Conductivity Detector
HAP:	Hazardous Air Polltant
HPLC:	High Performance Liquid Chromatography
IBR:	Incorporation by Reference
IC:	Ion Chromatography
ICAP:	Inductively Coupled Argon Plasma Emission Spectroscopy
ICP:	Inductively Coupled Plasma
IRIS:	Integrated Risk Information System
MBAS:	Methylene Blue Active Substances
MCL:	Maximum Comtaminant Limit
MDL:	Method Detection Limit
MF:	Membrane Filter
MFL:	Million Fibers per Liter
MPN:	Most Probable Number
MS:	Mass Spectrometry
NFPA:	National Fire Protection Association
NPDES:	National Pollutant Discharge Elimination System
NTA:	Nitrilotriacetic Acid

P:	Polyethylene
PM:	Particulate Matter
PQL:	Practical Quantitation Limits
RAC:	Reference Air Concentrations
RCRA:	Resource Conservation and Recovery Act
RsD:	Risk Specific Dose
SDWA	Safe Drinking Water Act
SASS	Source Assessment Sampling System
sv:	Semivolatile
T:	Titration
TAPPI:	Technical Association of the Pulp and Paper Industry
TCD:	Thermal Conductivity Detector
TCLP:	Toxicity Characteristic Leaching Procedure
TDE:	Tetrachlorodiphenylethane
TKN:	Total Kjeldahl Nitrogen
TLC:	Thin Layer Chromatography
TSCA:	Toxic Substances Control Act
UL:	Underwriter's Laboratories, Inc.
v	Volatile
VOC:	Volatile Organic Compounds
VOST:	Volatile Organic Sampling Train
WPCF:	Water Pollution Control Federation

REFERENCES

ASTM materials are available for purchase from at least one of the following addresses: American Society for Testing and Materials (ASTM), 1916 Race Street, Philadelphia, Pennsylvania 19103; or the University Microfilms International, 300 North Zeeb Road, Ann Arbor, MI 48106.

ASME materials are available for purchase from the American Society of Mechanical Engineers (ASME), 345 East 47th Street, New York, NY 10017.

(AOAC Method 9), "Official Methods of Analysis of the Association of Official Analytical Chemists," 11th Edition, 1970, pp. 11-12, IBR approved January 27, 1983 for 40CFR60.204(d)(2), 60.214(d)(2), 60.224(d)(2), 60.234(d)(2), 60.244(f)(2), Available from the Association of Official Analytical Chemists, 1111 North 19th Street, Suite 210, Arlington, VA 22209.

(API Publication 2517), "Evaporation Loss from External Floating Roof Tanks," Second Edition, February 1980, IBR approved January 27, 1983, for 40CFR60.111(i), 60.111a(f), 60.111a(f)(1) and 60.116b(e)(2)(i), Available for purchase from the American Petroleum Institute, 1220 L Street NW., Washington, DC 20005.

(ASME Interim Supplement 19.5 on Instruments and Apparatus), "Application, Part II of Fluid Meters," 6th Edition (1971). IBR Approved for 40CFR60.58a(h), Available for purchase from the American Society of Mechanical Engineers (ASME), 345 East 47th Street, New York, NY 10017.

(ASME PTC 4.1.), "Power Test Codes: Test Code for Steam Generating Units (1972)," IBR Approved for 40CFR60.46b and 60.58a(h), Available for purchase from the American Society of Mechanical Engineers (ASME), 345 East 47th Street, New York, NY 10017.

(ASME QRO-1-1989), "Standard for the Qualification and Certification of Resource Recovery Facility Operators," IBR Approved for 40CFR60.56a, Available for purchase from the American Society of Mechanical Engineers (ASME), 345 East 47th Street, New York, NY 10017.

(ASTM A100-69), "(Reapproved 1974), Standard Specification for Ferrosilicon," IBR approved January 27, 1983 for 40CFR60.261.

(ASTM A101-73), "Standard Specification for Ferrochromium," IBR approved January 27, 1983 for 40CFR60.261.

(ASTM A482-76), "Standard Specification for Ferrochromesilicon," IBR approved January 27, 1983 for 40CFR60.261.

(ASTM A483-64), "(Reapproved 1974), Standard Specification for Silicomanganese," IBR approved January 27, 1983 for 40CFR60.261.

(ASTM A495-76), "Standard Specification for Calcium-Silicon and Calcium Manganese-Silicon," IBR approved January 27, 1983 for 40CFR60.261.

(ASTM A99-76), "Standard Specification for Ferromanganese," IBR approved January 27, 1983 for 40CFR60.261.

(ASTM D1072-80), "Standard Method for Total Sulfur in Fuel Gases," IBR approved July 31, 1984 for 40CFR60.335(b)(2).

(ASTM D1137-53), "(Reapproved 1975), Standard Method for Analysis of Natural Gases and Related Types of Gaseous Mixtures by the Mass Spectrometer," IBR approved January 27, 1983 for 40CFR60.45(f)(5)(i).

(ASTM D1193-77), "Standard Specification for Reagent Water," for appendix A to part 60, Method 6, par. 3.1.1; Method 7, par. 3.2.2; Method 7C, par. 3.1.1; Method 7D, par. 3.1.1; Method 8, par. 3.1.3; Method 12, par. 4.1.3; Method 25D, par. 3.2.2.4; Method 26A, par. 3.1.1.

(ASTM D1266-87), "Standard Test Method for Sulfur in Petroleum Products (Lamp Method)," IBR approved August 17, 1989, for 40CFR60.106(j)(2).

(ASTM D129-64), "(reapproved 1978), Standard Test Method for Sulfur in Petroleum Products (General Bomb Method)," IBR approved for appendix A to part 60, Method 19.

(ASTM D129-64), "(Reapproved 1978), Standard Test Method for Sulfur in Petroleum Products (General Bomb Method)," IBR approved August 17, 1989, for 40CFR60.106(j)(2).

(ASTM D1475-60), "(Reapproved 1980), Standard Test Method for Density of Paint, Varnish, Lacquer, and Related Products," IBR approved January 27, 1983 for 40CFR60.435(d)(1), appendix A to part 60, Method 24, par. 2.1, and Method 24A, par. 2.2.

(ASTM D1552-83), "Standard Test Method for Sulfur in Petroleum Products (High-Temperature Method)," IBR approved August 17, 1989, for 40CFR60.106(j)(2).

(ASTM D1552-83), "Standard Test Method for Sulfur in Petroleum Products (High Temperature Method)," IBR approved for appendix A to part 60, Method 19.

(ASTM D1826-77), "Standard Test Method for Calorific Value of Gases in Natural Gas Range by Continuous Recording Calorimeter," IBR approved January 27, 1983, for 40CFR60.45(f)(5)(ii); 60.46(g); 60.296(f); appendix A to part 60, Method 19.

(ASTM D1835-86), "Standard Specification for Liquefied Petroleum (LP) Gases," to be approved for 40CFR60.41b.

(ASTM D1835-86), "Standard Specification for Liquefied Petroleum (LP) Gases," IBR approved for 40CFR60.41b; 60.41c.

(ASTM D1945-64), "(Reapproved 1976), Standard Method for Analysis of Natural Gas by Gas Chromatography," IBR approved January 27, 1983 for 40CFR60.45(f)(5)(i).

(ASTM D1946-77), "Standard Method for Analysis of Reformed Gas by Gas Chromatography," IBR approved for 40CFR60.45(f)(5)(i), 60.18(f), 60.614(d)(2)(ii), 60.614(d)(4), 60.664(d)(2)(ii), 60.664(d)(4),60.564(f), 60.704(d)(2)(ii) and 60.704(d)(4).

(ASTM D1946-82), "ASTM Standard Method for Analysis of Reformed Gas by Gas Chromatography," ASTM Standard D1946-82, Available from American Society for Testing and Materials, 1916 Race Street, Philadelphia, PA 19103.

(ASTM D2013-72), "Standard Method of Preparing Coal Samples for Analysis," IBR approved January 27, 1983, for appendix A to part 60, Method 19.

(ASTM D2015-77), "Standard Test Method for Gross Calorific Value of Solid Fuel by the Adiabatic Bomb Calorimeter," IBR approved January 27, 1983 for 40CFR60.45(f)(5)(ii); 40CFR60.46(g); appendix A to part 60, Method 19.

(ASTM D2016-74 (Reapproved 1983)), "Standard Test Methods for Moisture Content of Wood * * * for appendix A," Method 28.

(ASTM D2234-76), "Standard Methods for Collection of a Gross Sample of Coal," IBR approved January 27, 1983, for appendix A to part 60, Method 19.

(ASTM D2267-88), "ASTM Standard Test Method for Aromatics in Light Naphthas and Aviation Gasolines by Gas Chromatography," ASTM Standard D2267-88, Available from American Society for Testing and Materials, 1916 Race Street, Philadelphia, PA 19103.

(ASTM D2369-81), "Standard Test Method for Volatile Content of Coatings," IBR approved January 27, 1983 for appendix A to part 60, Method 24.

(ASTM D2382-76), "Heat of Combustion of Hydrocarbon Fuels by Bomb Calorimeter [High-Precision Method]," IBR approved for 40CFR60.18(f), 60.485(g), 60.614(d)(4), 60.664(d)(4), 60.564(f), and 60.704(d)(4).

(ASTM D2382-83), "ASTM Standard Test Method for Heat of Combustion of Hydrocarbon Fuels by Bomb Calorimeter (High-Precision Method)," ASTM Standard D2382-83, Available from American Society for Testing and Materials, 1916 Race Street, Philadelphia, PA 19103 (source: 40CFR260.11.a).

(ASTM D240-76), "Standard Test Method for Heat of Combustion of Liquid Hydrocarbon Fuels by Bomb Calorimeter," IBR approved January 27, 1983, for 40CFR60.46(g); 60.296(f); appendix A to part 60, Method 19.

(ASTM D2504-67), "(Reapproved 1977), Noncondensable Gases in C3 and Lighter Hydrocarbon Products by Gas Chromatography," IBR approved for 40CFR60.485(g).

(ASTM D2584-68), "Standard Test Method for Ignition Loss of Cured Reinforced Resins," IBR approved February 25, 1985 for 40CFR60.685(e).

(ASTM D2622-87), "Standard Test Method for Sulfur in Petroleum Products by X-Ray Spectrometry," IBR approved August 17, 1989, for 40CFR60.106(j)(2).

(ASTM D270-65), "(Reapproved 1975), Standard Method of Sampling Petroleum and Petroleum Products," IBR approved January 27, 1983, for appendix A to part 60, Method 19.

(ASTM D2879-83), "Test Method for Vapor Pressure-Temperature Relationship and Initial Decomposition Temperature of Liquids by Isoteniscope," IBR approved April 8, 1987 for 40CFR60.485(e), 60.111b(f)(3), 60.116b(e)(3)(ii), and 60.116b(f)(2)(i).

(ASTM D2879-86), "ASTM Standard Test Method for Vapor Pressure-Temperature Relationship and Initial Decomposition Temperature of Liquids by Isoteriscope," ASTM Standard D2879-86, Available from American Society for Testing and Materials, 1916 Race Street, Philadelphia, PA 19103.

(ASTM D2880-78), "Standard Specification for Gas Turbine Fuel Oils," IBR approved January 27, 1983 for 40CFR60.111(b), 60.111a(b), 60.335(b)(2).

(ASTM D2908-74), "Standard Practice for Measuring Volatile Organic Matter in Water by Aqueous-Injection Gas Chromatography," IBR approved for 40CFR60.564(j).

(ASTM D2986-71), "(Reapproved 1978), Standard Method for Evaluation of Air, Assay Media by the Monodisperse DOP (Dioctyl Phthalate) Smoke Test," IBR approved January 27, 1983 for appendix A to part 60, Method 5, par. 3.1.1; Method 12, par. 4.1.1; Method 17, par. 3.1.1.

(ASTM D3031-81), "Standard Test Method for Total Sulfur in Natural Gas by Hydrogenation," IBR approved July 31, 1984 for 40CFR60.335(b)(2).

(ASTM D3173-73), "Standard Test Method for Moisture in the Analysis Sample of Coal and Coke," IBR approved January 27, 1983, for appendix A to part 60, Method 19.

(ASTM D3176-74), "Standard Method for Ultimate Analysis of Coal and Coke," IBR approved January 27, 1983, for 40CFR60.45(f)(5)(i); appendix A to part 60, Method 19.

(ASTM D3177-75), "Standard Test Methods for Total Sulfur in the Analysis Sample of Coal and Coke," IBR approved January 27, 1983, for appendix A to part 60, Method 19.

(ASTM D3178-73), "Standard Test Methods for Carbon and Hydrogen in the Analysis Sample of Coal and Coke," IBR approved January 27, 1983 for 40CFR60.45(f)(5)(i).

(ASTM D323-82), "Test Method for Vapor Pressure of Petroleum Products (Reid Method)," IBR approved April 8, 1987 for 40CFR60.111(1), 60.111a(g), 60.111b(g), and 60.116b(f)(2)(ii).

(ASTM D3246-81), "Standard Method for Sulfur in Petroleum Gas by Oxidative Microcoulometry," IBR approved July 31, 1984 for 40CFR60.335(b)(2).

(ASTM D3278-78), "ASTM Standard Test Methods for Flash Point of Liquids by Setaflash Closed Tester," ASTM Standard D3278-78, Available from American Society for Testing and Materials, 1916 Race Street, Philadelphia, PA 19103.

(ASTM D3370-76), "Standard Practices for Sampling Water," IBR approved for 40CFR60.564(j).

(ASTM D3431-80), "Standard Test Method for Trace Nitrogen in Liquid Petroleum Hydrocarbons (Microcoulometric Method)," IBR approved November 25, 1986, for appendix A to part 60, Method 19.

(ASTM D3792-79), "Standard Method for Water Content of Water-Reducible Paints by Direct Injection Into a Gas Chromatograph," IBR approved January 27, 1983 for appendix A to part 60, Method 24, par. 2.3.

(ASTM D388-77), "Standard Specification for Classification of Coals by Rank," incorporation by reference (IBR) approved for 40CFR60.41(f); 60.45(f)(4)(i), (ii), (vi); 60.41a; 60.41b; 60.41c; 60.25(b), (c).

(ASTM D396-78), "Standard Specification for Fuel Oils," IBR approved for 40CFR60.40b; 60.41b; 60.41c; 60.111(b); 60.111a(b)..

ACRONYMS AND REFERENCES

(ASTM D4017-81), "Standard Test Method for Water in Paints and Paint Materials by the Karl Fischer Titration Method," IBR approved January 27, 1983 for appendix A to part 60, Method 24, par. 2.4.

(ASTM D4057-81), "Standard Practice for Manual Sampling of Petroleum and Petroleum Products," IBR approved for appendix A to part 60, Method 19.

(ASTM D4084-82), "Standard Method for Analysis of Hydrogen Sulfide in Gaseous Fuels (Lead Acetate Reaction Rate Method)," IBR approved July 31, 1984 for 40CFR60.335(b)(2).

(ASTM D4239-85), "Standard Test Methods for Sulfur in the Analysis Sample of Coal and Coke Using High Temperature Tube Furnace Combustion Methods," IBR approved for appendix A to part 60, Method 19.

(ASTM D4442-84), "Standard Test Methods for Direct Moisture Content Measurement in Wood and Wood-base Materials * * * for appendix A," Method 28.

(ASTM D4457-85), "Test Method for Determination of Dichloromethane and 1,1,1-Trichloroethane in Paints and Coatings by Direct Injection into a Gas Chromatograph," IBR approved for appendix A, Method 24.

(ASTM D737-85), "Standard Test Method for Air Permeability of Textile Fabrics," IBR approved January 27, 1983 for 40CFR61.23(a).

(ASTM D86-78), "Distillation of Petroleum Products," IBR approved for 40CFR60.593(d), 40CFR60.633(h), and 40CFR60.562-2(d).

(ASTM D93-79 or D93-80, D93-80), "ASTM Standard Test Methods for Flash Point by Pensky-Martens Closed Tester," ASTM Standard D93-79 or D93-80, D93-80 is available from American Society for Testing and Materials, 1916 Race Street, Philadelphia, PA 19103.

(ASTM D975-78), "Standard Specification for Diesel Fuel Oils," IBR approved January 27, 1983 for 40CFR60.111(b), 60.111a(b).

(ASTM E168-67), "(Reapproved 1977), General Techniques of Infrared Quantitative Analysis," IBR approved for 40CFR60.485(d), 40CFR60.593(b), and 40CFR60.632(f).

(ASTM E168-88), "ASTM Standard Practices for General Techniques of Infrared Quantitative Analysis," ASTM Standard E168-88, Available from American Society for Testing and Materials, 1916 Race Street, Philadelphia, PA 19103 (source: 40CFR260.11.a).

(ASTM E169-63), "(Reapproved 1977), General Techniques of Ultraviolet Quantitative Analysis," IBR approved for 40CFR60.485(d), 40CFR60.593(b), and 40CFR60.632(f).

(ASTM E169-87), "ASTM Standard Practices for General Techniques of Ultraviolet-Visible Quantitative Analysis," ASTM Standard E169-87, Available from American Society for Testing and Materials, 1916 Race Street, Philadelphia, PA 19103.

(ASTM E260-73), "General Gas Chromatography Procedures," IBR approved for 40CFR60.485(d), 40CFR60.593(b), and 40CFR60.632(f).

(ASTM E260-85), "ASTM Standard Practice for Packed Column Gas Chromatography," ASTM Standard E260-85, Available from American Society for Testing and Materials, 1916 Race Street, Philadelphia, PA 19103.

(ASTM E926-88), "ASTM Standard Test Methods for Preparing Refuse-Derived Fuel (RDF) Samples for Analyses of Metals," ASTM Standard E926-88, Test Method C-Bomb, Acid Digestion Method, Available from American Society for Testing Materials, 1916 Race Street, Philadelphia, PA 19103.

(EPA-77/8), "Procedures Manual For Ground-water Monitoring At Solid Waste Disposal Facilities," EPA-530/SW-611, August 1977.

(EPA-79/3), "Methods for Chemical Analysis of Water and Wastes," EPA600-4-79-020, March 1979.

(EPA-81/12), "APTI Course 415: Control of Gaseous Emissions," EPA Publication EPA450-2-81-005, December 1981, Available from National Technical Information Service, 5285 Port Royal Road, Springfield, VA 2216.

(EPA-83/3), "Methods for Chemical Analysis of Water and Wastes," Environmental Protection Agency, Environmental Monitoring Systems Laboratory-Cincinnati (EMSL-CI), EPA600-4-79-020, March 1983.

(EPA-84/2), "Sampling and Analysis Methods for Hazardous Waste Combustion, Appendix C," Prepared by ADL, EPA600-8-84-002, PB84-155845, February 1984.

8-7

(EPA-89/6), "Hazardous Waste Incineration Measurement Guidance Manual, Volume III of the Hazardous Waste Incineration Guidance Series," Prepared by MRI, EPA625-6-89-021, June 1989.

(EPA-91/7), "Methods for the Determination of Organic Compounds in Drinking Water," EPA600-4-88-039, July 1991.

(EPA-92/10), "Screening Procedures for Estimating the Air Quality Impact of Stationary Sources," Revised, October 1992, EPA Publication No. EPA450-R-92-019, Environmental Protection Agency, Research Triangle Park, NC (source: 40CFR260.11.a).

(EPA-93/1), "The ABCs of Risk Assessment," EPA Journal, EPA175-N-93-0014, Volume 19, Number 1, January/February/March 1993.

(EPA-93/8), "Methods for the Determination of Inorganic Substances in environmental Samples," EPA600-R-93-100, August 1993.

(NFPA-81), "Flammable and Combustible Liquids Code (1977 or 1981)," Available from the National Fire Protection Association (NFPA), 470 Atlantic Avenue, Boston, MA 02210.

(SW-846), "Test Methods for Evaluating Solid Waste, Physical/Chemical Methods," EPA Publication SW-846 [Third Edition (November, 1986), as amended by Updates I (July, 1992), II (September, 1994), IIA (August, 1993), and IIB (January, 1995)]. The Third Edition of SW-846 and Updates I, II, IIA, and IIB (document number 955-001-00000-1) are available from the Superintendent of Documents, U.S. Government Printing Office, Washington, DC 20402, (202) 512-1800. Copies may be inspected at the Library, U.S. Environmental Protection Agency, 401 M Street, SW, Washington, DC 20460.

(TAPPI Method T624 os-68), IBR approved January 27, 1983 for 40CFR60.285(d)(4), Available for purchase from the Technical Association of the Pulp and Paper Industry (TAPPI), Dunwoody Park, Atlanta, GA 30341.

(UL 103), "Sixth Edition revised as of September 3, 1986, Standard for Chimneys, Factory-built, Residential Type and Building Heating Appliance," Available for purchase from the following address: Underwriter's Laboratories, Inc. (UL), 333 Pfingsten Road, Northbrook, IL 60062.

(West Coast Lumber Standard Grading Rules No. 16), "Pages 5-21 and 90 and 91, September 3, 1970, revised 1984," Available for purchase from the following address: West Coast Lumber Inspection Bureau, 6980 SW. Barnes Road, Portland, OR 97223.

(WPCF Method 209A), "Total Residue Dried at 103-105 °C, in Standard Methods for the Examination of Water and Wastewater," 15th Edition, 1980, IBR approved February 25, 1985 for 40CFR60.683(b), Available for purchase from the Water Pollution Control Federation (WPCF), 2626 Pennsylvania Avenue NW., Washington, DC 20037.

Environmental and Health/Safety References

Total Quality for Safety and Health Professionals
F. David Pierce, a CSP and a CIH, shows you how to apply concepts - proven and successful - to your safety management program to achieve increased productivity, lowered costs, reduced inventories, improved quality, increased profits, and raised employee morale.
Hardcover, 244 pages, June '95, ISBN: 0-86587-462-X **$59**

Pollution Prevention Strategies and Technologies
This book is an indispensible guide to understanding pollution prevention policies and regulatory initiatives designed to reduce wastes. *Hardcover, Index, 484 pages, Oct '95, ISBN: 0-86587-480-8* **$79**

Environmental Audits, 7th Edition
This is the most comprehensive manual available on environmental audits! Completely updated, it provides you with all the step-by-step guidance you need on how to begin - and manage - a successful audit program for your facility. Includes over 50 pages of exercises that cover Audit Verification, Interviewing Skills, Management Assessments, Report Writing, and Audit Conferences.
Softcover, approx. 500 pages, Mar '96, ISBN: 0-86587-525-1 **$79**

Fate and Transport of Organic Chemicals in the Environment: A Practical Guide, 2nd Edition
This easy-to-use guide provides a unique tool to help you predict the fate and transport of chemicals in air, water, soil, flora and fauna.
Softcover, Index, 308 pages, Aug '95, ISBN: 0-86587-470-0 **$49**

Environmental Law Handbook, 13th Edition
Includes changes in major federal environmental laws. Details how those laws affect the regulated community.
Hardcover, Index, 550 pages, Mar '95, ISBN: 0-86587-450-6 **$79**

"So You're the Safety Director!" *An Introduction to Loss Control and Safety Management*
This book concentrates on your role in evaluating, managing, and controlling your company's losses and handling the OSHA compliance process.
Softcover, Index, 186 pages, Oct '95, ISBN: 0-86587-481-6 **$45**

Safety Made Easy: A Checklist Approach to OSHA Compliance
This new book provides a simple way of understanding your requirements under the complex maze of the Occupational Safety and Health Administration regulations. The authors have created safety and health checklists for compliance organized alphabetically by topic.
Softcover, 192 pages, June '95, ISBN: 0-86587-463-8 **$45**

Understanding Workers' Compensation: A Guide for Safety and Health Professionals
This book is designed to help you understand how the workers' comp system works, and provides a basic understanding of injury prevention, types of injuries, and cost containment strategies.
Softcover, 250 pages, July '95, ISBN: 0-86587-464-6 **$45**

Environmental Telephone Directory, 1996 Edition
All new directory contains *Electronic Mail Addresses and World-Wide Web Sites* for Federal Agencies and the most phone numbers and addresses of key U.S. Government and Environmental Contacts.
Softcover, 256 pages, Nov '95, ISBN: 0-86587-504-9 **$67**

Environmental Guide to the Internet, 2nd Edition
From environmental engineering to hazardous waste compliance issues, you'll have no problem finding it with this easy-to-use guide. Includes over 250 all new World Wide Web Sites.
Softcover, 236 pages, Feb '96, ISBN: 0-86587-517-0 **$49**

Ergonomic Problems in the Workplace: A Guide to Effective Management
The valuable insights you'll gain from this new book will help you develop and implement your own successful ergonomics program.
Softcover, 256 pages, Sept '95, ISBN: 0-86587-474-3 **$59**

Exposure Factors Handbook, Review Draft
The U.S. Environmental Protection Agency uses this document to develop pesticide tolerance levels, assess industrial chemical risks, and to undertake Superfund site assessments and drinking water health assessments.
Softcover, 866 pages, Nov '95, ISBN: 0-86587-509-X **$125**

Environmental and Health/Safety References (Cont'd)

Sampling, Analysis & Monitoring Methods: A Guide to EPA Requirements
This book provides a guide for determining which chemicals have sampling, analysis, and monitoring requirements under U.S. environmental laws and regulations, and where those testing and sampling methods can be found.
Softcover, 256 pages, Sept '95, ISBN: 0-86587-477-8 **$65**

OSHA Technical Manual, 4th Edition
This inspection manual is used nationwide by the U.S. Occupational Safety and Health Administration's inspectors in checking industry compliance with OSHA requirements. *Softcover, 400 pages, Feb '96, ISBN: 0-86587-511-1* **$85**

Environmental Statutes, 1996 Edition
The complete text of each statute as currently amended is included, with a detailed Table of Contents for quick reference.
Hardcover, 1,200 pages, Mar '96, ISBN: 0-86587-521-9 **$69**
Softcover, 1,200 pages, Mar '96, ISBN: 0-86587-522-7 **$59**
Also Available in Disk Format!
(Disks include the Folio search Engine)
• *3.5" Floppy Disks for Windows, (#4058)* **$135**
• *Statutes Package: hardcover book w/Windows disk (#4059)* **$204**

Emergency Planning & Management: *Ensuring Your Company's Survival in the Event of a Disaster*
This book will help you assess your exposure to disasters and prepare emergency response, preparedness, and recovery plans for your facilities, both to comply with OSHA and EPA requirements and to reduce the risk of losses to your company.
Softcover, 192 pages, Nov '95, ISBN: 0-86587-505-7 **$59**

Product Side of Pollution Prevention: Evaluating the Potential for Safe Substitutes
This report focuses on safe substitutes for products that contain or use toxic chemicals in their manufacturing process.
Softcover, 240 pages, Sept '95, ISBN: 0-86587-479-4 **$69**

GI Environmental Database CD ROM
Our four most popular environmental references — Directory of Environmental Information Sources; Environmental Telephone Directory; Environmental Regulatory Glossary; and Environmental Acronyms — are now included on one CD ROM — a powerful compliance tool for instant information retrieval on your computer. *Single CD ROM, Windows™, Nov '95, Product Code #4073* **$149**

Chemical Information Manual, 3rd Edition (Book and Disk Format)
This database contains essential data for over 1,400 chemical substances. The following information is available to you: proper identification synonyms, OSHA exposure limits, description and physical properties; carcinogenic status, health effects and toxicology, sampling and analysis.
Softcover, 400 pages, Aug '95, ISBN: 0-86587-469-7 **$99**
Also available on disk!
3.5" Floppy Disk for Windows, #4070 **$99**

Federal Facility Pollution Prevention: Planning Guide and Tools for Compliance
This U.S. Environmental Protection Agency guide presents pollution prevention tools and provides a step-by-step approach to develop a pollution prevention program plan for federal facilities.
Softcover, 240 pages, Aug '95 ISBN: 0-86587-476-X **$75**

Internet and the Law: Legal Fundamentals for the Internet User
Explains the principles of laws of copyright, trademark, trade secret, patent, libel/defamation and related issues, and the basic principles of licensing.
Softcover, 264 pages, Dec '95 ISBN: 0-86587-506-5 **$75**

Environmental Regulatory Glossary, 6th Edition
This glossary defines and standardizes more than 4,000 terms, abbreviations and acronyms, compiled directly from the environmental statutes or the U.S. Code of Federal Regulations.
Hardcover, 544 pages, June '93, ISBN: 0-86587-353-4 **$72**

Environmental Science and Technology Handbook
This is the first book to bridge the gap between the latest environmental science and technology available, and compliance with today's complex regulations.
Hardcover, 389 pages, Jan '94, ISBN: 0-86587-362-3 **$79**

Health Effects of Toxic Substances
This comprehensive book provides you with an excellent understanding of the toxicology and industrial hygiene of hazardous materials.
Softcover, Index, 256 pages, Aug '95, ISBN: 0-86587-471-9 **$39**

Principles of Environmental, Health and Safety Management
This book provides you with information and advice on how your company can develop a comprehensive environmental management program.
Softcover, Aug '95, 320 pages, ISBN: 0-86587-478-5 **$59**